ALFRED BENZON SYMPOSIUM 24

Viral Carcinogenesis

VIRAL CARCINOGENESIS

FUNCTIONAL ASPECTS

Proceedings of the Alfred Benzon Symposium 24 held at the premises of the Royal Danish Academy of Sciences and Letters. Copenhagen 15–19 June 1986

EDITED BY

NIELS OLE KJELDGAARD
JES FORCHHAMMER

Published by
Munksgaard, Copenhagen

Distributed in North and South America by
Raven Press, New York

Distributed in Japan by
Nankodo, Tokyo

Printed in Denmark by P. J. Schmidts Bogtrykkeri, Vojens
ISBN 87-16-06491-7
ISSN 0105-3639

Published simultaneously in the U.S.A. by Raven Press
ISBN 0-88167-150-9
LCCCN 86-42572

Contents

III Transcriptional Control and Modifications

IV Gene Activation by Integration

V GENE ACTIVATION BY TRANSLOCATION

VI TRANSFORMATION BY BPV

VII CYTOSKELETON AND ACTIVATION OF C-ONC

VIII Signal Pathways

List of Participants

D. Baltimore
Whitehead Institute for Biomedical Research
Nine Cambridge Center
Cambridge, Massachusetts 02142
U.S.A.

A. Berns
Antoni van Leeuwenhoekhuis
The Netherlands Cancer Institute
Plesmanlaan 121
NL-1066 CX Amsterdam
The Netherlands

D. G. Blair
Laboratory of Molecular Oncology
National Cancer Institute
Frederick, Maryland 21701
U.S.A.

R. Bravo
European Molecular Biology Laboratory
Postfach 10.2209
D-6900 Heidelberg
F.R.G.

J. S. Brugge
Department of Microbiology
State University of New York
Stony Brook, New York 11794
U.S.A.

J. E. Celis
Institute of Medical Biochemistry
University of Aarhus
DK-8000 Aarhus C
Denmark

S. Cory
The Walter and Eliza Hall Institute of Medical Research
P. O. Royal Melbourne Hospital
Victoria 3050
Australia

F. Cuzin
Centre de Biochimie
Université de Nice
Parc Valrose
F-06034 Nice CEDEX
France

W. S. Dynan
Department of Chemistry and Biochemstry
University of Colorado at Boulder
Campus Box 215
Boulder, Colorado 80309
U.S.A.

A. J. van der Eb
Sylvius Laboratories
Department of Medical Biochemistry
P. O. Box 9503
NL-2300 RA Leiden
The Netherlands

J. Forchhammer
The Fibiger Institute
70 Nordre Frihavnsgade
DK-2100 Copenhagen Ø
Denmark

R. C. Gallo
Laboratory of Tumor Cell Biology
National Cancer Institute
Building 37, Room 6A-09
Bethesda, Maryland 20205
U.S.A.

D. Hanahan
Cold Spring Harbor Laboratory
P. O. Box 100
Cold Spring Harbor, New York 11724
U.S.A.

W. A. Haseltine
Dana-Farber Cancer Institute
44 Binney Street
Boston, Massachusetts 02115
U.S.A.

H. zur Hausen
Deutsches Krebsforschungszentrum
Im Neuheimer Feld 280
D-6900 Heidelberg
F.R.G.

J. Hess Thaysen
Department of Medicine P
Rigshospitalet
9 Blegdamsvej
DK-2100 Copenhagen Ø
Denmark

N. O. Kjeldgaard
Institute of Molecular Biology
130 C. F. Møllers Allé
DK-8000 Aarhus C
Denmark

K. Leppard
Department of Molecular Biology
Princeton University
Princeton, New Jersey 08544
U.S.A.

D. R. Lowy
Laboratory of Cellular Oncology
National Cancer Institute
Building 37, Room 1B-26
Bethesda, Maryland 20205
U.S.A.

P. Matthias
Institut für Molekülar-Biologie II
der Universität Zürich
CH-8093 Zürich
Switzerland

B. Norrild
Institute of Medical Microbiology
University of Copenhagen
30 Juliane Maries Vej
DK-2100 Copenhagen Ø
Denmark

R. Nusse
Antoni van Leeuwenhoekhuis
The Netherlands Cancer Institute
Plesmanlaan 121
NL-1066 CX Amsterdam
The Netherlands

F. Skou Pedersen
Institute of Molecular Biology and Plant Physiology
University of Aarhus
130 C. F. Møllers Allé
DK-8000 Aarhus C
Denmark

U. R. Rapp
Laboratory of Viral Carcinogenesis
National Cancer Institute
Building 560, Room 21–77
Frederick, Maryland 21701
U.S.A.

P. Rigby
Laboratory of Eukaryotic Molecular Genetics
National Institute for Medical Research
The Ridgeway, Mill Hill
London NW7 1AA
England

B. Roizman
Marjorie B. Kovler Viral Oncology Laboratories
University of Chicago
910 East 58th Street
Chicago, Illinois 60637
U.S.A.

J. Schlessinger
Department of Molecular Biology
Meloy Laboratories
4 Research Court
Rockville, Maryland 20850
U.S.A.

C. J. Sherr
Department of Tumor Cell Biology
St. Jude Children's Research Hospital
332 North Lauderdale, P.O. Box 318
Memphis, Tennessee 38101
U.S.A.

J. Sümegi
Department of Tumor Biology
Karolinska Institutet
Box 60400
S-104 01 Stockholm
Sweden

B. Vennström
European Molecular Biology Laboratory
Postfach 10.2209
D-6900 Heidelberg
F.R.G.

R. A. Weinberg
Whitehead Institute for Biomedical Research
Nine Cambridge Center
Cambridge, Massachusetts 02142
U.S.A.

B. Westermark
Department of Pathology
University Hospital
University of Uppsala
S-751 85 Uppsala
Sweden

B. M. Willumsen
Institute of Microbiology
University of Copenhagen
2A Øster Farimagsgade
DK-1353 Copenhagen K
Denmark

F. Wong-Staal
Laboratory of Tumor Cell Biology
National Cancer Institute
Building 37, Room 6A-09
Bethesda, Maryland 20205
U.S.A.

M. Yaniv
Unité des Virus Oncogènes
Département de Biologie Moléculaire
Institut Pasteur
25, rue du Dr. Roux
F-75724 Paris CEDEX 15
France

Introduction

N. O. Kjeldgaard[1] *& J. Forchhammer*[2]

It is now generally accepted that malignant diseases are caused by multiple damage events to the cells. This concept was first based on analysis of the age-specific incidence of cancers (Armitage & Doll 1961) and many carcinomas in humans show an exponential increase with age (Cook *et al.* 1969). Different estimates of the number of events leading to malignancies range between three and seven. Obviously, such estimates do not tell us anything about the individual events, whether they are always identical in the same malignancy or whether the events always occur in an obligatory order. The estimates also give no information as to whether all these events affects cellular functions in the progenitors to the eventual tumor cells or if some of the events are related to other cells or integral cell functions of the organism.

Recent progress in the study of the molecular biology of tumor viruses has identified several functional viral genes involved in tumor induction, and studies of acutely transforming retroviruses have revealed the presence of a number of oncogenes derived from cellular genes having defined functions in the normal cell (Bishop 1985). The activation of cellular proto-oncogenes in malignant cells is observed not only in viral-induced tumors, but also in tumors arising "spontaneously" or after chemical induction. These discoveries have focused emphasis on the dysregulation of normal cellular functions as entities in the train of events leading to cancer.

In planning the Symposium we felt that it would be important to place strong emphasis on the virus as a genetic element, which imposes changes in some regulatory circuits of the cell.

As a guiding principle, we wanted to follow a scheme where we start out from

1) Department of Molecular Biology and Plant Physiology, Aarhus University, DK-8000 Aarhus C, Denmark.

2) Department of Molecular Oncology, The Fibiger Institute, The Danish Cancer Society, DK-2100 Copenhagen Ø, Denmark.

VIRAL CARCINOGENESIS, Alfred Benzon Symposium 24,
Editors: N. O. Kjeldgaard, J. Forchhammer, Munksgaard, Copenhagen 1987.

the nucleus, analyzing the manner in which the virus affects the transcription of the cellular DNA, either by an action in *cis* on neighboring genes or by an action in *trans* on the expression of genes through regulatory factors. The dysregulation affects the expression of known or as yet unknown proto-oncogenes which subsequently exert their effects on cellular growth through signal transducing functions in the cellular membranes, in the cytoplasm and, eventually, inside the nucleus.

The Affectors:

cis-affectors	enhancers promotors	
trans-affectors	chromosomal locations, chromosomal translocations regulatory proteins	Nucleus

The Effectors:

mitogenic	viral & cellular protein kinases	Cytoplasm
products	signal transduction growth factors receptors	Membranes

We hope that the proceedings of the Symposium will reflect these ideas and illuminate some of the multiple events involved in carcinogenesis.

REFERENCES

Armitage, P. & Doll, R. (1961) Stochastic models for carcinogenesis. In: *Proc. Fourth Berkeley Symp. on Mathematical Statistical and Probability,* ed. Neyman, J., pp. 19–38. University of California Press, Berkeley and Los Angeles, California.

Bishop, J. M. (1985) Viral oncogenes. *Cell 42,* 23–38.

Cook, P. J., Coll, R. & Fellingham, S. A. (1969) A mathematical model for age distribution of cancer in man. *Int. J. Cancer 4,* 93–112.

I. Enhancers and Transcriptional Factors

Transcriptional Control and Oncogenicity of Murine Leukemia Viruses

F. S. Pedersen[1], *M. Etzerodt*[1], *S. Lovmand*[1], *H. Y. Dai*[1], *A. J. Bækgaard*[1], *J. Sørensen*[1], *P. Jørgensen*[1], *N. O. Kjeldgaard*[1], *J. Schmidt*[2], *C. Leib-Mösch*[2], *A. Luz*[2] & *V. Erfle*[2]

Replication competent murine retroviruses can induce various types of tumors upon inoculation into newborn mice (Teich *et al.* 1982, 1985). These viruses do not carry host-derived oncogenes, and a critical step in the oncogenic process seems to be proviral integration in specific regions of the host genome (Teich *et al.* 1985).

Individual viruses differ in their oncogenic potency and in their specificity of tumor induction. Well-known examples of such viruses are the Moloney murine leukemia virus, that induces mainly T-cell lymphomas, and Friend murine leukemia virus that induces erythroid leukemia. Analysis of recombinants between these two viruses has demonstrated that the specificity of tumor induction is determined by structures in the U3 region of the LTR containing the transcriptional enhancer (Chatis *et al.* 1983, 1984).

For some inbred strains of mice production of endogenous retroviruses leads to induction of neoplastic diseases at high frequencies. In the AKR strain the animals develop thymic lymphomas at very high frequency within the first year of life. In these animals, the endogenous, ecotropic virus, Akv, in combination with numerous endogenous proviral sequences gives rise to a complex family of viruses, showing variation in oncogenicity as well as in a number of other biological properties (Coffin 1982, Stoye & Coffin 1985). Preleukemic and leukemic AKR mice therefore provide a rich source of closely related viruses for genetic studies of the interaction of the viral genomes with their host. Genetic

[1]Department of Molecular Biology and Plant Physiology, University of Aarhus, DK-8000 Århus C, Denmark, and [2]Abteilung für Pathologie, Gesellschaft für Strahlen- und Umweltforschung, D-8042 Neuherberg, Federal Republic of Germany.

VIRAL CARCINOGENESIS, Alfred Benzon Symposium 24,
Editors: N. O. Kjeldgaard, J. Forchhammer, Munksgaard, Copenhagen 1987.

TABLE I

Oncogenic properties of AKR viruses. The viruses SL3-2 and SL3-3 were isolated from a lymphoid cell line established from a spontaneous AKR tumor (Pedersen et al. *1981). The Akv virus was derived by transfection of the molecular clone, λCl. 623 (Lowy* et al. *1980). All viruses were cultivated on NIH 3T3 cells. The oncogenic activity of each virus was determined by inoculation of 0.1 ml cell-free supernatant intraperitoneally into newborn mice. The SL3-2- and SL3-3-induced diseases were, in all cases, thymic lymphomas, whereas the Akv-induced diseases in NMRI/Nhg mice included B- and T-cell lymphomas as well as myelogenous and non-differentiated leukemias. The data for disease induction in C_3H_f/Bi mice are taken from Pedersen* et al. *(1981). In brackets are given the minimum and maximum days of survival of diseased animals as well as the number of diseased animals out of the total number of animals injected*

Virus	Induction in $C3H_f$/Bi, days	Induction in NMRI/Nhg, days
SL3-2	99 (84–114, 8/8)	167 (82–445, 6/7)
SL3-3	100 (73–117, 9/9)	118 (91–164, 9/9)
Akv	no induction	272 (180–489, 29/31)
Control littermates	no induction	no induction

mapping studies, using recombinants between the endogenous virus, Akv, and its more oncogenic derivatives have localized a major oncogenic determinant in the transcriptional enhancer of the LTR, whereas regions of the viral *env* gene also have some effect (DesGroseillers *et al.* 1983, 1984, Lenz *et al.* 1984, Holland *et al.* 1985).

The discovery of the significance of the U3 sequences for the oncogenicity led to the hypothesis that the cell type specificity of the transcriptional enhancer determines tumor induction. In support of this hypothesis, Celander & Haseltine (1984) found that the transcriptional enhancer of the potent lymphomagenic AKR virus, SL3-3, in contrast to the Akv enhancer element showed a strong preference for T-cells. This cell specificity implies that different viruses are adapted to use different cell type-specific factors for transciptional enhancement.

The studies presented in this communication were undertaken to study the function of *cis*-acting control elements in the viral LTRs and their interaction with cellular *trans*-factors in determination of transcriptional activity and cell type specificity.

ONCOGENIC PROPERTIES

AKR viruses

Table I shows the oncogenic properties of some of the more potent AKR viruses and of the endogenous Akv virus. The SL3-2 and SL3-3 viruses are derived from a cell line established from a lymphoma of an AKR mouse (Pedersen *et al.* 1981,

TABLE II

Disease pattern of osteoma-derived viruses in NMRI/Nhg mice. The OA MuLV complex represents viruses produced by C_3H fibroblasts infected with cell-free osteoma extracts (Schmidt et al. *1984). A molecular clone, OA MuLV Type I was derived from rat cells infected by the OA MuLV complex (Leib-Mösch* et al. *1986). NIH 3T3 cultures producing this virus were derived by transfection. Newborn animals were injected intraperitoneally with 0.1 ml cell-free supernatant. Data are taken from Leib-Mösch* et al. *(1986)*

Inoculum	Osteoma Animals	Osteoma Mean age, days	Osteopetrosis Animals	Osteopetrosis Mean age, days	Lymphoma Animals	Lymphoma Mean age, days
OA MuLV complex	4/22	389±33	14/22	393±75	19/22	347±77
OA $MuLV_R$ Type I	2/10	532±65	3/10	464±197	7/10	545±56
Control	0/50	–	0/50	–	4/50	365±157

1982). The table shows that these viruses induce lymphomas of the T-cell type in $C3H_f$/Bi and NMRI/Nhg mice with a latency period of 2–5 months. The Akv virus was previously found to be non-oncogenic, as shown in Table I for the $C3H_f$/Bi strain. Our more recent results, however, using mice of other strains, have led to the discovery that Akv virus induces hematopoietic tumors in some strains. In the NMRI/Nhg strain, as shown in Table I, the latency periods for Akv virus are longer than for the SL3-2 and SL3-3 viruses. Also, the Akv virus shows a broader disease specificity, yielding T- and B-cell lymphomas as well as myelogenous and non-differentiated leukemias.

Viruses derived from bone tumors

In our search for viruses of novel tissue specificities, we have begun an investigation of retroviruses associated with bone tumors. One source of such viruses are the spontaneous benign bone tumors that are observed at high frequency in some inbred strains of mice (Teich *et al.* 1982). The 101/Nhg strain shows an osteoma incidence of 70% in old female mice. These osteomas produce retroviruses (Schmidt *et al.* 1984) and we have obtained infectious molecular clones of viruses derived from this material. The disease pattern induced by the osteoma-derived viruses in NMRI/Nhg mice is shown in Table II. It can be seen that the virus induces not only osteomas but also osteopetrosis and malignant lymphomas with latency periods of 8–18 months.

NUCLEOTIDE SEQUENCE ANALYSIS OF U3 REGIONS

Since the U3 region of the LTR has been shown to carry determinants of the oncogenic and transcriptional specificities, it is of interest to compare the structure of this region for different virus isolates. For our studies of both the lymphoma-associated and the bone tumor-associated viruses, we use the Akv nucleotide sequence as a prototype (Etzerodt *et al.* 1984).

Lymphoma-associated viruses

Previous studies have demonstrated the presence of repeated elements in the region of the murine leukemia virus LTRs associated with enhancer function (Weiss *et al.* 1985). Figure 1 shows a comparison of the nucleotide sequence in this region for Akv, SL3-2, SL3-3, and Gross A/NIH virus, a serially transmitted lymphomagenic AKR virus (Buchhagen *et al.* 1980, Villemur *et al.* 1983). Whereas the Akv virus contains a perfect 99 bp non-overlapping, tandem repeat in this region, the SL3-2 and SL3-3 viruses show a more complex pattern of repeated elements with repetitions of some parts 2 times and other parts 3 times. The Gross A virus contains a somewhat shorter sequence with only minor repetitions. As has been found previously by Lenz *et al.* (1984) for SL3-3- and by Villemur *et al.* (1983) for Gross A, the regions can be considered as consisting of a small number of repeat elements occurring in different numbers and relative order. Our recent results show that the SL3-2 nucleotide sequence fits into this general pattern. As shown in Figures 1 and 2, the repeat elements show only minor variation between the individual viruses. It should also be noted that a 16 nucleotide element found about 70 nucleotides upstream of the repeat area is found in complete form with one or two point differences in the repeat segment region of both the SL3-2 and the Gross A/NIH sequences. Part of this sequence element is also found repeated in all the viruses shown.

Characteristic for these thymic lymphoma-associated viruses relative to Akv are point differences in the B and C elements as shown in Figure 2.

Bone tumor-associated viruses

Nucleotide sequence analysis of the LTR of the osteoma-derived virus, OA MuLV$_R$ Type I (Leib-Mösch *et al.* 1986) showed that this region is very similar to Akv virus. Also for this virus, the main difference is found in the repeat segment area associated with enhancer function. This sequence contains a 46 bp imperfect tandem repeat element (D, E, and A, Figure 1b).

Comparison of the nucleotide sequence with other previously determined nucleotide sequences of murine leukemia virus LTRs led to the discovery that the OA MuLV$_R$ type I sequence is closely related to the sequence of FBJ MuLV, a virus isolated from a spontaneous osteosarcoma of a CF-1 mouse (Finkel *et al.* 1966,

Van Beveren *et al.* 1983). The FBJ MuLV is distinguished from the OA virus by having 3 copies of the 46 bp repeat (Figure 1b). also, the point differences from

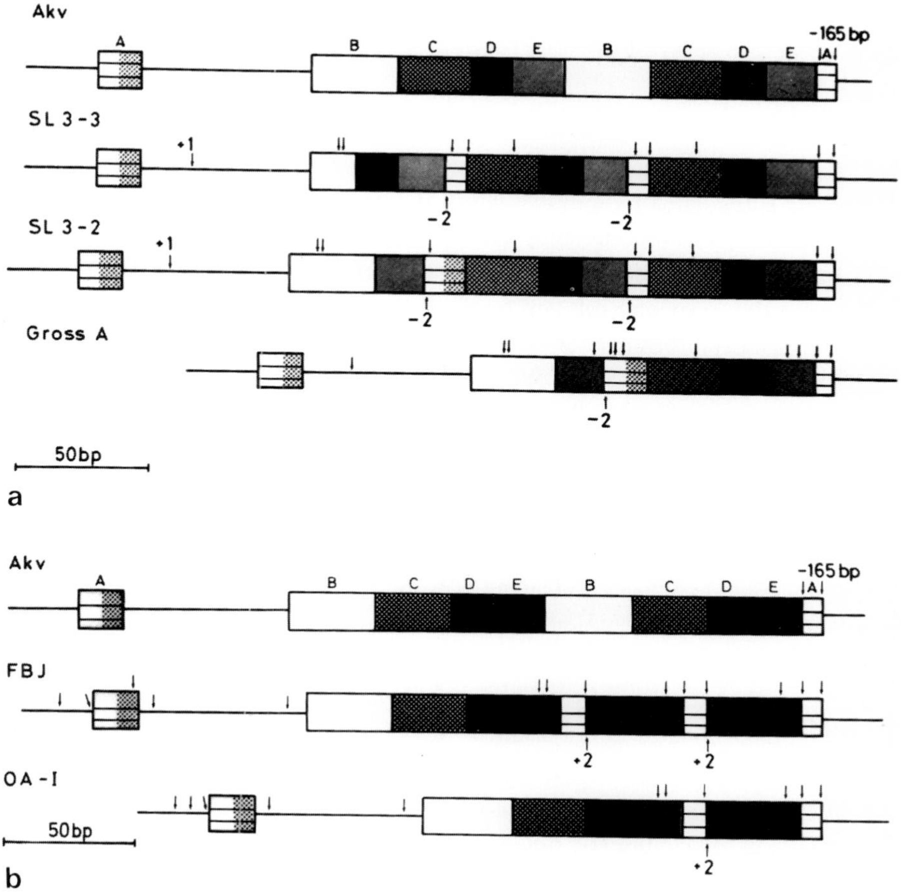

Fig. 1. Repeat segment structures in U3 regions of murine leukemia viruses. Panel a) Akv and T-cell lymphoma-derived viruses. Panel b) Akv and bone tumor-derived viruses. The 165 base pairs adjacent to the cap-site are not included in the figure. These nucleotides, as well as the R and U5 regions, are identical for all viruses. A, B, C, D, and E represent repeat segments. The arrows indicate point differences relative to Akv virus. The +1 and −2 marks indicate that an extra nucleotide is present or 2 nucleotides are missing, respectively, relative to the Akv prototype. The SL3-2 nucleotide sequence was determined by direct dideoxy sequencing of RNA using chemically synthesized oligodeoxynucleotides as primers for reverse transcriptase and was subsequently confirmed by analysis of molecular clones of cDNA copies of the viral RNAs. The remianing nucleotide sequences were published previously as listed. Akv (Van Beveren *et al.* 1982, Etzerodt *et al.* 1984, Herr 1984), SL3-3 (Lenz *et al.* 1984), Gross A/NIH (Villemur *et al.* 1983), FBJ MuLV (Van Beveren *et al.* 1983), OA MuLV$_R$ Type I (Leib-Mösch *et al.* 1984).

the Akv sequence found in the E segments are shared between OA MuLV and FBJ MuLV as shown in Figure 2.

ANALYSIS OF TRANSCRIPTIONAL CONTROL REGIONS

What is the biological significance of the variant segment structure of the murine leukemia virus enhancer region? We propose that the segments represent recognition sites for regulatory nuclear proteins. This proposal is in accordance with the recent findings that the transcriptional activity of a number of eucaryotic promoters is regulated by multiple interactions with nuclear proteins (Dynan & Tjian 1985). For the present study such a model would imply that the transcriptional activity of a virus in a given cell type is determined by the pattern of these proteins in this particular cell and by the exact segment structure of the virus.

To further explore this possibility we want to localize the functional regions of the different LTR elements. We have therefore developed a general strategy for construction of random deletion and duplication mutants of these regions.

Strategy of in vitro *mutagenesis*

The method is illustrated in Figure 3a. The *E. coli* plasmids paL and ptLPCA both contain one copy of Akv LTR. The plasmid paL also contains a functional ampicillin resistance gene and a deleted non-functional tetracycline resistance gene. The plasmid ptLPCA contains a functional tetracycline resistance gene and a functional chloramphenicol acetyl transferase gene (CAT gene). The CAT gene is inserted in a eukaryotic transcriptional unit with the LTR serving as transcriptional initiator and a segment of the SV40 early region containing the polyadenylation signal. The *Pst*I recognition site in the U3 region and the *Kpn*I site in the R region are unique in both plasmids.

Generation of deletion and duplication mutants is accomplished in the following manner:

a) paL is digested with *Kpn*I, then with the exonuclease *Bal* 31, and finally with *Sal*I that cleaves once in the *tet* gene. The large fragments carrying variable deletions from the *Kpn*I site and upstream into the U3 region are gel-purified.
b) ptLPCA is digested with *Pst*I, then with *Bal* 31, and finally with *Sal*I. The large fragments containing variable deletions from the *Pst*I site and downstream in the U3 region are gel-purified.
c) The two gel-purified DNA samples are combined, ligated, and ampicillin- and tetracycline-resistant colonies selected after transformation of bacteria.

The structure of the mutated region in the plasmids is determined by direct

NUCLEOTIDE SEQUENCE OF U3 SEGMENTS:

Akv; SL3-3; SL3-2; Gross A; FBJ; OA-I

Segment B

CAGAGAGGCTGGAAAGTACCGGGACTAGGGCCAA	
CAGAGAGGCTAAAAAGTACCGGGACTAGGGCCAA	SL3-2
CAGAGAGGCTAAAAAGTA	SL3-3

Segment C

ACAGGATATCTGTGGTCAAGCACTAGGGC	
ACAGGATATCTGTGGTTAAGCACTAGGGC	SL3-3 1,2; SL3-2 1,2; GrossA

Segment D

CCCGGCCCAGGGCCAA

NUCLEOTIDE SEQUENCE OF U3 SEGMENTS:

Akv; SL3-3; SL3-2; Gross A; FBJ; OA-I

Segment E

GAACAGATGGTCCCCAGAAA	
GAACAGACGGTTCCCAGAAA	GrossA 1
GAACAGATGATTCCCAGAAA	FBJ 1; OA-I 1
GAACAGATGGTTCCCAGAAA	FBJ 2,3; OA-I 2
GAACAGATGGTCCCCAGA	SL3-3 1,2; SL3-2 1,2
GAACAGATGGTCTCCAGA	GrossA 2

Segment A

CAGCTAACTGCAGTAACG	
CAGCTAACTGCAGTAATG	FBJ 1
CCGCTAACTGCAGTAACG	SL3-2 2
CCACTCACTGCAGTAACG	GrossA 2
CCGCTAACG	SL3-3 2,3
CAGCTAAA	FBJ 2; OA-I 2
TAGCTAAA	terminals; FBJ 3

Fig. 2. Nucleotide sequence of repeat segment structures. For each segment the first sequence correponds to Akv. Only the nucleotide sequences of those segments differing from the Akv prototype are listed.

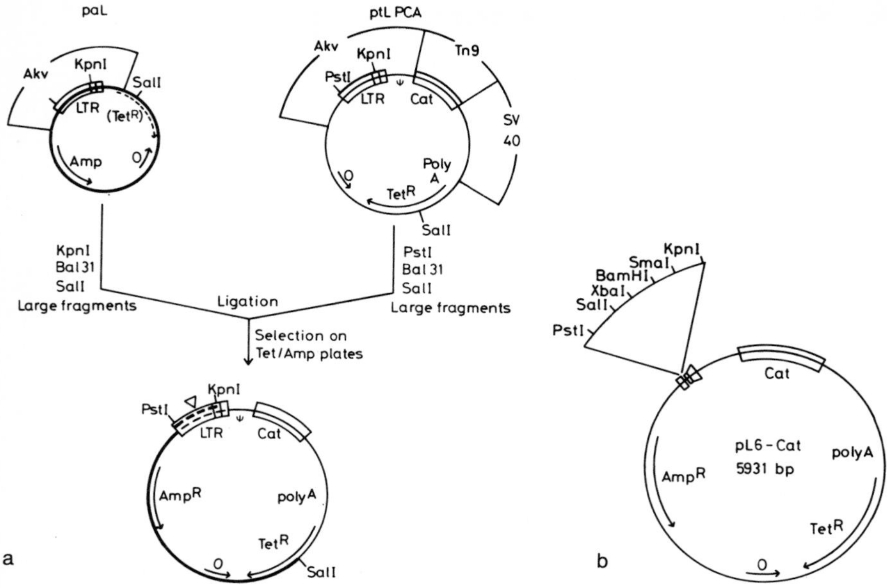

Fig. 3. a) Construction of deletion and duplication mutants of the LTR. The principles of the method are given in the text. A detailed description of the plasmids and the mutagenesis strategy will be presented elsewhere. b) The structure of plasmid pL6-Cat.

dideoxysequencing using a panel of synthetic primers. The transcriptional activity of the different plasmids in transient expression assays can be assessed by measurements of chloramphenicol acetyl transferase levels.

A plasmid, pL6-cat (Figure 3b) contains a polylinker region inserted between *Pst*I and *Kpn*I. This plasmid is used as a negative control in transient expression studies and as transient expression vector for insertion of variant U3 sequences.

Analysis of deletions and duplications of the Akv LTR

A panel of mutants of the Akv LTR was chosen for transient expression studies in NIH 3T3 fibroblast cells (Figure 4a). The panel includes two short duplication mutants and a series of longer deletion mutants. Whereas the two duplication mutants yield transient expression values similar to the non-mutated construct, the deletion mutants gives values ranging from full activity to 3–5%, which is background levels.

Interpretation of data of this sort should be performed with some caution since the extension of the deletions vary at both ends. Also, verification of the use of

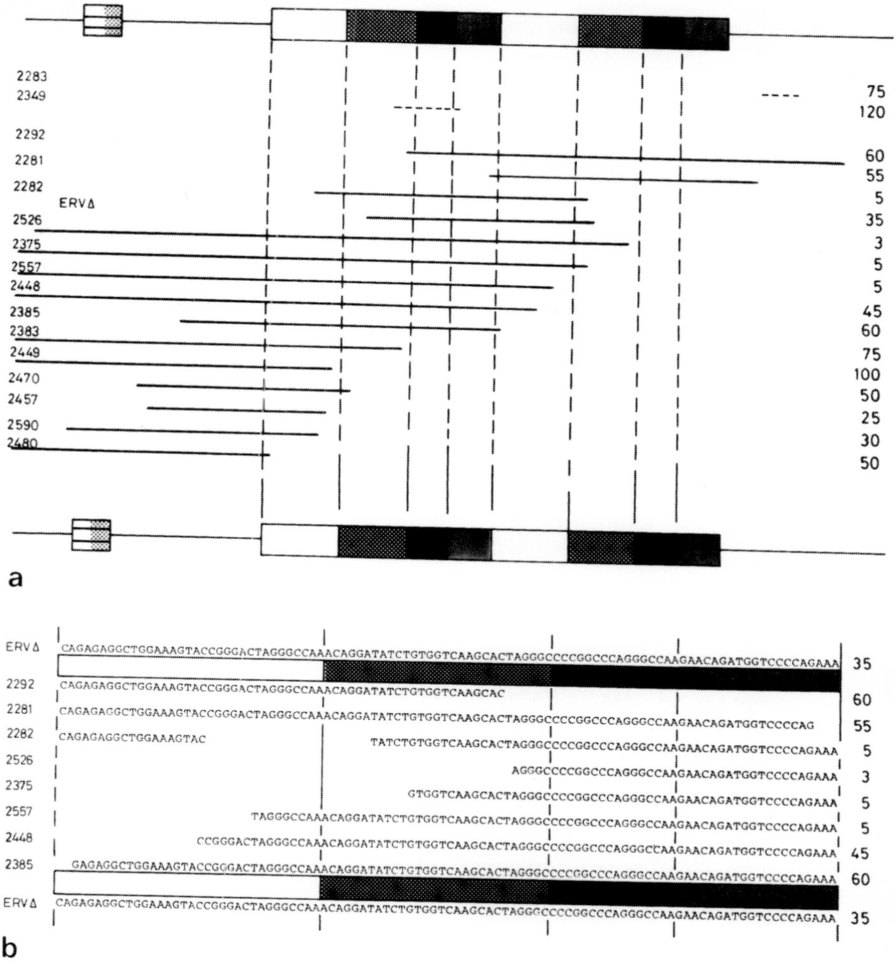

Fig. 4. Transient expression activity of deletion and duplication mutants of Akv LTR in NIH 3T3 cells. The deletion mutants (——) and duplication mutants (----) were generated by the method illustrated in Figure 3, with the exception of ERV*Δ*, in which one of the two 99 bp repeats was removed by digestion with EcoRV and religation. The transient expression activity in NIH3T3 cells was determined by $CaPO_4$ transfection and measurement of chloramphenicol acetyl transferase activity in extracts of cells harvested 48 h after transfection essentially as described by Gorman *et al.* (1982). The values, representing the average of at least 6 transfections are given relative to the values for the intact Akv LTR (=100). The background values obtained by transfection of pL6-cat DNA, carrier DNA or from untransfected cultures, all give values ranging from 1 to 5. Panel a) shows the structure for the mutants relative to the intact Akv U3 region. Panel b) shows the structure of mutants missing more than one copy of the 99 bp repeat. The structure of these mutants is shown relative to a U3 region carrying one copy of the Akv 99 bp repeat.

the correct transcriptional initiation site will be required before final conclusions can be drawn from these data.

From inspection of the results (Figure 4b) we note the following:

a) *Duplication mutants.* Two mutants, 2283 and 2349, containing duplications of short regions show little or no variation from non-mutated Akv LTR.

b) *Deletions starting upstream and extending into the segment repeat area.* All deletions retaining one complete copy of the repeat segment give significant expression values. However, differences between some of these mutants (e.g. 2457 giving 25% versus 2449 giving 100%) may point to effects of the upstream sequences or to effects of the exact structure of the repeat segments retained. This phenomena has not been analyzed further.

c) *Deletions starting downstream and going into the segment repeat area.* Two mutants missing more than one copy of the repeat (2292, 2281) have retained activity.

d) *Deletions missing more than one copy of the repeat.* All the low transient expression values are obtained with plasmids of this structure. Comparison of values from mutants of this structure is shown in Figure 4b, using an Akv LTR with one repeat as prototype. The values can be considered as falling into two groups. One group is represented by the mutant containing one perfect repeat copy (ERVΔ) and deletion mutants 2292, 2281, 2448, and 2385, all giving around 50% of the value obtained for the intact LTR. The second group of deletion mutants (2282, 2526, 2375, and 2557) all give values close to the background.

The results show that an Akv LTR element carrying only one copy of the 99 bp element (B, C, D, E) still gives significant transient expression values in NIH 3T3 cells. Within this structure both E, D, and part of C, seem dispensable in this assay (mutant 2292). Similarly the upstream part of element B is not absolutely required (mutant 2448).

The 39 bp segment (part of B, part of C) between these two deletion mutants seems to contain a positive transcriptional element. All mutants missing part of this segment (2282, 2526, 2375, 2557) give background levels of expression, whereas all mutants retaining this segment (2292, 2281, 2448, 2385) give full activity. The difference between mutants 2448 and 2557 may indicate a critical function of the 7 bp sequence, CCGGGAC, in element B. It has not yet been analyzed, however, if the 39 bp segment is sufficient for activity, i.e. if a mutant carrying both the deletion of mutant 2448 and the deletion of mutant 2292 is transcriptionally active.

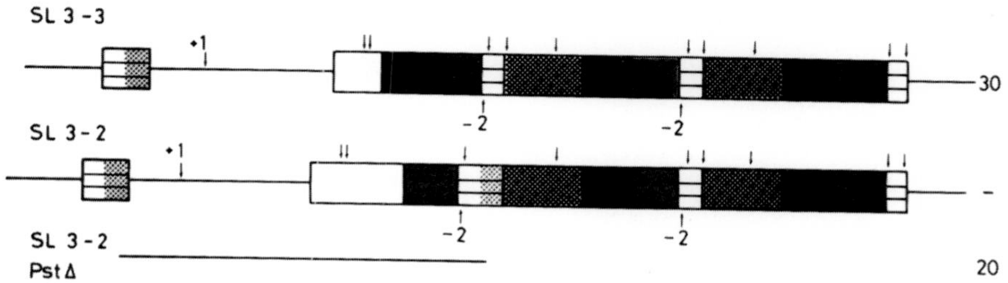

Fig. 5. Transient expression activity of LTR elements of thymic lymphoma-derived virus in NIH 3T3 cells. The *Pst*I-*Kpn*I fragments of SL3-3 (from a molecular clone kindly provided by Dr. J. Lenz) and SL3-2 were inserted into pL6-Cat to give the transient expression vectors with the indicated LTR structures. The measurements were performed as described in the legend to Figure 4. The extent of the deletion in SL3-2 PstΔ is given by the horizontal line.

Analysis of variant LTR elements

Identification of nucleotides within the 39 bp segment of the Akv LTR as being critical for transient expression in NIH 3T3 cells led us to ask if a similar segment is also critical for expression of the variant viruses in these cells. Inspection of the repeat segment structures in the T lymphoma-derived viruses shows that this segment is not found in intact form in these viruses: The SL3-3 structure contains only the upstream part of the B element followed by elements D and E rather than by C. The SL3-2 and Gross A structures contain a complete B element followed by element E.

In our initial expression studies of these elements in NIH3T3 cells we find, in accordance with Celander & Haseltine (1984), that the SL3-3 LTR gives a value of about 30% of the Akv LTR (Figure 5). We also find that a mutant of the SL3-2 LTR, named SL3-2 PstΔ, carrying a deletion that covers the B element and adjacent sequences gives about 20% of the Akv value.

Taken together with the deletion analysis, these data imply that the 39 bp segment (and in particular the 7 bp sequence CCGGGAC) containing nucleotides indispensable for the Akv LTR driven expression is not required for expression from the variant LTR elements, SL3-3 and SL3-2. These LTR elements must therefore contain positive *cis*-acting transcriptional control elements of a structure different from those used in Akv. These elements may represent recognition sites for different regulatory proteins in NIH 3T3 cells or variant recognition sites for the same protein. It is of immediate interest to analyze the activity of the *cis*-elements in different cell types and to characterize the host cell factors responsible for this activity. We expect that such studies will contribute significantly to our understanding of mammalian gene regulation as well as retroviral oncogenesis.

REFERENCES

Buchhagen, D. L., Pedersen, F. S., Crowther, R. L. & Haseltine, W. A. (1980) Most sequence differences between the genomes of the AKV virus and the *in vitro* passaged, leukemogenic Gross A virus are located near the 3′ terminus. *Proc. Natl. Acad. Sci. (USA) 77*, 4354–64.

Celander, D. & Haseltine, W. A. (1984) Tissue-specific transcription preference as a determinant of cell tropism and leukaemogenic potential of murine retroviruses. *Nature 312*, 159–63.

Chatis, P. A., Holland, C. A., Hartley, J. W., Rowe, W. P. & Hopkins, N. (1983) Role for the 3′ end of the genome in determining disease specificity of Friend and Moloney murine leukemia viruses. *Proc. Natl. Acad. Sci. (USA) 80*, 4408–11.

Chatis, P. A., Holland, C. A., Silver, J. E., Frederickson, T. N., Hopkins, N. & Hartley, J. W. (1984) A 3′ end fragment encompassing the transcriptional enhancers of nondefective Friend virus confers erythroleukemogenicity on Moloney leukemia virus. *J. Virol. 52*, 248–54.

Coffin, J. (1982) Endogenous viruses. In: *RNA tumor viruses*, eds. Weiss, R., Teich, N., Varmus, H. & Coffin, J., pp. 1109–1204. Cold Spring Harbor Laboratory, Cold Spring Harbor, New York.

DesGroseillers, L., Rassart, E. & Jolicoeur, P. (1983) Thymotropism of murine leukemia virus is conferred by its long terminal repeat. *Proc. Natl. Acad. Sci. (USA) 80*, 4203–7.

DesGroseillers, J., Villemur, R. & Jolicoeur, P. (1984) The high leukemogenic potential of Gross passage A murine leukemia virus maps in the region of the genome corresponding to the long terminal repeat and to the 3′ end of *env*. *J. Virol. 47*, 24–32.

Dynan, W. S. & Tjian, R. (1985) Control of eucaryotic messenger RNA synthesis by sequence-specific DNA-binding proteins. *Nature 316*, 774–8.

Etzerodt, M., Mikkelsen, T., Pedersen, F. S., Kjeldgard, N. O. & Jørgensen, P. (1984) The nucleotide sequence of the Akv murine leukemia virus genome. *Virology 134*, 196–207.

Finkel, M. P., Biskis, B. O. & Jinkins, P. B. (1966) Virus induction of osteosarcomas in mice. *Science 151*, 698–701.

Gorman, C. M., Moffat, L. F. & Howard, B. H. (1982) Recombinant genomes which express chloramphenicol acetyltransferase in mammalian cells. *Mol. Cell. Biol. 2*, 1044–51.

Herr, W. (1984) Nucleotide sequence of AKV murine leukemia virus. *J. Virol. 49*, 471–8.

Holland, C. A., Hartley, J. W., Rowe, W. P. & Hopkins, N. (1985) At least four viral genes contribute to the leukomegenicity of murine retrovirus MCF 247 in AKR mice. *J. Virol. 53*, 158–65.

Leib-Mösch, C., Schmidt, J., Etzerodt, M., Pedersen, F. S., Hehlmann, R. & Erfle, V. (1986) Oncogenic retrovirus from spontaneous murine osteomas II. Molecular cloning and genomic characterization. *Virology 150*, 96–105.

Lenz, J., Celander, D., Crowther, R. L., Patarca, R., Perkins, D. W. & Haseltine, W. A. (1984) Determination of the leukemogenicity of a murine retrovirus by sequences within the long terminal repeat. *Nature 295*, 467–70.

Lowy, D. R., Rands, E., Chattopadhyay, S. K., Garon, S. K. & Hager, G. L. (1980) Molecular cloning of infectious integrated murine leukemia virus DNA from infected mouse cells. *Proc. Natl. Acad. Sci. (USA) 77*, 614–8.

Pedersen, F. S., Crowther, R. L., Hays, E. F., Nowinski, R. C. & Haseltine, W. A. (1982) Structure of retroviral RNAs produced by cell lines derived from spontaneous lymphomas of AKR mice. *J. Virol. 41*, 18–29.

Pedersen, F. S., Crowther, R. L., Tenney, D. Y., Reimold, A. & Haseltine, W. A. (1981) Novel leukemogenic retroviruses isolated from a tumour cell line derived from a spontaneous AKR tumour. *Nature 292*, 167–70.

Schmidt, J., Erfle, V., Pedersen, F. S., Rohmer, H., Schetters, H., Marquart, K.-H. & Luz, A. (1984) Oncogenic retroviruses from spontaneous murine osteomas I. Isolation and biological characterization. *J. Gen. Virol. 65*, 2237–48.

Stoye, J. & Coffin, J. (1985) Endogenous viruses. In: *RNA tumor viruses 2/Supplements and appendixes*, eds. Weiss, R., Teich, N., Varmus, H. & Coffin, J., pp. 337–405. Cold Spring Harbor Laboratory, Cold Spring Harbor, New York.

Teich, N., Wyke, J., Kaplan, P. (1985) Pathogenesis of retrovirus-induced disease. In: *RNA tumor viruses 2/Supplements and appendixes*, eds. Weiss, R., Teich, N., Varmus, H. & Coffin, J., pp. 187–248. Cold Spring Harbor Laboratory, Cold Spring Harbor, New York.

Teich, N., Wyke, J., Mak, T., Bernstein, A. & Hardy, W. (1982) Pathogenesis of retrovirus-induced disease. In: *RNA tumor viruses*, eds. Weiss, R., Teich, N., Varmus, H. & Coffin, J., pp. 785–998. Cold Spring Harbor Laboratory, Cold Spring Harbor, New York.

Van Beveren, C., Rands, E., Chattopadhyay, S. K., Lowy, D. R. & Verma, I. M. (1982) Long terminal repeat of murine retroviral DNAs: Sequence analysis, host-proviral junctions, and preintegration site. *J. Virol. 41*, 542–56.

Van Beveren, C., Van Straaten, F., Curran, T., Müller, R. & Verma, I. M. (1983) Analysis of FBJ-MuSV provirus and c-fos (mouse) gene reveals that viral and cellular fos gene products have different carboxy termini. *Cell 32*, 1241–55.

Villemur, R., Rassart, E., DesGroseillers, L. & Jolicoeur, P. (1983) Molecular cloning of viral DNA from leukemogenic Gross passage A murine leukemia virus and nucleotide sequence of its long terminal repeat. *J. Virol. 45*, 539–46.

Weiss, R., Teich, N., Varmus, H. & Coffin, J., eds. (1985) *RNA tumor viruses 2/Supplements and appendixes*. Appendix A, pp. 765–805. Cold Spring Harbor Laboratory, Cold Spring Harbor, New York.

DISCUSSION

VENNSTRÖM: I have a question regarding your osteopetrosis viruses. In the avian system there are a number of osteopetrosis viruses which do not seem to integrate at a common locus, and they carry no oncogenes. Rather, in these tumors there is a large number of unintegrated circular DNAs that seem to replicate. Is that also the case with the murine viruses?

SKOU PEDERSEN: We do not have any results from our work yet. It is, strictly speaking, hard to work with bones. We are doing various things trying to immortalize the cells from the osteomas in culture to look at their viral pattern. We also look at various tissue culture systems, hopefully enabling us to dissect these phenomena. But I am not able to give you an answer.

GALLO: They are not really tumors, are they?

VENNSTRÖM: Not really, strictly speaking.

GALLO: Even less than strictly speaking; are they really tumors?

SKOU PEDERSEN: No, not the osteopetroses, but here we are also talking about osteomas which are benign bone tumors.

GALLO: How often do you find specific integrations in mouse tumors that are not laboratory-induced. Let's say in any wild type animal tumors that are induced by virus or even with viral-induced tumors in laboratory experiments? Do you routinely find specific integration areas?

SKOU PEDERSEN: We have not done such an analysis, but looking at the literature, it is obvious that the analysis of murine system was somewhat delayed relative to the avian and other systems because of the difficulties in hybridization analysis of integrating viruses. I think it is still not clear whether some negative cases may be due to variant viruses that we have not been able to detect with specific hybridization analyses. I think we should be cautious about this interpretation.

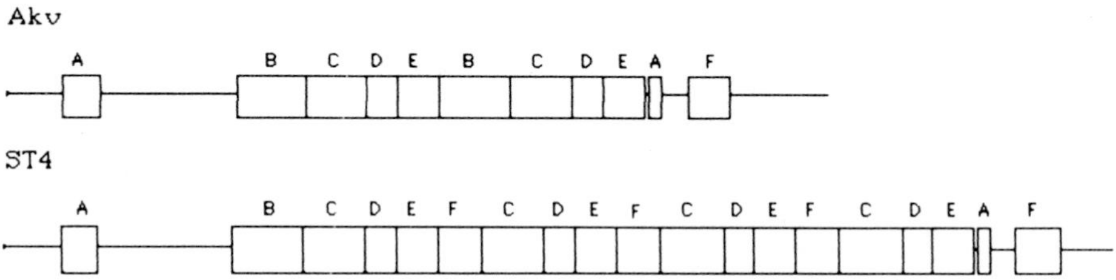

Fig. A.

RAPP: In the cases in the mouse where the virus has been found to activate an oncogene, can you recall what the sequence of the LTR looked like, after the virus had actually activated an oncogene?

SKOU PEDERSEN: There are different examples. Some have MCF-like structure, which is somewhat mutated relative to the sequences I have shown you here. There are also examples of ecotropic viral sequences. There is one example (Corcoran *et al.* 1984, *Cell 37*, 113) of a structure having a triplicate repeat structure (Fig. A). In fact, that virus has a repeat segment containing three copies of an element (F), that in Akv is found downstream of the repeat area. So, it seems as if, in analogy with the A element located upstream of the repeat area, there are also elements from further downstream that in some cases may be used to generate these viruses. What it actually means for the transcription phenotype of these elements is not clear to me in this case.

RAPP: So, in these cases where the LTR of an integrated virus has activated an oncogene, that LTR has not been taken out to ask the question as to what its biological activity would be when it would be presented to the animal as part of a virus again?

SKOU PEDERSEN: I have not done it, but some transient expression analyses have been carried out. I am not aware of any complete viruses being reconstituted, and that question asked.

WILLUMSEN: Presumably, these AKR viruses are generated by recombination of endogenous sequences. What is the current thought on at what level this recombination takes place? Is it on the RNA level during reverse transcription, or is it at the DNA level? Are there any hard data addressing this problem?

SKOU PEDERSEN: This has been a matter of controversy for a long time, even doing the experiments in culture, where you have a chance of following the process more closely. It is even more difficult to study in the animal. My belief is that to get a recombination with the endogenous provirus sequence requires that these endogenous elements are transcriptionally activated. But that does not tell you if it will occur by co-packaging of RNA or on some other stage of the replication process.

CORY: Maybe I could briefly summarize our results with AKR T-lymphomas. Some 25% of the tumors carried a viral insertion near the *myc* oncogene. We isolated one of those and sequenced it, and that is the one that Dr. Pedersen was talking to you about. In all the other cases, as judged by restriction endonuclease analysis, the inserts near the *myc* gene seemed to be of the recombinant MCF type. Another subset of AKR lymphomas ($\sim$10%) which we analyzed involved insertion of a recombinant MCF virus in a locus which we have termed *pvt*-t. The locus most frequently involved in the variant 6 is translocation in mouse plasmacytomas.

BERNS: It is still not clear to me, in different diseases induced by OA virus, if there are different LTRs related to the different diseases. Because, as you mentioned, the incoming virus might easily be changed, and the LTR might be mutilated. If you look at the different tumor tissues, would you find a LTR which is different from the ingoing virus?

SKOU PEDERSEN: I completely agree with yout that one should not interpret that data to mean that in the tumors you will necessarily find exactly the same structure of the LTR. For the OA virus the osteomas are very difficult to work with, because they are difficult to take the DNA out of. We have done some restriction analyses of the viruses in the lymphomas, and they seem to be similar to incoming virus, but not to the point of resolution where one can tell about minor differences in LTR structures or otherwise. For instance in the Akv virus-induced tumors which I showed you, do the kinds of structures you find in tumors represent recombination selection, or do they represent Akv virus still

```
       10        20        30        40        50        60        70        80
AATGAAAGACCCCTTCATAAGGCTTAGCCAGCTAACTGCAGTAACGCCATTTTGCAAGGCATGGAAAATACCAGAGCTG

       90       100       110       120       130       140       150       160
ATGTTCTCAGAAAAACAAGAACAAGGAAGTACAGAGAGGCTGGAAAGTACCGGGACTAGGGCCAAACAGGATATCTGTGG

      170       180       190       200       210       220       230       240
TCAAGCACTAGGGCCCCGGCCCAGGGCCAAGAACAGATGGTCCCCAGAAATAGCTAAAACAACAACAGTTTCAAGAGACC

      250       260       270       280       290       300       310       320
CAGAAACTGTCTCAAGGTTCCCCAGATGACCGGGGATCAACCCCAAGCCTCATTTAAACTAACCAATCAGCTCGCTTCTC

      330       340       350       360       370       380   ↓   390       400
GCTTCTGTACCCGCGCTTATTGCTGCCCAGCTCTATAAAAAGGGTAAGAACCCCACACTCGGCGCGCCAGTCCTCCGATA

      410       420       430       440       450       460       470       480
GACTGAGTCGCCCGGGTACCCGTGTATCCAATAAAGCCTTTTGCTGTTGCATCCGAATCGTGGTCTCGCTGATCCTTGGG

      490       500       510       520
AGGGTCTCCTCAGAGTGATTGACTGCCCAGCCTGGGGGTCTTTCATT
```

Fig. B.

being present? If the former was the case, that might be very useful to try to select for elements of a somewhat different structure and possible specificity.

BERNS: In the analysis of tumors similar to what Suzanne Cory just mentioned we also see that if you have activated oncogenes by proviral insertion you generally see a difference in the LTR. In most cases a duplication. For one or another reason there is preference in the activation event for an altered LTR and in general with Moloney there is a duplicated or triplicated enhancer sequence. This fits with the idea that you need this alteration for the activation.

MATTHIAS: Did you actually look for known consensus sequences within these U3 regions?

SKOU PEDERSEN: The figure B shows the occurrence of consensus sequences of transcriptional control elements within the LTR of Akv and MuLV. The sequence shown contains only one copy of the 99 bp tandem repeat (indicated by the bracket).

The boxes represent the following homologies:

a) Glucocorticoid responsive elements (consensus AG$^{A}/_{T}$CA(G)$^{A}/_{T}$), positions 100–107, 190–198;
b) Adenovirus 5 core enhancer I (consensus $^{A}/_{C}$GGAAGTG$^{A}/_{G}$) positions 106–112;
c) SV40 core enhancer (consensus GTGG$^{A}/_{T}$$^{A}/_{T}$$^{A}/_{T}$G) positions 121–127, 157–165.

The CCAAT box (pos. 303–307) and the TATA box (pos. 354–360) are underlined and the cap-site (pos. 384) marked by an arrow.

The 39 bp fragment identified by the deletion analysis covers position 129–168. This region contains an element homologous to the SV40 core enhancer, which may be of relevance for our observations. On the other hand, detailed inspection of our data led us to point to nucleotides 129–136 as important. These may possibly be part of another *cis*-element.

GALLO: To try to put all this together and go back to the biology of development of the malignancy, could one speculate that maybe, if you replicate enough, it does not matter what kind of enhancer you have, as you replicate and have random integrations, eventually one may occur somewhere near the right cellular gene important for proliferation, but in your case you have a more efficient mechanism. Integration does not have to occur so presicely, therefore there can be less replication?

SKOU PEDERSEN: Yes. This would fit my point of view.

GALLO: It would be of interest to see substantial data correlating incidence of malignancy, distance of the provirus integration site to the gene suspected to be important to cell growth, and the nature of the enhancer sequence.

SKOU PEDERSEN: I do not think that there is data to do this kind of calculation from the actual measurements at the moment. But that is a kind of working hypothesis which I like.

YANIV: I think that there is more and more evidence that this kind of change will modify the host range of the virus, the type of cells that the virus will infect.

GALLO: Is that true here in these cases? Are the cell type or the state of differentiation of the cell that is infected altered in any way?

SKOU PEDERSEN: There are some data on the site of integration that may tell something about the type of cell and the kind of selective process. The loci activated in lymphomas induced by some viruses seem to be different from these activated in spontaneous AKR lymphomas, so somewhat different types of viruses may actually lead to activation of different cellular genes. This may be taken as one indication of differences in either cell type or distance of activation or possibly other parameters also.

YANIV: I think Ostertag's lab (personal communication) has nice data on selecting variants of retroviruses that will infect, for example, embryonal cells. He showed that deletions and rearrangements occur in the LTR of these viruses.

Determinants of Cell Type Specificity within the Immunoglobulin Heavy Chain Gene Enhancer

Patrick Matthias, Thomas Gerster, Markus Thali, Dirk Bohmann[+], Walter Keller[+], Werner Zürcher, Joe Jiricny* & Walter Schaffner*

Analysis of the DNA sequences regulating the rate of transcription of eukaryotic genes has revealed two classes of cis-acting regulatory sequences, namely promoters and enhancers. Promoters are located within about 100 bp upstream of the mRNA start site and usually contain the so-called TATA box, as well as one or two upstream elements (reviewed in Dynan & Tjian 1985). The relative location as well as the spacing of the various elements of the promoter are important for efficient transcription to take place; for instance, some upstream elements can be inverted without deleterious effect on transcription (Everett *et al.* 1983) but they cannot be moved away from the rest of the promoter (E. Serfling, M. Jasin and W. Schaffner, unpublished).

A second set of DNA sequences has been identified which can increase the transcription of linked genes up to 2 orders of magnitude, and this essentially independently of their position (relative to the promoter) or orientation. These DNA sequences, which can be moved several kb upstream or downstream of the gene's promoter and still act, have been called enhancer sequences and were first identified in the genome of simian virus 40 (SV40); (Banerji *et al.* 1981, Moreau *et al.* 1981). Subsequently, enhancers were also discovered in several other viruses as well as in the vicinity of various cellular genes (for review, see Yaniv 1984, Schaffner 1985). In its natural context the enhancer is usually found 150–300 bp upstream of the transcription start site, but it can also be located 3' to the initiation site (such as in the Ig genes, Banerji *et al.* 1983, Gillies *et al.* 1983, or

Institut für Molekularbiologie II, der Universität Zürich, CH-8093 Zürich, Switzerland, [+]Deutsches Krebsforschungszentrum Im Neuenheimer Feld, D-69 Heidelberg, FRG, and *Friedrich-Miescher Institut, P.O. Box 2543, CH-4002 Basel, Switzerland.

VIRAL CARCINOGENESIS, Alfred Benzon Symposium 24,
Editors: N. O. Kjeldgaard, J. Forchhammer, Munksgaard, Copenhagen 1987.

in bovine papilloma virus, Lusky *et al.* 1983). Cellular genes can also harbor enhancer elements which often confer inducibility or tissue-specificity to the gene they control (reviewed by Serfling *et al.* 1985).

Intensive investigation during the past few years suggests that enhancer and promoter elements are, at least in some cases, overlapping physically and functionally. Enhancers seem to have a modular structure, being composed of an array of short conserved sequence motifs ("modules" or "elements"; Serfling *et al.* 1985), which are binding specific trans-acting factors (Wildeman *et al.* 1984). Inducibility and tissue specificity are most probably governed by the differential availability and activity of these regulatory factors. As a model system for the enhancer mechanism and the tissue specificity we have chosen to study the mouse Ig heavy chain gene enhancer. Ig genes consist of multiple gene segments. In the course of B cell differentiation a functional Ig heavy chain gene is assembled by recombining one from several hundred variable (V) segments to a diversity (D) and a joining (J) segment. This VDJ exon is itself separated from the constant (C) region by a large intron (reviewed by Tonegawa 1983). While germ-line V regions are not expressed at an appreciable level, rearranged V genes are highly transcribed. An obvious explanation for this is that the large intron between J and C actually contains a strong transcriptional enhancer (Banerji *et al.* 1983, Gillies *et al.* 1983, Neuberger 1983). DNA rearrangements bring the V region promoter under the control of this enhancer. This immunoglobulin enhancer turned out to be highly cell-type specific. Test genes linked to a 1 kb XbaI fragment from the J-C intron were active only when transfected into lymphoid cells of the B lineage, but not when transfected into fibroblast or epithelial cells (Banerji *et al.* 1983, Gillies *et al.* 1983). By means of deletions and resections within the XbaI fragment, an HinfI fragment (224 bp) could be identified which retained full activity and still had the same tissue specificity as the larger fragment.

We are focusing on the heavy chain gene enhancer as a model system for the enhancer mechanism and for tissue specificity. The main questions we want to address are:

– Can the Ig enhancer be further dissected to identify at the DNA level the units of enhancer activity?

– Are there modules which are constitutive (i.e., active in a wide variety of cells) and modules which are truly tissue-specific (active only in B cells)?

– What are the trans-acting factors interacting with the various enhancer modules?

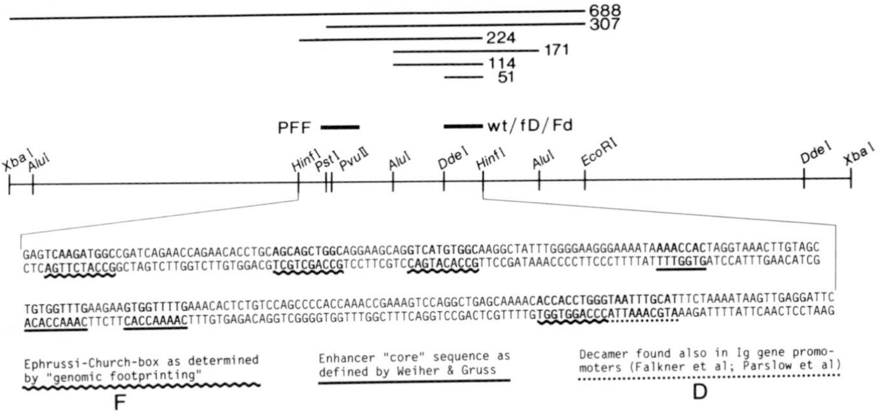

Fig. 1. Map of the XbaI fragment containing the IgH enhancer. The fragments tested for activity are depicted by solid bars in the upper part of the Figure. The number beside each fragment indicates its length in base pairs. These numbers are used throughout the text to designate the different fragments. The thicker solid bars labelled PFF, wt, fD or Fd represent the synthetic oligonucleotides tested. The nucleotide sequence of the HinfI fragment is presented below the map, and some potentially important sequence elements are indicated.

RESULTS

To identify the regions within the Ig enhancer which contribute to tissue-specific or constitutive expression, we have linked several restriction fragments to an enhancerless SV40 T-antigen gene and measured their enhancing activity by RNA mapping after transfection into various cell lines. In a first series of experiments we have shown that three segments, namely a 307 bp PstI-EcoRI, a 224 bp HinfI, or a 171 bp AluI fragment all were tissue-specific in their activity (not shown). These fragments are overlapping and hence define a central region of the Ig enhancer as being involved in the tissue specificity (see Figure 1). Close inspection of the sequence presented in Figure 1 reveals that the overlapping region, as well as the area upstream of it (i.e., up to the HinfI site), contains several conserved sequence motifs which may be important, such as:

(i) three enhancer "core" sequences GTGGA/TA/TA/TG (Weiher *et al.* 1983) in the middle of the HinfI fragment.

(ii) a decanucleotide sequence TAATTTGCAT (Falkner *et al.* 1984, Parslow *et al.* 1984), which was shown to be an important component of the heavy or light chain gene promoters (Mason *et al.* 1985, Falkner *et al.* 1984). These promoters show by themselves a strong preference for B cells (Picard & Schaffner 1985, Foster *et al.* 1985, Mason *et al.* 1985).

(iii) four lymphoid cell-specific boxes identified by DMS protection experiments performed *in vivo* ("genomic footprinting"; Ephrussi *et al.* 1985).

Next, we tested even smaller fragments such as the 114 bp AluI-HinfI fragment (which is contained in all three fragments previously mentioned, see Figure 1), which also turned out to be tissue-specific (not shown). This AluI-HinfI fragment can conveniently be cut into two halves by DdeI; the resulting 66 bp AluI-DdeI fragment contains two enhancer core sequences, while the 51 bp DdeI-HinfI fragment contains a lymphoid-footprint box directly abutting a decanucleotide sequence (Figure 1). These fragments were inserted as single or multiple copies into the EcoRI site downstream of a rabbit β-globin test gene and the activity of the constructs was evaluated in transient expression assays. Figure 2 shows the results with the 51 bp DdeI-HinfI enhancer fragment. Various cell lines were transfected with a 5:1 mixture of the test plasmid and a truncated β-globin gene driven by the SV40 enhancer as an internal reference (REFΔ). Forty-eight hours after transfection, cytoplasmic RNA was prepared and the relative quantities of β-globin transcripts from the test and reference genes were determined by an RNAase protection assay (Melton *et al.* 1984). Correctly initiated RNA from the reference gene gives one protected band, approximately 170 nucleotides in length, while correctly initiated RNA from the test gene produces two protected fragments. This is due to the fact that the SP6 RNA probe used is derived from the genomic DNA and extends up to the BamHI site within the second exon of the gene (see Figure legend for details). As can be seen (Figure 2), three to six copies of the 51 bp fragment have a very strong enhancing activity in B cells (X63Ag8), approximately equal to that of the SV40 enhancer. The multimers of the DdeI-HinfI fragments also contain interspersed polylinker sequences which increase the overall enhancing activity (see Figure 2 legend and compare Figures 2 and 5). The same constructs are completely inactive in HeLa cells, a line which normally does not express Ig genes. Figure 2 also reveals that, in the permissive cell line (B cell), a higher number of copies of the DdeI-HinfI fragment results in a greater enhancer activity (lanes 3–10). This is in agreement with our previous findings that a certain length is required to elicit a strong enhancing effect (natural enhancers are usually 200–300 bp long, reviewed in Schaffner 1985). Therefore, short restriction fragments or synthetic oligonucleotides have to be arranged in the form of multiple tandem copies in order to obtain sufficient activity.

To identify the protein(s) interacting with DNA sequences within this 51 bp DdeI-HinfI fragment, *in vitro* experiments were performed using a band shift assay (Garner & Revzin 1981, Singh *et al.* 1986). End-labelled DNA of the 51

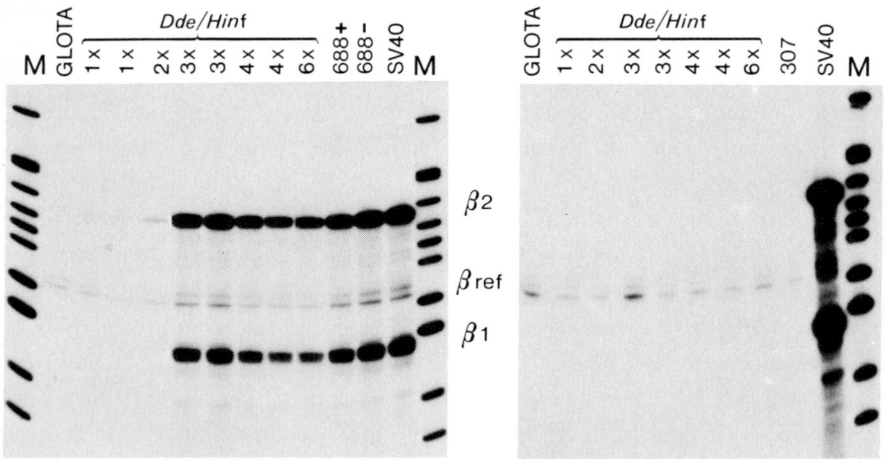

Fig. 2. Cell type specificity of the 51 bp DedI-HinfI fragment. Transient expression assays were done in X63Ag8 (myeloma, left panel) and HeLa (cervixcarcinoma, right panel) cells. Different DNA constructions were transfected into cultured cells by the DEAE-Dextran technique. Cytoplasmic RNA was isolated after 42 h and mapped with a SP6-generated β-globin-specific RNA probe. All transfected recombinants are based on the vector GLOTA containing the rabbit β-globin gene and the SV40 early region. 3' of the globin gene, different fragments of the IgH enhancer were inserted (a 51 bp DedI/HinfI, a 688 bp XbaI/EcoRI and a 307 bp PstI/EcoRI fragment) as well as the SV40 enhancer. When two lanes are labelled identically or with + and −, it indicates that two plasmids with inserts in opposite orientation were tested. The experiment shows that oligomers of the 51 bp fragment act as a strong enhancer which is tissue-specific, i.e. transcripts are only abundant in X63Ag8 but not in HeLa cells. The oligomers with more than two copies of the 51 bp DdeI-HinfI DNA fragments contain, on each side, polylinker sequences which contribute to the enhancing effect, but do not alter the tissue specificity. Band 1 corresponds to correctly initiated transcripts spanning from the cap site to the end of the first exon, band 2 to RNA extending from the beginning of the second exon to the BamHI site of the rabbit β-globin gene. REF RNA is derived from a truncated globin gene driven by the SV40 enhancer, a reference plasmid that is mixed into all transfected DNAs as an internal standard.

bp fragment was incubated with a nuclear extract from either BJA-B (a B cell), Molt-4 (a T cell) or HeLa (a cervix carcinoma). From DNA transfection data we knew that the IgH enhancer is active in the former two lines (although it is not active in several other T cell lines) and is inactive in the latter one. The results of such a band shift experiment are shown in Figure 3. The first three lanes represent negative controls with an end-labelled DNA fragment from the plasmid pSP64; no shift is observed in any extract under the experimental conditions used. Lanes 5, 6 and 7 correspond to the incubation of the DdeI-HinfI fragment in HeLa, Molt-4 or BJA-B extract. While one retarded band is common to all

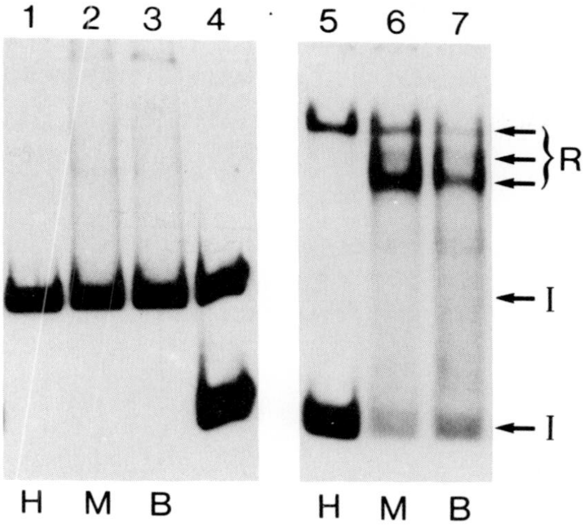

Fig. 3. Binding of a factor to the DdeI-HinfI enhancer fragment. 5' end-labelled DNA of the pSP64 EcoRI-PvuII fragment (lane 1–3) or of the Ig enhancer 51 bp DdeI-HinfI fragment (lane 5–7) were incubated with nuclear protein extracts (4 μg protein). Lanes 1 and 5, HeLa; lane 2 and 6, Molt-4; lane 3 and 7, BJA-B. After incubation, the complexes were resolved on a 4% polyacrylamide gel run at 4°C. Lane 4 contains approximately 3000 CPM of each fragment, without any extract addition. I and R indicate the position of input and retarded fragments, respectively.

three extracts, two faster migrating bands are found uniquely in Molt-4 or BJA-B extract. These could correspond to activator protein(s) responsible for the cell type specificity of the DdeI-HinfI fragment.

We next carried out competition experiments in conjunction with the band shift assay to define which sequences within the 51 bp DNA fragment are binding the factor(s). In this case the 114 bp AluI-HinfI end-labelled DNA fragment was used and incubated with the protein extract along with an excess of various unlabelled competitor DNAs. Gel purified natural DNA fragments or double-stranded synthetic oligonucleotides were used as competitors. Figure 4 displays the result of such an experiment performed with BJA-B extract. When the whole 224 bp HinfI fragment from the Ig enhancer is used as a competitor, the retarded bands are efficiently competed out (lane 8). By contrast, no competition is observed when an EcoRI-PvuII DNA fragment from pSP64 is used (lane 9). The wild-type oligonucleotide (lane 3) and the one having a mutation within the lymphoid footprint box (fD, lane 4) both compete efficiently. Another oligonucleotide, mutated within the decanucleotide sequence, does not compete anymore

A

```
                            F            D
wt     5'CTGAGCAAAACACCACCTGGGTAATTTGCATTTCTAAAATAAGTCGA 3'
fD     5'-------------AC--G----------------------------- 3'
Fd     5'------------------------G-T--G----------------- 3'
            F                       F
PFF    5'GCAGCAGCTGGCAGGAAGCAGGTCATGTGGCAAGGCTATTTGGG 3'
                          D
SphI   5'CTGCAGATGCATGCTTTGCATACTT 3'
```

B

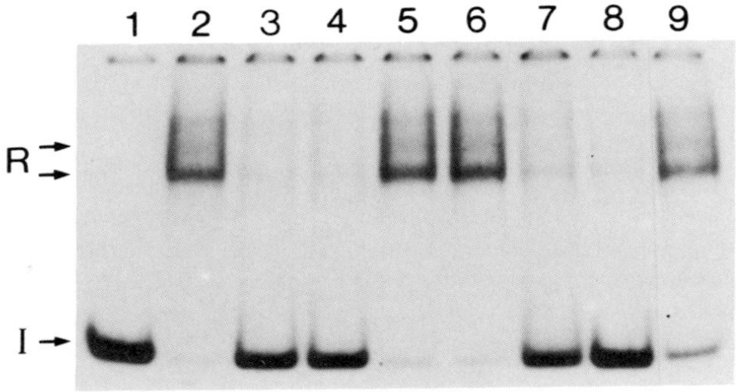

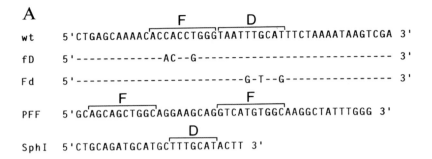

Fig. 4. A. Sequence of the oligonucleotides used for competition: the three upper oligonucleotides (wt, fD, Fd) correspond to the DdeI-HinfI fragment of the Ig enhancer, except that at the very end of the fragment one nucleotide has been changed to generate a SalI site. The wt oligonucleotide has no other change, the fD oligonucleotide has, additionally, three base changes in the lymphoid footprint box, and the Fd oligonucleotide has, additionally, three base changes in the decanucleotide sequence. The PFF oligonucleotide comprises nucleotides 377 to 421 from the Ig enhancer, and contains two lymphoid-specific boxes, as identified by *in vivo* DMS protection experiments (Ephrussi *et al.* 1985). The SV40 oligonucleotide comprises nucleotides 196 to 220 from SV40 and contains the direct repeats found near the SphI site (Zenke *et al.* 1986); overlapping with this sequence is an almost perfect match to the Ig decanucleotide. All oligonucleotides were annealed to their complementary strand before use.
B. Competition binding analysis in nuclear BJA-B extract. Binding assays were done as described in Figure 3 (legend), except that the 114 bp AluI-HinfI was used (see Figure 1 for map). This fragment gives the very same band shift as the 51bp DdeI-HinfI fragment. The various lanes additionally contain 1000 fmoles each of: wild-type oligonucleotide, lane 3; fD oligonucleotide, lane 4; Fd oligonucleotide, lane 5; PFF oligonucleotide, lane 6; SV40 oligonucleotide, lane 7; 1600 fmoles of the HinfI IgH enhancer fragment, lane 8; 2000 fmoles of the pSP64 EcoRI-PvuII fragment, lane 9. 1000 fmoles of the oligonucleotides correspond to approximately a 60-fold molar excess. Lane 2 is a reaction without competitor DNA and lane 1 contains only input DNA fragment, but no extract.

(Fd, lane 5), nor does an oligonucleotide from another region of the IgH enhancer near the PstI site which contains two lymphoid footprint boxes (PFF, lane 6). Conversely, an oligonucleotide corresponding to a segment of the SV40 enhancer (nucleotide 196 to 220), which contains an almost perfect homology to the decanucleotide sequence, does compete efficiently for the band-shift (lane 7). Note that all retarded bands are being competed out, not only the B cell-specific ones. Taken together, these data identify the decanucleotide sequence as a binding site for a specific protein factor. Similar competition experiments with HeLa extract led to the same conclusion, namely that the decanucleotide sequence is also essential for the DNA: protein interaction in these nonpermissive cells (not shown).

To assess the biological relevance of the observed DNA-protein interactions, the various oligonucleotides which had been used in competition assays were tested for activity *in vivo*. Multiple copies of each type of oligonucleotide (wt, fD, Fd) were inserted downstream of the rabbit β-globin gene. These constructs were transfected into B cells together with a reference plasmid, and cytoplasmic RNA was analyzed after 2 days, as described above. Figure 5 presents the results of one such experiment. Four copies of the wild-type oligonucleotide give a stimulation in X63 cells equivalent to 20% of that obtained with the SV40 enhancer (lanes wt), while four copies of the oligonucleotide containing the mutated decanucleotide sequence have a drastically reduced activity (lane Fd). Mutation of the lymphoid footprint sequence results in only a moderate reduction in activity (lanes fD). This result suggests that both the footprint and the decanucleotide sequence contribute to *in vivo* activity, but that the decanucleotide sequence is the dominant element.

DISCUSSION

The data presented in this paper indicate that a short segment (51 bp) from within the Ig enhancer still displays a strict cell specificity in that it behaves like the complete enhancer when tested in B or HeLa cells. When cloned downstream of the rabbit β-globin gene, three to six copies of this short DNA segment (with interspersed polylinker sequences) are approximately as active as the whole SV40 enhancer, but only in B cells. In HeLa cells, no activity is observed. When the oligonucleotides are multimerized an increase in activity is also observed (not shown), although much less pronounced than if polylinker sequences are present. In this case, four copies of the wild type sequence (without polylinker DNA) have approximately 20% of the activity of the SV40 enhancer. It is striking that

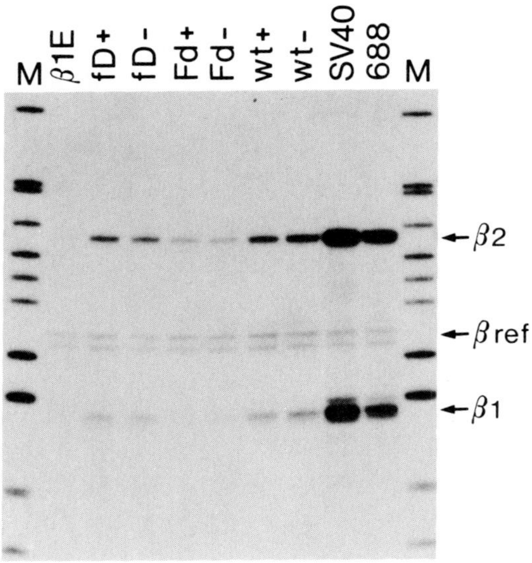

Fig. 5. Mutation of the decanucleotide sequence is a down mutation *in vivo*. Tandem repeats of four copies of oligonucleotides fD, FD and wt were cloned into vector β-1E in both orientations. β-1E contains a 4.7 kb fragments of the rabbit β-globin gene in plasmid pJC-1 with a single EcoRI site 3' of the globin genes into which the tetramers of the oligonucleotides were inserted. These constructs were transfected into X63Ag8 myeloma cells and the cytoplasmic RNA analyzed by RNAase protection mapping. The result shows that mutations in the lymphoid footprint region have only a minor, if any, effect on globin transcription. However, the mutation in the decanucleotide region shows a reduction of enhancement, although transcription stimulation is not entirely abolished.

the polylinker sequences can augment the activity of the DdeI-HinfI fragment without increasing the basal activity in HeLa cells.

When radiolabelled DNA of the DdeI-HinfI fragment is incubated in nuclear protein extracts, a complex pattern of interactions is detected, which is clearly cell-type specific. From the three types of DNA-protein complexes observed in the gel retardation assay, two are specific for B cells, while the third one is common to both B and HeLa cells. Binding competition experiments done with various wild type or mutant synthetic oligonucleotides have identified the decanucleotide sequence motif as a binding site for a nuclear factor. Furthermore, the oligonucleotide containing a mutated decanucleotide sequence showed a considerably reduced enhancer activity when cloned downstream of the β-globin gene and tested *in vivo*.

A possible explanation for the partially common and partially specific gel retardation pattern seen in the various extracts could be that a ubiquitous protein factor binds to the decanucleotide in every tissue, and that some auxiliary factors bind to the left or to the right of this motif in lymphoid cells only. As a first attempt to answer this question, we performed gel retardation assays directly with the mutated oligonucleotides. The pattern we observed was as follows: the oligonucleotide with a mutated lymphoid footprint box gave the wild type pattern of interactions, while mutation of the decanucleotide sequence entirely abolished complex formation, both in B and HeLa extracts (data not shown). Furthermore, *in vivo* testing of these oligonucleotides showed that the former mutation had only a weak effect on enhancing activity, while the latter was a strong down mutation (Figure 5). This indicates that the decanucleotide is the dominant element and that only a minor contribution comes from the footprint box. The lack of a detectable band shift with the footprint sequence in crude protein extracts could mean that the cognate protein does not bind under the conditions used, or that it is present at a too low concentration.

Together, these results confirm and extend previously published data and indicate that the decanucleotide sequence is not only an important component of the Ig gene promoters (Falkner *et al.* 1984, Bergman *et al.* 1984, Mason *et al.* 1985), but also of the Ig heavy chain gene enhancer. A factor binding to a similar recognition sequence in the promoters of both the heavy and kappa light chain gene and also in the heavy chain enhancer has recently been identified and called IgNF-A (Singh *et al.* 1986). However, IgNF-A did not seem to be tissue-specific, as the same band shift pattern was observed with both B and HeLa cell extracts (Singh *et al.* 1986). Hence, it remains to be established whether IgNF-A and the factor we have detected are one and the same protein.

Testing the appropriate oligonucleotides both *in vivo* and *in vitro* will hopefully answer the question of whether the decanucleotide sequence is by itself a truly cell type-specific enhancer module, and to what extent its specificity is influenced by neighbouring enhancer modules.

MATERIAL AND METHODS

Plasmid constructions were done by standard procedures according to Maniatis *et al.* (1982). Cell transfections by the DEAE-dextran technique and RNA extraction were done according to Banerji *et al.* (1983). RNA mapping was done essentially following the protocol of Melton *et al.* (1984) except that only 8 μg/ml RNAase A and 40 μ/ml RNAase Tl were used. Gel retardation assays were done according to Singh *et al.* (1986) with the following modifications: the binding reaction was performed in a buffer containing 10 mM Hepes pH 7.9, 10% Glycerol, 110 mM KCl, 4 mM $MgCl_2$, 0.1 mM EDTA, 0.25 mM DTT, 4 mM Spermidine. After binding at room temperature for 10 min, the samples were put on ice and then loaded on a 4% polyacrylamide gel which was run at 4° C. Synthetic oligonucleotides were provided by J. Jiricny and W. Zücher, Friedrich Miescher Institute, Basel. They were annealed before use and 5'end-labelled when appropriate.

REFERENCES

Banerji, J., Olson, L. & Schaffner, W. (1983) A lymphocyte-specific cellular enhancer is located downstream of the joining region in immunoglobulin heavy chain genes. *Cell 33,* 728–40.

Banerji, J., Rusconi, S. & Schaffner, W. (1981) Expression of a β-globin gene is enhanced by remote SV40 DNA sequences. *Cell 27,* 299–308.

Bergman, Y., Rice, D., Grosschedl, R. & Baltimore, D. (1984) Two regulatory elements for immunoglobulin kappa light chain gene expression. *Proc. Natl. Acad. Sci. (USA) 81,* 7041–5.

Dynan, W. S. & Tjian, R. (1985) Control of eukaryotic messenger RNA synthesis of sequence-specific DNA-binding proteins. *Nature 316,* 774–8.

Ephrussi, A., Church, G. M., Tonegawa, S. & Gilbert, W. (1985) B lineage-specific interactions of an immunoglobulin enhancer with factors in vivo. Science 227, 134–40.

Everett, R. D., Baty, D. & Chambon, P. (1983) The repeated GC-rich motifs upstream of the TATA box are an important element of the SV40 early promoter. *Nucl. Acids. Res. 11,* 2447–64.

Falkner, F. G., Neumann, E. & Zachau, H. G. (1984) Tissue specificity of the initiation of immunoglobulin kappa gene transcription. *Hoppe-Seyler's Z. Physiol. Chem. 365,* 1331–43.

Foster, J., Stafford, J. & Queen, C. (1985) An immunoglobulin promoter displays cell-type specificity independently of the enhancer. *Nature 315,* 423–25.

Garner M. M. & Revzin, A. (1981) A gel electrophoresis method for quantifying the binding of proteins to specific DNA regions: application to components of the E. coli lactose operon regulatory system. *Nucl. Acids Res. 9,* 3047–60.

Gillies, S. D., Morrison, S. L., Oi, V. T. & Tonegawa, S. (1983) A tissue-specific transcription enhancer element is located in the major intron of rearranged immunoglobulin heavy chain gene. *Cell 33,* 717–28.

Lusky, M., Berg, L., Weiher, H. & Botchan, M. (1983) Bovine papilloma virus contains an activator of gene expression at the distal end of the early transcription unit. *Mol. Cell. Biol. 3,* 1108–22.

Maniatis, T., Fritsch, E. F. & Sambrook, J. (1982) *Molecular cloning: A laboratory manual.* Cold Spring Harbor Laboratory, New York.

Mason, J. O., Williams, G. T. & Neuberger, M. S. (1985) Transcription cell type specificity conferred by an immunoglobulin V gene promoter that includes a functional consensus sequence. *Cell 41,* 479–87.

Melton, D. A., Krieg, P. A., Rebagliati, M. R., Maniatis, T., Zinn, K. & Green, M. R. (1984) Efficient in vitro synthesis of biologically active RNA and RNA hybridization probes from plasmids containing a bacteriophage SP6 promoter. *Nucl. Acids Res. 12,* 7035–56.

Moreau, P., Hen, R., Wasylyk, B., Everett, R., Gaub, M. P. & Chambon, P. (1981) The SV40 72 base pair repeat has a striking effect on gene expression both in SV40 and other chimeric recombinants. *Nucl. Acids Res. 9,* 6047–67.

Neuberger, M. S. (1983) Expression and regulation of immunoglobulin heavy chain gene transfected into lymphoid cells. *EMBO J. 2,* 1373–8.

Parslow, T. G., Blair, D. L., Murphy, W. J. & Granner, D. K. (1984) Structure of the 5' ends of immunoglobulin genes: A novel conserved sequence. *Proc. Natl. Acad. Sci. USA 81,* 2650–4.

Picard, D. & Schaffner, W. (1985) Cell-type preference of immunoglobulin κ and λ gene promoters. *EMBO J 4,* 2831–38.

Schaffner, W. (1985) Introduction to *Eukaryotic Transcription, Current Communications in Molecular Biology,* ed. Gluzman, Y., pp. 1–18. Cold Spring Harbor Laboratory Press, Cold Spring Harbor New York.

Serfling, E., Jasin, M. & Schaffner, W. (1985) Enhancers and eukaryotic gene transcription. *Trends in Genetics 1,* 224–30.

Singh, H., Sen, R., Baltimore, D. & Sharp, P. A. (1986) A nuclear factor that binds to a conserved sequence motif in transcriptional control elements of immunoglobulin genes. *Nature 319,* 154–8.

Tonegawa, S. (1983) Somatic generation of antibody diversity. *Nature 302,* 575–600.

Weiher, H., König, M. & Gruss, P. (1983) Multiple point mutations affecting the simian virus 40 enhancer. *Science 219,* 626–31.

Wildeman, A. G., Sassone-Corsi, P., Grundström, T., Zenke, F. & Chambon, P. (1984) Stimulation of in vitro transcription from the SV40 early promoter by the enhancer involves a specific trans-acting factor. *EMBO J. 3,* 3129–33.

Yaniv, M. (1984) Regulation of eukaryotic gene expression by trans-activating proteins and cis-acting DNA elements. *Biol. Cell. 50,* 203–11.

Zenke, M., Grundström, T., Matthes, H., Wintzerith, M., Schatz, C., Wildeman, A. & Chambon, P. (1986) Multiple sequence motifs are involved in SV40 enhancer function. *EMBO J. 5,* 387–97.

DISCUSSION

YANIV: Did you try to take the element from the U2 promoter or from the histone H2B promoter to see if out of its context it becomes a tissue-specific enhancer?

MATTHIAS: No, we have not done that yet. We know that fragments from the histone H2B promoter compete very efficiently for band shift. We have analyzed individually different elements from the SV40 enhancer, and they have various cell type specificities. This is work which has been done by Sabine Schirm, and the major conclusion is that short oligonucleotides from within the SV40 enhancer do not behave like the whole SV40 enhancer. For example, if she takes an oligo covering the SphI site region, it is more active in B or some T cells than in fibroblasts. And the SphI region of the SV40 enhancer contains a very good homology to the decamer sequence. On the other hand, if she tests a short oligo containing the SV40 core sequence, she finds it active in essentially every cell tested. It was known from the work done in Pierre Chambon's lab that the core sequence is a crucial part in the SV40 enhancer as a whole and, surprisingly, the cell line where the core element is least active for SV1, which is a natural host for SV40. On the basis of these and other data, our hypothesis is that if one looks at the whole enhancer it may have cell type-specificity A, but if one now takes subsegments of it, they can have a different cell type-specificity, B, C, or D. But once they are put together in the same arrangement as *in vivo*, then one gets cell type-specificity A again.

RIGBY: I will admit to a certain amount of confusion. Your data show that very short segments from within the IgH enhancer retain cell type-specificity?

MATTHIAS: Yes.

RIGBY: But if I properly understand the paper of Wasylyk (*EMBO J.* 1986, *5*, 553), in his hands the entire enhancer works in all cell types and the importing thing is the flanking sequences which repress in non-B cells.

MATTHIAS: We do not get exactly those data, and I should say that there is one major difference between Wasylyk's and our experiments. We are testing everything for enhancer activity in a 3′ position, downstream of the beta globin gene.

Wasylyk has hooked his fragment directly to the TATA box of the conalbumin promoter. If we think of the Ig enhancer, and the deca-nucleotide sequence in the HinfI fragment, this is one of the areas where sterile transcripts start. So, this is by definition an area which also has promoter properties. We have done constructions similar to those of Wasylyk and we do not completely agree with his results. We can also see some activity in non-B cells, for example, in LTK, 3T6, but not in 3T3 or HeLa. But we detect activity only if we have the IgH enhancer directly to the promoter. As soon as we move the whole fragment to the 3′ side, then it is completely dead. So, for us it is a little bit questionable whether this is a real enhancer effect.

KJELDGAARD: So, you feel it is important to test the 3′ position?

MATTHIAS: It is needed to make sure.

YANIV: But Wasylyk was studying by S1 mapping transcripts initiated from the conalbumin promoter.

MATTHIAS: I agree. But one could say that it is a different mechanism to activate transcription from a 3′ position than to activate it from a place where you are sure direct contact can happen between enhancer and promoter.

CORY: I would like to explore with you for a moment the question of B versus T lymphoid specificity. As you just mentioned, some T-cell lines do express these sterile RNA transcripts and this is also true of some myeloid cell lines. I have always thought this reflected the activity of the Ig heavy chain enhancer in these particular cell lines. You said that you did not find the enhancer fragment to be active in BW S147. That is not a cell line, I believe, where sterile μ transcripts are found. Have you tested T cell lines which do not contain sterile μ transcripts to see if such cell lines contain these factors?

MATTHIAS: No, we have not tested specifically such cell lines. But in the light of the data by Bob Perry recently in *Nature*, it need not be that the enhancer is active. One could argue that only the deca-nucleotide sequence needs to be active, because it is known to be a promoter component. So one could say that if that region is active, it is enough to get sterile transcripts.

CORY: I am very interested in this, because in our Eμ-myc transgenic mice we expected to find some T-cell lymphomas and maybe even some myeloid malignancies, and we did not find them. The fact that we only found B-cell tumors argues for a greater tissue specificity of the enhancer than we expected from our interpretation of the sterile μ transcript data.

MATTHIAS: Normally in T cell the Ig enhancer is not active, but we knew from the published literature that MOLT4 cells activate the enhancer. That was done by Peter Gruss and we checked this by DNA transfection experiments, and that is true, the enhancer is active is these cells. Also, the pattern we observe in bandshift assays is of the B-cell type. We have recently obtained a B-cell line where the enhancer is not active *in vivo*. We have done transfection experiments, comparing the SV40 enhancer and the Ig enhancer, and the SV40 is doing fine while the Ig is completely dead. We are very eager to test that cell line, a human cell line, for its pattern in bandshifts.

YANIV: Do you have head-to-tail polymers or random ligation when you use oligonucleotides?

MATTHIAS: They are head-to-tail, yes. We synthetize oligos with sticky ends to make sure that they ligate in an ordered fashion.

YANIV: The fact that you can add a linker and get increased activation indicates that stereospecificity of the binding site is not so important or that something is binding to the linker and increasing its activity.

MATTHIAS: It could be either, but we observe that also with motives from the SV40 enhancer. We have not checked extensively, but it does not look as if the polylinker by itself is an enhancer.

LOWY: Would you be able to give us the current concepts of how an enhancer that is located either downstream or a long distance upstream from a promoter might function?

MATTHIAS: The possibility we favor mostly is that the DNA sort of makes a loop and thereby allows the enhancer and the promoter to come in direct vicinity. Then, their respective factors may interact by making direct protein:protein contacts. Although this model seems to be the most probable, there is one drawback with it, namely that one should expect some trans-enhancement to occur. Theoretically, plasmid molecules containing either only an enhancer or only a promoter should be able to interact. But we have observed that (only) *in vivo* so far. One would need good *in vitro* systems to show that. A second model would be the so-called entry site model. In this scheme the enhancer somehow attracts the RNA polymerase which then scans the DNA in search of a promoter. One could envisage that, while scanning, the polymerase already transcribes, maybe short unstable RNAs. We have tried to test that by doing a construction where we have the enhancer far away from the promoter, and where we have terminators to the left and to the right of that unit. We then measured the transcription through the intervening segment by run-on assay and using M13 probes. The result is that we could not detect transcription of the intervening DNA. So even if polymerase scans this DNA, it does not seem to transcribe it.

DYNAN: With the loop model, one would expect there would be a minimum distance at which the enhancer could work, the minimum being the smallest segment that could be formed into a loop. As far as I know such a minimum is not observed. This is evidence against the loop model. As for the entry site model, there is some data, at least *in vitro*, using beta ^{32}P labelled nucleoside triphosphates, suggesting that the 5′ end of the RNA product does correspond to the actual site where interaction first occurs. One can incorporate beta ^{32}P label into the 5′ or SV40 or polyoma late RNAs, and I think experiments have been done with the globin gene as well. Is this not evidence against your entry site model?

MATTHIAS: Yes, that is true, but you could still argue that polymerase scans the whole area and, while scanning, it transcribes short RNA segments for instance, and when the polymerase recognizes the TATA-box, then it really starts and puts the Cap and everything. Again, it seems that it is not the case. Regarding the distance, we have again done a careful analysis of the effect of the distance, and we still do not observe a distance effect with the beta-globin promoter.

YANIV: This is contrary to the data from Chambon's laboratory. (Wasylyk & Chambon *NAR* 1984 *12*, 5589–5608).

MATTHIAS: Yes, but he uses always plasmid sequences between the enhancer and the promoter. We have tested plasmid sequences from pBR and there are some sequences which are very deleterious to the enhancer effect. I think that this may explain the discrepancies, at least partly. We have now clearly put the enhancer from position −60 or so, down to −1.5 kb, and we see essentially no decrease in activity.

RIGBY: Perhaps more of a comment on the *trans*-enhancer effect of the type postulated? It is in fact quite well documented in *Drosophila*, at the white locus for example.

MATTHIAS: You mean the transversion?

RIGBY: Yes. So, there are *in vivo* data that actually support that.

HANAHAN: What has happened to the ideas that perhaps these enhancers were involved in compartmentalization or localizing transcribed sequences in the nuclear matrix, which would also explain the lack of binding?

MATTHIAS: It is not anymore a current model. The only thing I could mention is that in the kappa gene locus there is indeed an attachment site, but next to the enhancer and not overlapping with it.

Transcription from Unconventional Promoters: Dihydrofolate Reductase and SV40 Late

William S. Dynan

Our ideas about the sequences controlling transcriptional initiation in mammalian cells have been heavily influenced by the study of a relatively small group of genes. These include histones, globins, immunoglobulins, ovalbumin and conalbumin, and a variety of viral transcription units. Because these mRNAs are expressed at high levels, the promoters which direct their synthesis were naturally the first to be cloned, sequenced, and studied in detail. Sequence comparisons within this group of promoters, supported by mutational and biochemical evidence, have led to the development of a detailed and accepted view of how transcriptional control regions are organized (see reviews by Breathnach & Chambon 1981, Dynan & Tjian 1985).

Recently, however, there have been many reports describing the identification and sequencing of genes encoding less abundant messages, such as those for housekeeping enzymes and those involved in growth control (McGrogan *et al.* 1985, Reynolds *et al.* 1984, Melton *et al.* 1986, review by Dynan 1986a). It has become clear that the 5' flanking regions of these genes lack many of the elements that had hither to been thought to be general features of promoter organization.

The most obvious difference is the lack of a TATA box. This consensus sequence was the first to be recognized in mammalian promoters, and is present in virtually all of the more than 60 promoters compared by Breathnach & Chambon in their early and comprehensive review (1981). The TATA box region has been shown to play a role in both promoter strength and in fixing the start site for transcription, and appears to be the binding site for one of the multiple protein factors required for transcriptional initiation (Parker & Topol 1984, Sawadogo & Roeder

Department of Chemistry & Biochemistry, Campus Box 215, University of Colorado, Boulder, CO 80309, U.S.A.

VIRAL CARCINOGENESIS, Alfred Benzon Symposium 24,
Editors: N. O. Kjeldgaard, J. Forchhammer, Munksgaard, Copenhagen 1987.

1985). While there had been a few promoters recognized early – on as being able to function without a TATA box, it is only recently that it has become apparent that there is a major group of cellular promoters for which this moiety seems to be totally dispensable.

The second distinctive feature of the new group of promoters is their upstream region. This is invariably GC-rich, sometimes conspicuously so. The GC-rich region may be as long as several hundred base pairs, and in every case contains actual or proposed binding sites for the transcription factor Sp 1. There is a statistical excess of the dinucleotide C_pG, and in those promoters that have been examined directly, these sites are markedly undermethylated (see review by Bird 1986).

In the studies reported here, we report several preliminary experiments and observations relating to what we call unconventional promoters, those which lack a TATA box and which have an extended GC-rich upstream region. It is important to recognize that 'unconventional' is a historical designation, chosen to emphasize that these promoters are different from those first characterized in mammalian cells. Unconventional does not imply uncommon; in fact, the frequency of this form of organization among newly cloned housekeeping and other low-expression genes suggests that these may be the numerically predominant class of promoter in mammalian cells.

DIHYDROFOLATE REDUCTASE

The gene for dihydrofolate reductase is certainly one of the most well-characterized of the new group and in some ways may be regarded as the prototype. DHFR catalyzes the production of tetrahydrofolate, which is required for *de novo* synthesis of thymidylate, purines, and glycine. The level of DHFR mRNA varies during the cell cycle, being maximal at the onset of DNA synthesis (Farnham & Schimke 1985). The structural gene has been cloned in mouse, hamster, and human species, and a detailed comparative analysis of all three genes has been published (Mitchell 1986). In each species, there are two GC-rich clusters, one immediately upstream of the major transcriptional initiation site, and the other 400 to 500 bp upstream. In mouse, it has been shown that either of these regions can function individually as a transcriptional promoter (McGrogan et al. 1985).

Both of the GC-rich regions of the mouse DHFR 5' flanking sequence were observed to contain segments matching the consensus recognition sequence (Kadonage *et al.* 1986) for the transcription factor Sp1. Sp1 was recognized originally as a host-cell factor that binds the Simian Virus 40 promoter region and activates transcription (Dynan & Tjian 1983a, 1983b). Sp1 has been isolated from human tissue culture (HeLa) cells (Dynan & Tjian 1983a, Kadonaga & Tjian 1986) and the availability of these isolated fractions enabled us to test directly the proposal that Sp1 interacted with the DHFR promoter region.

Partially purified Sp1 preparations were incubated with DHFR promoter DNA fragments under conditions previously established as allowing binding to SV40. The Sp1-DNA complexes were treated with controlled amounts of DNase I to allow visualization of a "DNase footprint" (Galas & Schmitz 1978) with results as shown in Figure 1. There are four distinct regions of protection, each encompassing about 25 bp on either strand of the DNA, and each centered on the "GC box" moiety now recognized as the fundamental recognition unit of factor binding.

Subsequent studies showed that the binding, as measured by DNase footprinting, could be specifically and effectively competed for by SV40 DNA, thus establishing that the protection was almost certainly brought about by the same protein factor (Dynan *et al.* 1986). Similar footprinting studies were undertaken with the more distal cluster of GC-rich DNA in the mouse DHFR flanking region. These revealed a similar protected area in the approximate position predicted from the sequence homology (results not shown).

In vitro transcription experiments with a HeLa cell nuclear extract showed that the DHFR promoter is active in this system, and that initiation occurred predominantly at the same nucleotide position as *in vitro,* some 55 bp upstream of the translational start (Dynan *et al.* 1986, Sazer & Schimke 1986). Further experiments with a fractionated transcription system showed that transcription of DHFR was strongly dependent on the presence of an Sp1 containing fraction. These results suggest that, as with SV40, the binding of Sp1 in the upstream region influences the remaining components of the transcriptional machinery to initiate transcription.

Two facts are notable. The rate of RNA synthesis under *in vitro* conditions was of the same order of magnitude as that with the adenovirus major late promoter, often regarded as a benchmark "strong promoter". The adenovirus promoter requires its TATA box for *in vitro* RNA synthesis, whereas the DHFR promoter achieves a reasonable rate of initiation in the apparent absence of this element. The other fact of interest is that most transcripts mapped to a single site, laying to rest the notion that initiation in the absence of a TATA box is necessarily heterogeneous and dispersed.

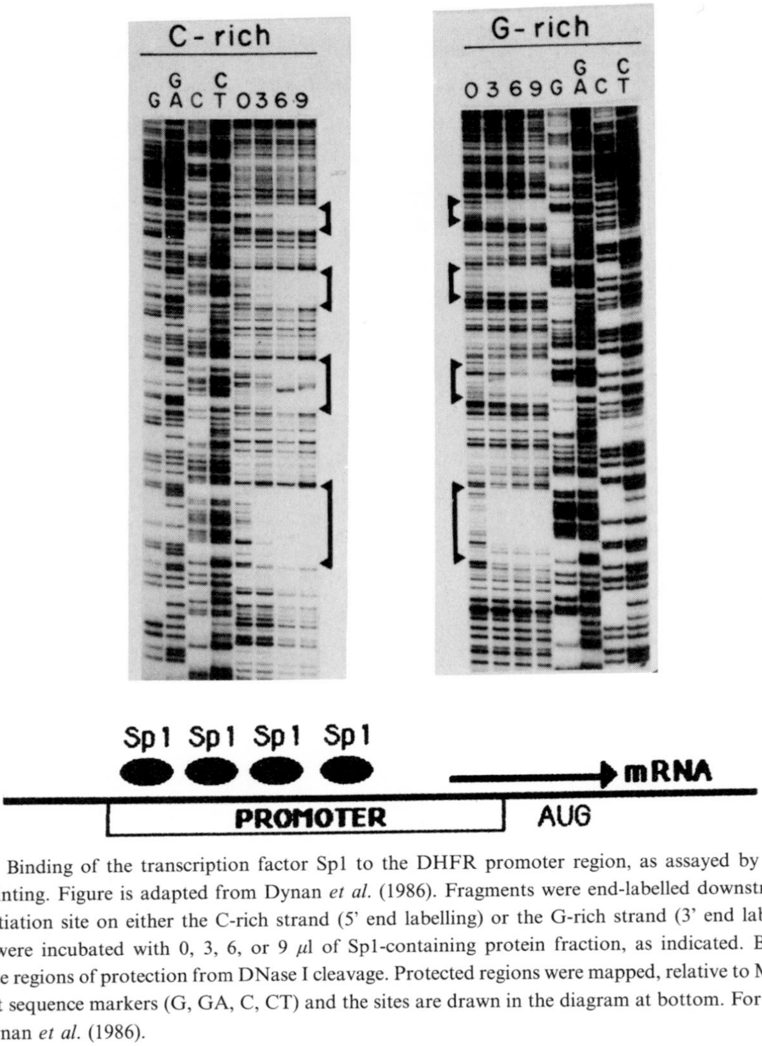

Fig. 1. Binding of the transcription factor Sp1 to the DHFR promoter region, as assayed by DNase footprinting. Figure is adapted from Dynan *et al.* (1986). Fragments were end-labelled downstream of the initiation site on either the C-rich strand (5' end labelling) or the G-rich strand (3' end labelling). They were incubated with 0, 3, 6, or 9 μl of Sp1-containing protein fraction, as indicated. Brackets indicate regions of protection from DNase I cleavage. Protected regions were mapped, relative to Maxam-Gilbert sequence markers (G, GA, C, CT) and the sites are drawn in the diagram at bottom. For details, see Dynan *et al.* (1986).

It has not yet been shown whether activity of the DHFR promoter region requires sequence elements in addition to the Sp1 binding regions. One possibility is that the conserved sequences between the multiple Sp1 sites in mouse DHFR play some role. Another possibility is that there is a functional region between the most proximal Sp1 site and the major initiation site. Sequences in this region are conserved between all three mammalian species, and a consensus has been

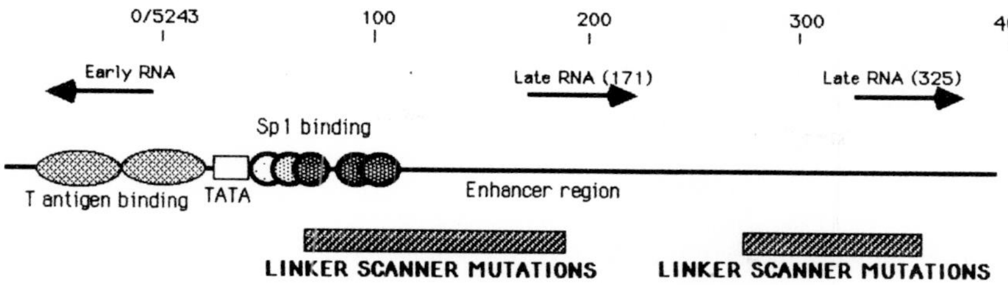

Fig. 2. The SV40 transcription control region. For more information, see text.

proposed (Mitchell 1986). Interestingly, this sequence is located in the place where a TATA box would be expected. A final possibility is that there is a functional element immediately downstream of the major transcriptional initiation site, where again there is strong conservation between the three mammalian genes.

With DHFR, there are no direct experiments bearing on the existence of these additional functional elements, and the only evidence comes from inspection and comparison of DNA sequences. As will be seen in the next section, however, there is some recent evidence for the importance of sequences at these positions in another non-TATA box regulatory region, the Simian Virus 40 late promoter.

SV40 LATE

Simian Virus 40 has two transcription units, one expressed early during lytic infection, the other late, after the onset of viral DNA replication. The early promoter (Figure 2) follows the conventional pattern: an AT rich "TATA" region, an upstream element containing multiple binding sites for the host cell transcription factor Sp1, and further upstream, an enhancer. The late promoter is unconventional, and in some respects resembles DHFR. There is no TATA box, and the GC-rich repeats of the Sp1 binding region play a prominent role in late expression, at least under some conditions.

In the present set of experiments, we set out to identify specific transcription factors and binding sites involved in late expression, using the same general approach as had been successful in demonstrating the role of Sp1 in early gene expression. There are multiple late starts (for review see Acheson 1980), and we picked two late starts to study in detail. One at nucleotide 325, corresponds to the major *in vivo* site of initiation; the other, at nucleotide 171, corresponds to a minor *in vivo* site but is of special interest because it is the late initiation site closest to the Sp1 binding region.

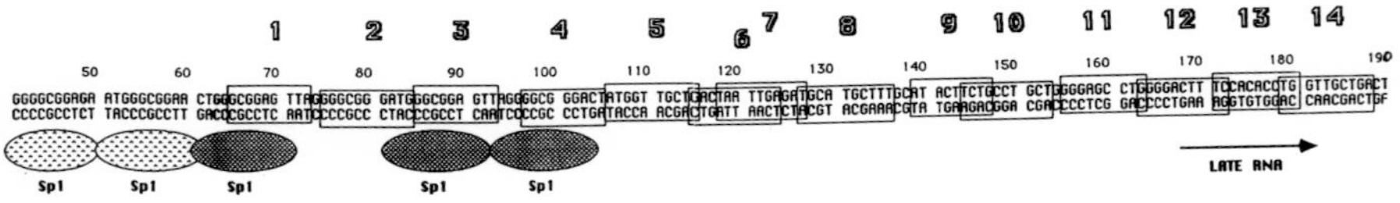

Fig. 3. Linker scanner mutations. Map represents a portion of the region shown in Figure 2. Boxes indicate the positions of mutations LS1–LS14. In each case, the sequence within the box has been substituted with the linker sequence "CGAGATCTCG".

A series of clustered point mutations was constructed covering the regions indicated in Figure 2. The choice of these regions for saturation mutagenesis was based on previous studies by other workers (Fromm & Berg 1982, Brady *et al.* 1982, Rio & Tjian 1983, Hansen & Sharp 1984, Vigneron 1984) and also on analogies with other mammalian promoters. We believed that the region at the left was most likely to contain elements contributing to transcription from the 171 site, and that both regions were likely to contain elements contributing to transcription from the 325 site.

Mutants were constructed by standard *in vitro* techniques. Closely spaced deletion mutants covering both regions from both directions were constructed and sequenced by facile methods (Haltiner *et al.* 1985) and recombined according to the "linker scanner" scheme first described by McKnight & Kingsbury (1982). Recombination according to this scheme results in the substitution of a defined sequence (CGAGATCTCG) for 10 bp of viral DNA. In almost every case, parental deletion mutants could be matched so as to preserve exactly the spacing of sequences flanking the substituted region.

We have completed a preliminary analysis of the 14 linker scanner mutants in the group at left in Figure 2. These map in the Sp1 binding and enhancer regions of the viral genome, and their precise positions are shown in Figure 3. All mutants were tested for their ability to support initiation of late transcription at the 171 site. The test system was an *in vitro* reaction containing HeLa cell nuclear extract.

Results were as follows:

i) Mutants LS1, LS2, LS3, LS4, all had decreased levels of late transcription. These mutations disrupt portions of the Sp1 binding region, and their effect was predicted from previous studies. Recombination of LS 1 and LS 4, eliminating all three strong binding sites for Sp1, resulted in a severe decrease in late transcription.

ii) Mutants LS5, LS6 and LS7 also had decreased levels of transcription. These mutations lie entirely outside the Sp1 binding region and do not affect binding of this factor, as measured by DNase footprinting. We believe these mutations affect the binding of a different factor, the footprint of which is shown in Figure 4. This factor is apparently separable from Sp1 (see Dynan & Tjian 1983b).

iii) Mutant LS9 showed severely decreased transcription from the 171 site. This mutation encompasses the region 24 to 32 bp upstream from the start, and thus covers the region where the TATA box would lie in a conventional promoter.

iv) Mutant LS10 activated a cryptic start within the substituted sequence.

v) Mutant LS14 severely decreased transcription, although the substitutions map entirely downstream of the initiation site. This result with LS14 is consistent with the long-established finding that the late promoter is totally inactivated by deletion of the sequence downstream of nucleotide 160 (Rio *et al.* 1980).

vi) Other LS mutations (8, 11, 12, 13) gave essentially wild-type levels of transcription. LS8 is especially significant. Results from other laboratories (Herr & Clarke 1986, Zenke *et al.* 1986) led to the strong prediction that this mutation will destroy enhancer function, yet we seen no effect on the level of late transcription.

DISCUSSION

Our results indicate that both DHFR and the SV40 late promoter are activated by Sp1-binding DNA sequences. There are numerous additional parallels between the two transcription units, including lack of methylation at CpG dinucleotides, and the presence of DNase hypersensitive sites flanking the GC-rich segments (Jongstra *et al.* 1984, Shimada & Nienhuis 1985, Mitchell *et al.* 1986).

Significantly, both control regions are bidirectional. This has been shown for both mouse and hamster DHFR promoters (Farnham & Schimke 1985, Mitchell 1986), where there is an opposite strand transcript of unknown function, and obviously has long been recognized for SV40. It is a property shared also by an SV40-homologous promoter region obtained from the monkey genome (Saffer & Singer 1984). It has been suggested that bidirectionality is an inherent property of the Sp 1-binding elements; despite their asymmettry, for example, they can be inverted and retain their ability to promote early viral transcription (Everett *et al.* 1983). Now that a large number of cellular housekeeping promoters have been identified that contain GC-rich 5' flanking regions, it will be interesting to see how many are associated with opposite-strand transcripts.

We have referred to the "apparent" absence of a TATA box in SV40 late and DHFR. In fact, we have little information about what structural features of the DNA are recognized in the TATA region by the transcriptional machinery, and

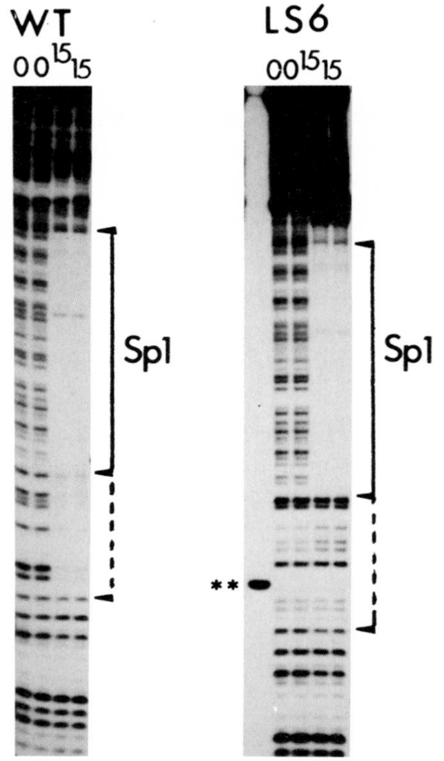

Fig. 4. Protection of DNA fragments by proteins in a HeLa cell nuclear extract. Left panel, wild type SV40; right panel, LS6. Fragments were prepared and subjected to DNase I footprinting as described (Dynan & Tjian 1983b) using 0 or 15 μl of HeLa nuclear extract, as indicated. Brackets mark the Sp1 binding region. The dashed line indicates the region occupied by a putative transcription factor required for late expression. The binding site of the late factor corresponds approximately to the region covered by LS5, LS6, and LS7 (Figure 3). Note that the factor is unable to bind to the LS6 probe.

there remains the outside possibility that the DNA sequences in these promoters interact with the same proteins as recognize the TATA box in conventional promoters, despite the dissimilarity in sequence. This question can only be evaluated by study of the proteins themselves at a biochemical level.

The work with DHFR and SV40 late raises interesting questions about the remaining components of the transcriptional machinery, those required in addition to Sp1, the "TATA factor" and similar DNA-binding proteins. Work carried out primarily with the adenovirus 2 major late promoter suggests that there are probably three of these transcription factors in addition to RNA polymerase II that are required for transcription. We and others have referred to these as "general" transcription factors, with the thought that they are involved in fundamental enzymatic aspects of initiation other than DNA sequence recognition, and that they will be required for initiation at all promoters. Clearly, this must be reexamined in light of the activity of unconventional promoters such as DHFR and SV40 late. There may be more than one mechanistic pathway for initiation, and we hope that the preliminary experiments described here will eventually enable us to investigate this possibility directly.

REFERENCES

Acheson, N. H. (1980) Lytic cycle of SV40 and polyoma virus. In: *DNA Tumor Viruses. 2d ed.*. ed. Tooze J., pp. 125–204. Cold Spring Harbor Laboratory, New York.

Bird, A. (1986) C_pG-rich islands and the function of DNA methylation. *Nature 321,* 209–13.

Brady, J., Radonovich, M., Vodkin, M., Natarajan, V., Thoren, M., Das, G., Janik, J. & Salzman, N. P. (1982) Site-specific base substitution and deletion mutations that enhance or suppress transcription of the SV40 major late RNA. *Cell 31,* 625–33.

Breathnach, R. & Chambon, P. (1981) Organization and expression of eucaryotic split genes coding for proteins. *Ann. Rev. Biochem 50,* 349–83.

Dynan, W. S. (1986) Promoters for housekeeping genes. *Trends In Genetics 2,* 196–7.

Dynan, W. S., Sazer, S. Tjian, R: & Schimke, R. (1986) Transcription factor Sp 1 recognizes a DNA sequence in the mouse dihydrofolate reductase promoter. *Nature 319,* 246–8.

Dynan, W. & Tjian, R. (1983a) Isolation of transcription factors that discriminate between different promoters recognized by RNA Polymerase II. *Cell 32,* 669–80.

Dynan, W. & Tjian, R. (1983b) The promoter-specific transcription factor Sp 1 binds to upstream sequences in the SV40 early promoter. *Cell 35,* 79–87.

Dynan, W. & Tjian, R. (1985) Control of eukaryotic messenger RNA synthesis by sequence-specific DNA-binding proteins. *Nature 316,* 774–8.

Everett, R. D., Baty, D. & Chambon, P. (1983) The repeated GC-rich motifs upstream from the TATA box are important elements of the SV40 early promoter. *Nucl. Acids. Res. 11,* 2447–64.

Farnham, P. J., Abrams, J. M. & Schimke, R. T. (1985) Opposite-strand RNAs from the 5' flanking region of the mouse dihydrofolate reductase gene. *Proc. Natl. Acad. Sci. USA 82.* 3978–82.

Farnham, P. J. & Schimke, R. T. (1986) Murine dihydrofolate reductase transcripts through the cell cycle. *Mol. Cell Biol. 6,* 365–71.

Fromm, M. & Berg, P. (1982) Deletion mapping of DNA regions required for SV40 early region promoter function in vivo. *J. Mol. Appl. Genet. 1,* 457–81.

Galas, D. & Schmitz, A. (1978) DNase footprinting: a simple method for the detection of protein-DNA binding specificity. *Nucl. Acids Res. 5,* 3157–70.

Haltiner, M. M., Kempe, T. & Tjian, R. (1985) A novel strategy for constructing clustered point mutations. *Nucl. Acids Res. 13,* 1015–25.

Herr, W. & Clarke, J. (1986) The SV40 enhancer is composed of multiple functional elements that can compensate for one another. *Cell 45,* 461–70.

Jongstra, J., Reudelhuber, T. L., Oudet, P., Benoist, C., Chae, C. B., Jeltsch, J. M., Mathis, D. & Chambon, P. (1984) Induction of altered chromatin structures by the SV40 enhancer and promoter elements. *Nature 308,* 708–14.

Kadonaga, J. T., Jones, K. A. & Tjian, R: (1986) *Trends in Biochem. Sci. 11,* 20–3.

Kadonaga, J. T. & Tjian, R. (1986) Affinity purification of sequence-specific DNA binding proteins. *Proc. Natl. Acad. Sci. USA 83,* 5889–5893.

McGrogan, M., Simonsen, C. C., Smouse, D. T., Farnham, P. J. & Schimke, R. T. (1985) Heterogeneity at the 5' termini of mouse dihydrofolate reductase mRNAs. *J. Biol. Chem. 260,* 2307–14.

McKnight, S. & Kingsbury, R. (1982) Transcriptional control signals of a eukaryotic protein-coding gene. *Science 217,* 316–24.

Melton, D. W., McEwan, C., McKie, A. B. & Reid, A. M. (1986) Expression of the mouse HPRT gene: deletional analysis of the promoter region of an X-chromosome linked housekeeping gene. *Cell 44,* 319–28.

Mitchell, P. J., Carothers, A. M., Han, J. H., Harding, J. D., Kas, E., Venolia, L. & Chasin, L. A. (1986) Multiple transcription start sites, DNase I-hypersensitive sites, and an opposite-strand exon in the 5' Region of the CHO dhfr gene. *Mol. Cell. Biol. 6,* 425–40.

Parker, C. S., & Topol, J. (1984) A drosophila RNA polymerase II transcription factor contains a promotor-region-specific DNA-binding activity. *Cell 36,* 357–69.

Reynolds, G. A., Basu, S. K., Osborne, T. F., Chin, D. J., Gil, G., Brown, M. S., Goldstein, J. L. & Luskey, K. L. (1984) HMG CoA reductase: A negatively regulated gene with unusual promoter and 5' untranslated regions. *Cell 38,* 275–85.

Rio, D., Robbins, A., Myers, R. & Tjian, R. (1980) Regulation of simian virus 40 early transcription by a purified tumor antigen. *Proc. Natl. Acad. Sci. USA. 77,* 5706–10.

Rio, D. & Tjian, R. (1984) Multiple control elements in the initiation of SV40 late transcription. *J. Mol. Appl. Genet. 2,* 423–35.

Sazer, S. & Schimke, R. T. (1986) *J. Biol. Chem. 261,* 4685–90.

Shimada, T. & Nienhuis, A. W. (1985) Only the promoter region of the constitutively expressed normal and amplified human dihydrofolate reductase gene is DNase I hypersensitive and undermethylated. *J. Biol. Chem. 260,* 2468–74.

Vigneron, M., Barrera-Saldaña, H. A., Baty, d., Everett, R. E. & Chambon, P. (1984) Effect of the 21-bp repeat upstream element on in vitro transcription from the early and late SV40 promoters. *EMBO J. 3.* 2373–82.

Zenke, M., Grundstrom, T., Matthes, H., Wintzerith, M., Schatz, C., Wildeman, A. & Chambon, P. (1986) Multiple sequence motifs are involved in SV40 enhancer function. *EMBO J. 55,* 387–97.

DISCUSSION

YANIV: Do you see any effect of these mutations on the major late start site at nucleotide 325?

DYNAN: You are asking about mutations LS1 through LS14. None of these has a large effect on the late start at nucleotide 325. There have been several reports from other groups of alterations in this far-upstream region having a large effect on the nucleotide 325 site. We would like to be able to understand the apparent discrepancy.

YANIV: Some people claim that if you mutate the Spl site and then check *in vivo* you increase late versus early transcripts. In your experiments I did not see anything like that.

DYNAN: No, the Spl site seems to be just as important for the late start at nucleotide 171 as for the early start. If you make a mutation that takes out all the strong Spl binding sites you lose the transcription from nucleotide 171 altogether. As for the late start at nucleotide 325, there have been reports both ways, either that the Spl-binding region is essential or that you actually stimulate *in vivo* transcription when you remove it. I cannot explain the discrepancy.

RIGBY: Do any of your scanning mutations affect the T antigen responsiveness?

DYNAN: I have not looked at that yet. As you know, if you add the purified T antigen *in vitro* you do not see a stimulation of late transcription (Rio et al. 1980 *Proc. Natl. Acad. Sci. 77,* 5706–5710). Presumably, the extracts lack a host cell factor that is required for T-antigen transactivation of late transcription.

HASELTINE: I have two questions. One is about Spl redirecting the specificity. Have you looked at that in any detail? And what are your thoughts regarding what you call the cryptic start site that appears once you make a deletion? How far away was it, and what do you know about it?

DYNAN: The cryptic start in mutant LS10 is in the linker itself. Its position corresponds to an RNA that is present as a minor start with some of the other constructs.

HASELTINE: So, you have created a start site. And to the redirection of specificity with Spl. Is there something limiting in your reactions, and if there is, what is it?

DYNAN: The system responds to added components as if no single component is limiting. If you add more of any component you get a stimulation. Your comment gets into the question of how Spl really works at the mechanistic level. We do not know the answer to that yet. Spl could influence binding of the RNA polymerase, or could affect polymerase isomerization, as is seen with some bacterial promoters. Presently, we are limited by the ability to isolate the components other than Spl. These are a little bit fragile, but in time we should be able to solve the technical problems.

YANIV: Concerning the distance between the Spl and the TATA-box proximal to the start site. In SV40 the Spl is close to the TATA-box in the early direction, in BK it is a little bit further upstream, and in some of the cellular promoters you have Spl 200–300 bp away. How do you think all this works?

DYNAN: In many of the promoters that have been looked at in detail the closest Spl site is something like 40–45 bp from the initiation site. We now know that this is true even in SV40; we just missed the two weakest sites in the early experiments. Similarly, in HSV thymidine kinase there is a very weak Spl site, I think about 40–45 bp from the transcriptional start site. The sites in DHFR are at a similar distance. I heard last night that Walter Schaffner's lab had done some experiments to move the Spl sites at a large distance from the transcriptional start site. Maybe Dr. Matthias can comment on the results of that experiment.

MATTHIAS: Edgar Serfling had inserted several copies of the Spl binding sites 3′ to the globin gene, but we did not see any enhancing effect. We had the idea that the 21 bp repeats could be functionally overlapping with an enhancer, but that does not seem to be the case.

ROIZMAN: When you have multiple Spl binding sites, are they always in the same orientation or can there be mixed, different orientations, and does the distance between them remain the same, or does it vary among promoters?

DYNAN: There are now 15 or 20 published promoter sequences that you can look at. Only a few of these have been tested experimentally, but in all of them you can see a good consensus sequence for Spl binding. By far the largest number have all the sites in the same orientation, but that orientation can differ from one promoter to another. In DHFR the G-rich strand reads in the sense of the coding region. In HPRT it is also the G-rich strand that reads in the sense of the coding region, but in APRT it is the C-rich strand. There are several examples with the herpes virus promoters, where the Sp sites have a mixed orientation. It is as if it really does not make much difference. We tried to figure this out by making some synthetic oligomers that could be put together either head-to-tail or head-to-head, but we did not get very far with that. The constructs we were making had no activity indicating that there is something about how Spl works that we do not understand.

ROIZMAN: What about the distance between individual Spl sites?

DYNAN: The distance varies. In SV40 they are as close as packing permits and they are all on the same face of the helix. The second transcription unit that we looked at was a monkey promoter of unknown function. It has three binding sites in a compact region. The middle site has been rotated around, so it is on opposite faces of the helix from the other two. Finally, we discovered a large number of promoters, such as DHFR and some of the herpes virus promoters where there is actually a clear unprotected region between the sites of Spl binding. So, as far as we know there is no pattern to the distance. For that matter, we do not really know what the upstream sites are doing. You could imagine that a site at a fixed distance from the promoter induces binding of RNA polymerase, but why there would be 5 or 3 additional sites further upstream is something that we do not yet know anything about.

RIGBY: A point that may be relevant to what you just said. Aaron Klug has argued (Rhodes & Klug 1986 *Cell 46,* 123–132) by analogy to the TFIIIA binding to the internal control region of RNA polymerase III promoters, in which its metal binding fingers look along the DNA at a certain periodicity that the important thing about Spl binding is not the consensus penta-nucleotide but the spacing of G residues; that there must be one G every half turn of the helix. Do any of the data that you have comment on that argument?

DYNAN: Every half turn?

RIGBY: Every half turn is important, yes. There is 10 1/2 bp to the turn of the helix and you have to have one G every 5 bases, but the sequence around those Gs does not matter. The rest of the sequence is only there to impose that spacing.

DYNAN: Certainly, if one looks at the known binding sites one would say that more of a consensus is present than simply a G every 5 bases.

RIGBY: And it would certainly explain why it does not matter which orientation you have.

DYNAN: In terms of the spacing between bases in the Spl binding site, I would add that there is no experimental evidence for what the helix looks like in these binding regions. Possibly, the DNA within the Spl sites is quite different from what we usually think of as B-form DNA. One sees odd chemical reactivities. One sees that the Gs in unoccupied sites are very much non-equivalent as far as the rate at which they are methylated by DMS. This is something that we are looking at, but we do not have any evidence on. We hope to measure binding to some chemically substituted oligonucleotides in conjunction with Marvin Caruther's lab.

LOWY: Have Spl binding sites been introduced into any genes that ordinarily do not contain them to see if the biology of these genes might change?

DYNAN: No, I cannot think of any examples.

LOWY: Are there any other features that are common to all the different genes in which Spl binding sites have been found?

DYNAN: There is no commonality to all the genes. There are a lot of genes for housekeeping enzymes, in fact almost all the ones that have been cloned that have GC-rich regions in the 5′ flanking regions. If you look at these sequences you can see areas that ought to be Spl binding sites. Unfortunately, there are also a number of miscellaneous genes, that you would not ordinarily call housekeeping genes, that have Spl sites as well. Metallothionein-IIA has one very nice site that is sitting among a lot of other regulatory elements.

MATTHIAS: Is there correlation between the strength of the binding and the function of the site?

DYNAN: We do not have numbers to the binding constants to different sites. There are clearly some promoters where some of the sites are very weak, and where the weak sites are important for transcription. That is, for example, clearly true in SV40, where mutants have been made that affect these very weak sites. These sites in SV40 are so weak that we could not even pick up the binding with the initial preparations of Spl.

MATTHIAS: If I recall correctly, in SV40 there is a Spl binding site where you do not get binding because of steric hinderance. Is that site important for transcription functionally? If one makes a mutation in there, does it increase the transcription *in vitro* or *in vivo?*

DYNAN: You are asking about GC box IV. You can mutate box IV without having much effect on transcription (Barrera-Saldana *et al.* 1985 *EMBO J. 4,* 3839–3849, Gidoni *et al.* 1985 *Science 230,* 511–517).

KJELDGAARD: Do you have information about mutations in the region of the second box in the vicinity of the start site of nucleotide 325?

DYNAN: The one clear effect we see is downstream. There are two mutations there that cause a very strong decrease in transcription. I think there is quite a bit of data in the older literature about possible downstream effects which we should go back to and examine in more detail.

Identification of Cellular Proteins that Interact with the Polyoma Virus Enhancer

M. H. Kryszke, J. Piette & M. Yaniv

Studies of DNA sequences that are required in cis for the transcriptional regulation of viral or cellular genes have revealed that crucial sequence elements can be located more than 100 base pairs upstream of the transcription initiation site. In many cases, these far upstream elements when cloned in both possible orientations, 5' or 3' to a transcription unit such as β-globin, can activate or enhance transcription from such a heterologous promoter (see recent review in Eukaryotic Transcription, Y. Gluzman (ed.), Cold Spring Harbor, 1985). These elements were defined as enhancers and they can be of both viral and cellular origin. Among viruses, enhancers were found in SV40, polyoma, adenovirus and papillomavirus, and they are present in the LTR of many retroviruses. Cellular enhancers are frequently but not exclusively associated with genes that are expressed only in specific cell types or tissues. The rearranged immunoglobulin heavy or light chain coding sequences include a lymphocyte-specific enhancer inside an intron (see Matthias *et al.,* this volume). The rat insulin gene transcription control element contains a tissue-specific enhancer which can activate transcription from the tk promoter only in endocrine cells of the pancreas (Edlund *et al.* 1985).

Soon after the discovery of enhancers, it became apparent that they can play an important role in events that occur during cell transformation. The integration of the Avian leukosis virus next to the cellular myc gene is associated with malignant lymphoma in the chicken. The viral LTR can either serve directly as a promoter for the cellular oncogene or, alternatively, it acts as an enhancer when the viral control elements are integrated 5' or 3' in both orientations away from the gene. Some chromosomal translocations in mouse plasmacytomas or human

Unité des virus oncogènes, VA 041149 du CNRS, Département de biologie moléculaire, Institut Pasteur, 25 rue du Dr. Roux, 75724 Paris Cedex 15, France.

VIRAL CARCINOGENESIS, Alfred Benzon Symposium 24,
Editors: N. O. Kjeldgaard, J. Forchhammer, Munksgaard, Copenhagen 1987.

Burkitt lymphomas bring the myc gene into the vicinity of the immunoglobulin enhancer (see Cory *et al.*, this volume). It is highly probable that other events concerning oncogene activation occur by proximal integration of viral enhancers or by chromosomal rearrangements that bring cellular enhancers next to oncogenes.

Studies of viral minichromosome structure or of the chromatin of the immunoglobulin domain have revealed that enhancers are frequently associated with a DNase I hypersensitive region. This segment of DNA, frequently devoid of nucleosomes, may be associated with cellular proteins that confer a typical pattern of protected and hypersensitive residues (Cereghini & Yaniv 1986). To try to understand the molecular mechanisms that are involved in enhancer function, we set out to study the cis DNA elements that constitute the polyoma virus enhancer and to search for cellular transacting proteins that are required for enhancer function.

RESULTS

The polyoma non-coding region contains at least two distinct enhancer elements with different cell specificities

Studies by several groups have shown that sequences located on the late side of the viral origin, more than 150 bp upstream of the early cap site, are crucial for early promoter function (Tyndall *et al.* 1981). A fragment of 246 bp located between the BclI and the PvuII sites on the late side of the origin was shown by de Villiers & Schaffner (1981) to exhibit enhancer activity when cloned next to the rabbit β-globin gene (see Figure 1). Herbomel *et al.* (1984) undertook to dissect this region by cloning different subfragments of this region 5' or 3' to a transcription unit including the chicken α-collagen promoter linked to the bacterial chloramphenicol acetyl transferase gene. Transfection of the different plasmids into mouse fibroblasts showed that two adjacent fragments, the BclI-PvuII fragment (109 bp) and the PvuII-PvuII fragment (138 bp) exhibit enhancer activity. The first one (element A) contains a core sequence homologous to the adenovirus 5 E1a enhancer as well as to the Ig heavy chain gene enhancer, and the second element (B) contains sequences homologous to the SV40, Ig and BPV1 enhancers. In fibroblasts the A enhancer was roughly 3- to 5-fold stronger than the B enhancer. A point mutation in the B enhancer of a polyoma mutant that was selected for growth in mouse embryonal carcinoma cells increased the activity of the B enhancer already in fibroblasts by 2-to 3-fold. When tested in embryonal carcinoma cells, the activity of the A enhancer decreased 3- to 4-fold as compared

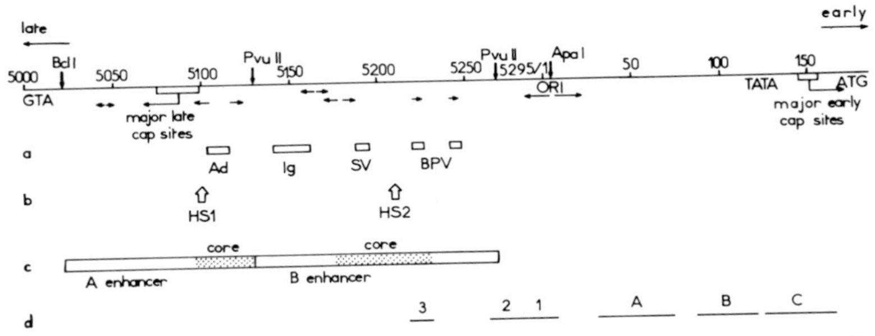

Fig. 1. The non-coding region of polyoma virus. The nucleotide numbering is according to Tyndall *et al.* (1981). The restriction sites for BclI, PvuII and ApaI are indicated by vertical arrows. Direct or inverted repeats are represented by horizontal arrows. The major late RNA cap sites (Cowie *et al.* 1981) and the major early cap sites (Cowie *et al.* 1982) are indicated by boxes with arrows. Also indicated are the replication origin (ORI) with the TATA box of the early promoter and the start codons for late and early genes.

a. Boxes indicate polyoma sequences homologous to the adenovirus 5 E1a enhancer (Ad) (Hearing & Shenk 1983), the immunoglobulin heavy chain gene enhancer (Ig) (Banerji *et al.* 1983), the SV40 enhancer (SV) (Weiher *et al.* 1985) and the BPV1 enhancer (BPV) (Weiher & Botchan 1984).

b. The two DNase I hypersensitive regions mapped in the chromatin by Herbomel *et al.* (1981) are designated by open arrows termed HS1 and HS2.

c. The A and B enhancers defined by Herbomel *et al.* (1984) are indicated by boxes. The dotted segments represent the enhancer core sequences.

d. The large T antigen binding sites are shown (Cowie & Kamen 1984).

to that measured in fibroblasts. On the contrary, the wild type B enhancer or the F9 variant of the B enhancer retained an activity identical to that observed in fibroblasts. Consequently, the F9 mutant enhancer exhibited a 3-fold higher activity (when placed 5' to the collagen promoter) than the A enhancer in PCC3 embryonal carcinoma cells. In the case of polyoma, replication of the viral DNA requires, in addition to the viral minimal origin, at least one enhancer element (Muller *et al.* 1983, de Villiers *et al.* 1984). DNA fragments containing either the A or the B enhancer element can fulfil this function (Luthman *et al.* 1982, Muller & Hassel 1986). In agreement with our results on the cell specificity of the polyoma enhancers, it was shown recently that only the B and not the A polyoma enhancer can activate viral DNA replication in the 2–4 cell stage mouse early embryos (Wirak *et al.* 1985).

Subcloning and deletion analysis suggested that each polyoma enhancer contains a crucial core element which by itself is not sufficient to supply full enhancer activity. Auxiliary sequences on either side of the core residues are also required

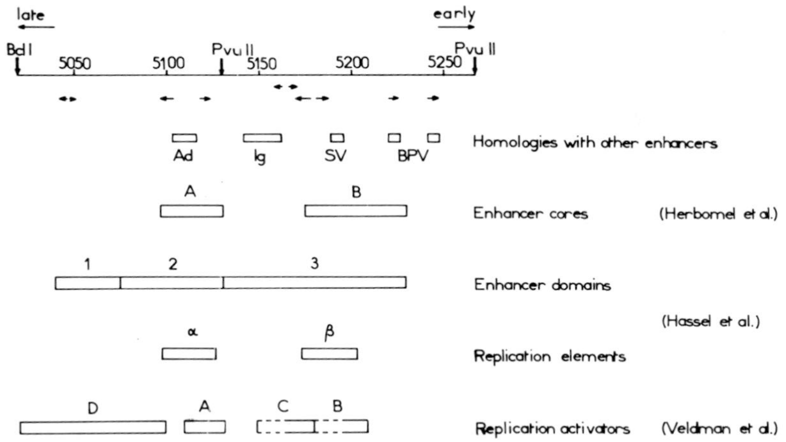

Fig. 2. Functional map of the polyoma virus enhancer region. As in Figure 1, the restriction sites for BclI and PvuII, the direct and inverted repeats, and the sequences homologous to other enhancers are indicated. Also represented are the A and B enhancer cores defined by Herbomel *et al.* (1984), the three enhancer domains described by Hassel *et al.* (1985), the elements involved in viral replication (Hassel *et al.* 1985) and the four functional domains of the enhancer, A, B, C and D, defined by Veldman *et al.* (1985).

for full enhancer function. The combined results of several groups suggest that the 246 bp BclI-PvuII fragment can be divided into four distinct domains: A, B, C and D as detailed in Figure 2. The A and B cores that were defined by Herbomel *et al.* (1984) are roughly identical to the α and β core elements defined by Muller & Hassel (1986) by deletion analysis as the crucial sequences required to activate replication when cloned next to the polyoma minimal origin. The conjunction of two consecutive elements, B+C, A+C or A+D is sufficient to supply enhancer function. Several repeats of the A or the B domain (the cores of the A and B enhancer elements) also supply such a function (Veldman *et al.* 1985, J. A. Hassel, personal communication).

Identification of three cellular proteins that interact with the polyoma enhancer

Several lines of evidence strongly suggest that cis acting enhancer sequences are the site of action of cellular factors. As previously mentioned, these sequences are excluded from normal nucleosomal structure and exhibit typical DNase I hypersensitive sites. Competition experiments have shown that excess DNA containing the crucial sequence element of an enhancer can compete with the homologous enhancer both *in vivo* and *in vitro* (Schöler & Gruss 1984, Sassone-

Corsi *et al.* 1985). For these reasons, we set out to search for proteins that interact specifically with the polyoma enhancer elements. Two experimental approaches were used: direct DNase I footprintings (Galas & Schmitz 1978) and gel retardation assays (Strauss & Varshavsky 1984). Nuclear extracts prepared from mouse 3T6 fibroblasts by salt wash were mixed with end-labelled DNA fragments in the presence of excess poly(dI-dC) or salmon sperm carrier DNA and treated briefly with DNase I. In parallel, end-labelled DNA was treated with nuclease in the absence of nuclear proteins. After deproteinization, the DNA samples were run on a sequencing gel alongside purine and pyrimidine sequence ladders. As shown in Figure 3, the addition of increasing amounts of nuclear proteins strongly protects two segments of viral DNA. A weak protection is observed along a third segment. These segments correspond to domains B-C, A and D of the polyoma enhancer. A dialysed extract which permitted the addition of higher protein concentrations to the assay and the maintenance of a lower ionic strength gave better protection of the D element. However the protection of the B-C element was lost after dialysis (Figure 3, b). The protection of the B-C domain is not continuous, it is interrupted by hypersensitive sites. Such hypersensitive sites or regions are also seen on the late side of domain A and on the early side of domain B. The protection is observed on both DNA strands, as summarized in Figure 6.

Another technique that we applied recently for the study of DNA-protein interactions in eucaryotes is the gel retardation assay. In this approach, a radioactively labelled fragment is mixed with excess non-specific carrier DNA before the addition of nuclear proteins. The mixture is then fractionated on a native polyacrylamide gel. The presence of proteins specifically interacting with the DNA fragment can give rise to a stable slow migrating band on the gel. We detected such a retarded band with DNA fragments containing the B-C enhancer domain (Figure 4a). No stable complex was detected under these conditions with DNA fragments containing the A or the D enhancer domain. To prove that the complex detected is sequence-specific, we proceeded with competition experiments. An excess of non-labelled DNA fragments containing the B-C domain abolished complex formation while an excess of fragments containing the A and D domains, the SV40 or the immunoglobulin enhancer did not interfere with the formation of the B-C complex (Figure 4b). We concluded that different proteins interact with the two different polyoma enhancer segments that we studied previously and that, despite some sequence homology between the polyoma and the SV40 or the Ig enhancer, the proteins complexed with the polyoma B-C

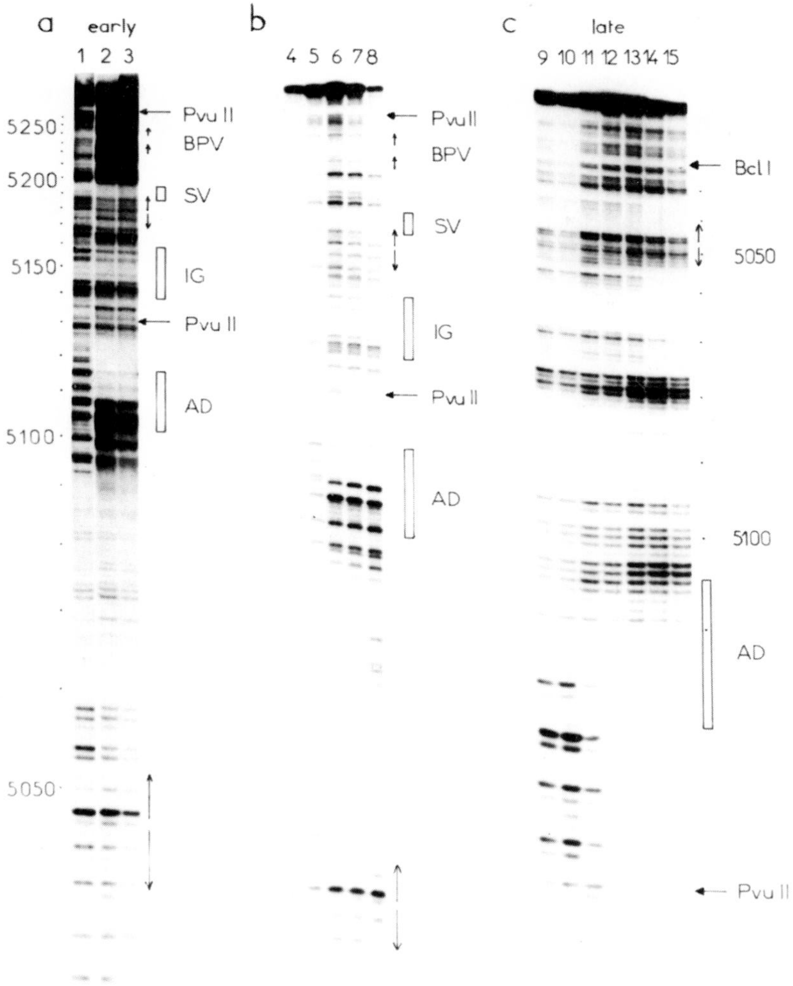

Fig. 3. Direct DNase I footprinting on the polyoma enhancer. Nuclear extracts from mouse 3T6 fibroblasts were prepared by washing isolated nuclei with a 0.4 M NaCl solution. They were incubated on ice with radioactively labelled DNA covering the whole enhancer region of polyoma, in the presence of poly(dI-dC)-poly(dI-dC) and treated by DNase I at 20°C. The mixture was then digested by proteinase K before phenol extraction, ethanol precipitation and electrophoresis on a sequencing gel. Panel a represents DNase I footprinting on the early strand. Lane 1 shows DNA digested in the absence of extract; lanes 2 and 3 show DNA digested in the presence of 18 μg proteins using two different DNase I concentrations. Panel b illustrates footprinting on the early strand using a dialysed nuclear extract. Lanes 4 and 5 contain DNA digested in the absence of extract; lanes 6, 7 and 8 DNA digested in the presence of 2.5, 5 and 10 μg proteins respectively.

Panel c shows DNase I footprinting on the late strand. Control DNA is shown on lanes 9 and 10. In lanes 11 to 15, 2.5, 5, 10, 20 or 30 μg proteins were used, respectively. Typical features of the DNA sequence are indicated, as symbolized in Figure 1.

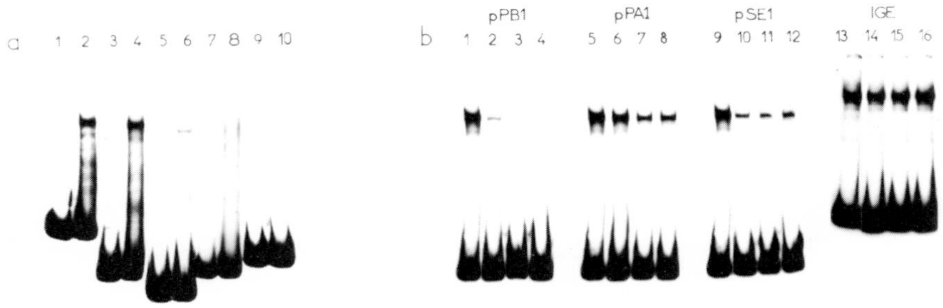

Fig. 4. Gel retardation assays and competition experiments. a. After incubation of the nuclear extract with radioactively labelled DNA in the presence of excess salmon sperm carrier DNA, the protein-DNA complex formed was separated from free DNA by electrophoresis on a native polyacrylamide gel. The 3T6 nuclear extract was tested for complex formation with DNA fragments containing the whole enhancer region of polyoma virus (1, 2), the B enhancer alone (3, 4), the B enhancer core sequence (5, 6), the A enhancer (7, 8) or only procaryotic sequences (9, 10). Lanes designated by odd numbers contain only free DNA incubated in the absence of extract, lanes designated by even numbers correspond to DNA incubated in the presence of nuclear extract.
b. The competition experiments were carried out as follows. The nuclear extract was incubated with a labelled DNA fragment containing the B enhancer of polyoma virus in the presence of excess pML2 carrier DNA. Increasing amounts of pML2 were replaced by cold competitor DNA containing the B enhancer of polyoma virus (pPB1, lanes 2–4), or the polyoma A enhancer (pPA1, lanes 6–8) or the SV40 enhancer (pSE1, lanes 10–12) or the Ig enhancer (IGE, lanes 14–16). The amounts of competitor DNA used were 0 (lanes 1, 5, 9 and 13), a 7-fold molar excess over probe (lane 14), a 20-fold molar excess (lanes 2, 6, 10, 15), a 40-fold molar excess (lane 16), a 60-fold molar excess (lanes 3, 7, 11) and a 100-fold molar excess (lanes 4, 8, 12).

domain are different from those interacting directly with the SV40 or the Ig enhancer.

Mapping of fine contacts in the B enhancer element

The gel retardation assay permits a relatively easy analysis of the minimal sequences required for complex formation. Using a series of Bal31 deletions, we could show that removal of sequences from the origin to nucleotide 5198 (S48) did not hamper complex formation. Deletion of another 10 bp segment (S49) abolished complex formation. Coming from the late side, deletion to nucleotide 5140 (E3) did not affect complex formation, while deletions to nucleotide 5165 (E1) or nucleotide 5173 (E17) greatly reduced or abolished complex formation, respectively. Replacement of the residues deleted in E1 with pBR322 sequences largely restored the yield of complex. This was not true for E17 or S49.

In gel retardation assays, only a small fraction of the total DNA is complexed with the specific factor. Brief DNase I treatment prior to application to the gel permits the identification of the phosphodiester bonds protected against DNase I cleavage in the final complex as shown in Figure 5a. The protected domain on both strands is roughly identical to that observed by direct DNase I footprinting. Both the B and C segments are protected with a hypersensitive site in between. A similar approach also permitted the mapping of close contacts between the protein and the bases or phosphate residues (Siebenlist & Gilbert 1980). Treatment of the polyoma DNA fragment with dimethylsulfate (DMS) before complex formation allowed the identification of sites of purine methylation that interfere with protein binding (methylation interference). As shown in Figure 5b and summarized in Figure 6, methylation of G or A residues in a stretch of 11 nucleotides interferes with complex formation. These residues consist of the early proximal part of the GC-rich palindrome and include only the first two G residues of the GTGTGGTTT Weiher and Gruss enhancer consensus sequence (Weiher *et al.* 1983). Methylation of purine residues in the symmetrical late proximal part of the palindrome did not interfere with complex formation even though these residues were protected against DNase I digestion. DMS protection experiments showed, in addition to the protection of the purine stretch, increased methylation of two G residues on the early strand. No close contacts were revealed by DMS interference or protection in the C domain homologous to the Ig enhancer.

DISCUSSION

Deletion and point mutation analysis in the viral context, as well as construction of chimeric plasmids and analysis of their biological activity permitted the definition of at least two distinct enhancer elements in the polyoma non-coding region. For their full activity, the A and B core enhancer elements require either auxiliary sequences located in the C and D domains or polymerization of a single core element. In this paper we describe the identification of three distinct mouse proteins that interact with different domains of this region.

PEB1 (polyoma enhancer binding protein 1): The most detailed study was undertaken with this factor. The possibility to isolate a stable complex between this factor and the B enhancer permitted the identification of the fine contacts between this protein and the DNA double helix. As judged by DNase I protection, this factor covers roughly 50 bp on both strands. This binding induces hypersensitive sites on the early proximal boundary and roughly in the middle of the protected domain. Altogether, this factor protects the Ig homology, the GC-rich palindrome and the Weiher and Gruss SV40 enhancer homology. Deletion of the

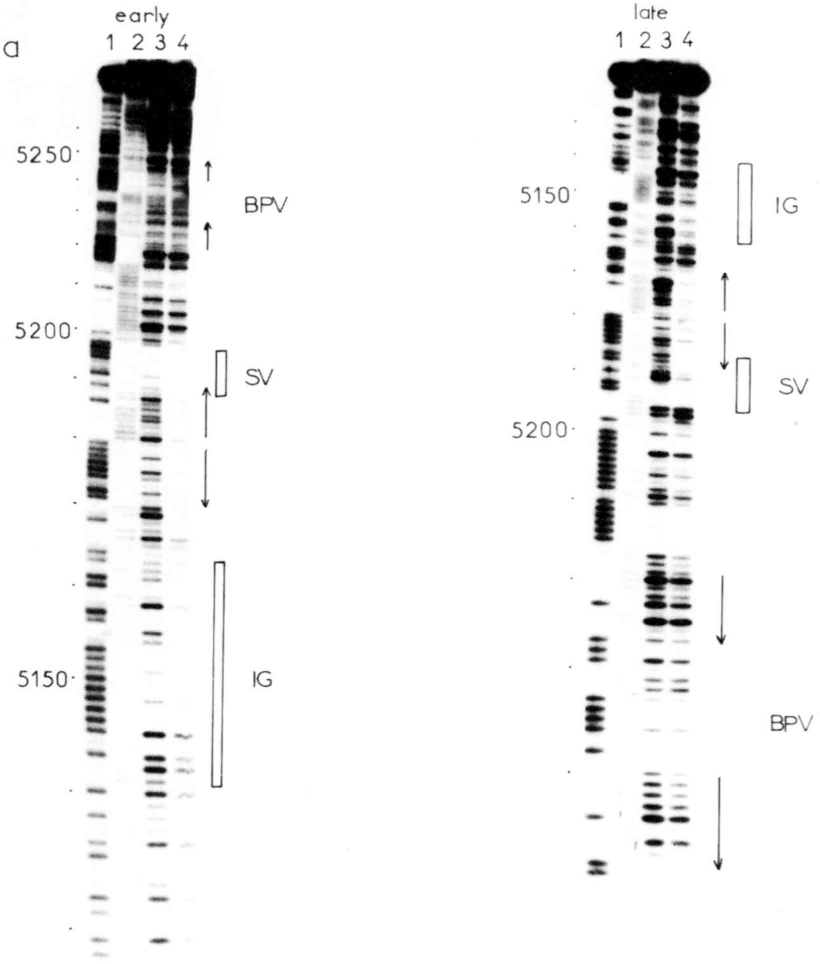

Fig. 5. Mapping of fine contacts within the B enhancer. a. DNase I footprinting. The nuclear extract was incubated with radioactively labelled DNA encompassing the B enhancer of polyoma virus. The mixture was then treated with DNase I before being loaded on a native polyacrylamide gel. Free DNA and complexed DNA were separated by electrophoresis, eluted, and their respective cleavage patterns were compared on a sequencing gel. The left panel illustrates footprinting on the early strand, the right panel footprinting on the late strand. Lanes 1 and 2 show G+A and C>T degradation products, lane 3, free DNA and lane 4, retarded complexed DNA. Characteristic features of the sequence are indicated as symbolized in Figure 1.

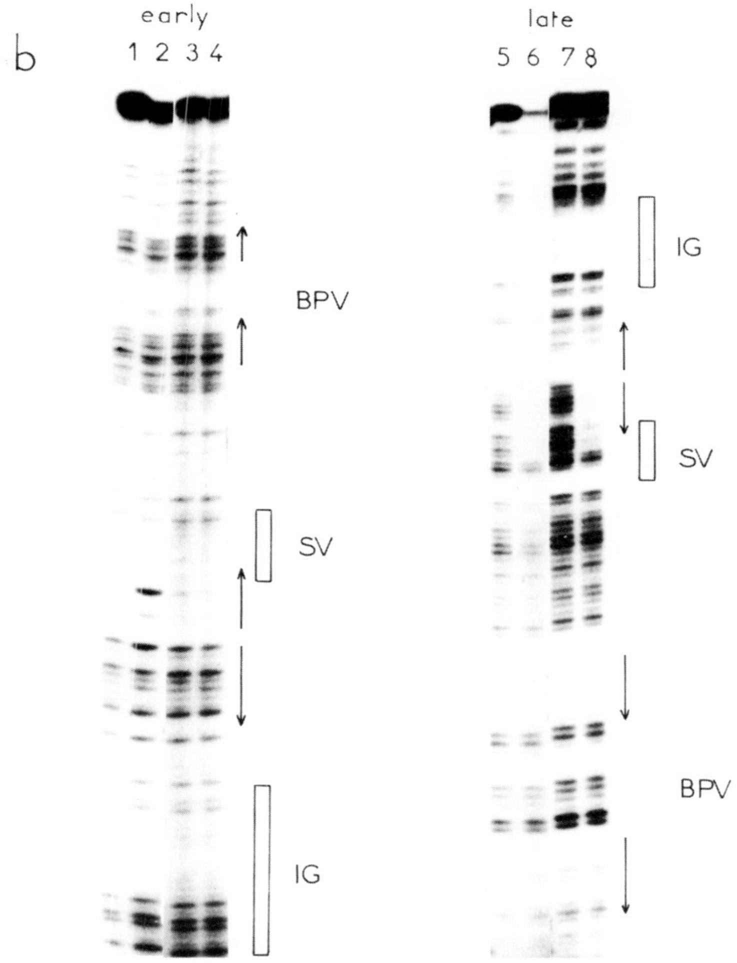

b. Methylation protection and interference. The DNA fragment containing the B enhancer of polyoma virus was treated with DMS either before (methylation interference experiments, lanes 3, 4, 7, 8) or after (protection experiments, lanes 1, 2, 5, 6) incubation with the nuclear extract. Separation and gel analysis were performed as described previously. The left panel shows the results obtained with the early strand, the right panel corresponds to the late strand. Free DNA appears in lanes 1, 2, 5 and 7, retarded complexed DNA in lanes 2, 4, 6 and 8. Typical features of the sequence are also indicated.

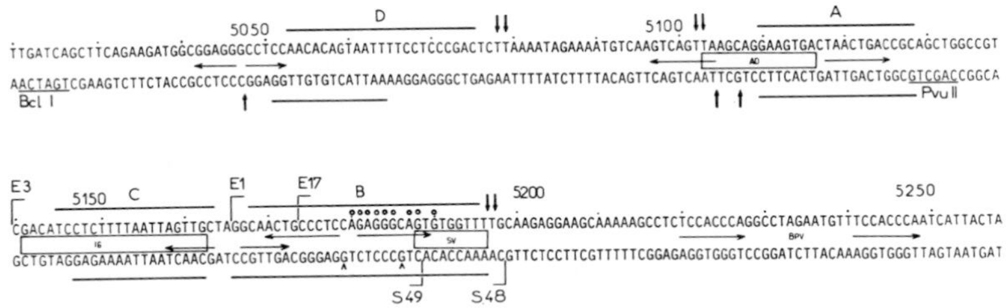

Fig. 6. Summary of the protection and interference experiments and position of the deletion endpoints on the DNA sequence. The sequences of both strands of the polyoma enhancer are represented with the typical features depicted previously. Sequences protected against DNase I cleavage are indicated by lines, sites hypersensitive to DNase I are schematized by vertical arrows. o symbolizes the sites of methylation interference and protection, ∧ sites hypersensitive to DMS. Deletion endpoints referred to in the text are indicated by hooks.

Ig homology (domain C) reduced but did not abolish complex formation. Its replacement by pBR322 sequences almost completely restored the ability of the B domain to form a stable complex. However these plasmid sequences were no longer protected against DNase I. The dimethylsulfate (DMS) interference and protection data confirm the primordial role of the origin proximal sequences of the B domains. We detected close contacts between the protein and G or A residues in a sequence of 11 bp covering the origin proximal part of the GC-rich palindrome and only part of the SV40 homology. It is tempting to conclude that this factor differs from the factors that interact directly with the Weiher and Gruss core (Wildeman *et al.* 1986, S. L. McKnight, personal communication). It is possible that the contacts with G residues outside of the palindrome are important since we do not observe any symmetry in DMS exclusion and protection experiments. The nature of the DNase I protection pattern, the DMS sensitivity and the possibility of replacing part of the protected domain with neutral sequences is somewhat reminiscent of the interaction of TFIIIA transcription factor with the 5S gene of Xenopus (Sakonju & Brown 1982). However, TFIIIA does not interact with the polyoma B enhancer, nor does our factor with the TFIIIA recognition domain on Xenopus 5S gene (J. Piette, unpublished observations). It is possible that PEB1 is a member of the class of DNA binding proteins that resemble TFIIIA by having repeated finger-like domains which bind Zn^{2+} ions and interact with DNA specifically (Miller *et al.* 1985).

PEA1: DNase I footprinting revealed strong binding of a cellular factor to the A or *a* enhancer core element (the A domain). The sequence homologous to the adenovirus E1a enhancer constitutes the boundary of the protected segment, with DNase I hypersensitive sites on its late side. As already mentioned, this complex was not visible in gel retardation assays. Perhaps the association constant is not sufficiently high to prevent dissociation of the DNA-protein complex during gel electrophoresis.

PED1: Direct DNase I footprinting revealed weak protection between residues 5051 and 5075. This protection was more pronounced with dialysed extracts, conditions which permitted the analysis of the binding in lower ionic strength and in the presence of higher protein concentrations. Both strands were protected to roughly the same extent. The protected area at least partially covers the D domain of the polyoma enhancer sequences that correspond to an auxiliary element or perhaps also have intrinsic enhancer function. Further characterization of the nature of this element and its precise boundary are required to confirm the possible correlation between the DNase I footprint we observed and the crucial sequences of this element.

The DNase I protection pattern that we observed *in vitro* with alternation of protected and hypersensitive sites is very similar to the DNase I hypersensitivity pattern observed in viral chromatin by Herbomel *et al.* (1981). The sequences highly protected *in vivo* correspond to the B and C domains, the hypersensitive sites observed *in vivo* correspond to the sites seen *in vitro* in the origin proximal part of the B element and in the late proximal part of the A element. All these observations strongly argue that the protection pattern we see *in vitro* is relevant to the events occurring *in vivo*. It is still possible that additional cellular proteins directly interact with the polyoma enhancer region or, alternatively, that they interact with the protein moieties of the complexes we described here. Evidence for the binding of still another protein to the polyoma enhancer was presented by Fujimura (1986) who characterized a protein which footprints on the boundary between the B and C domains. We have also detected such a protein in extracts prepared from slow growing cells; its concentration increased after cells reached confluency. Such a factor may be involved in the repression of enhancer function and not in its activation. In certain experiments, we also observed binding to a CCAAT box-like sequence located between the two repeats homologous to the BPV1 enhancer (M.-H. Kryszke, unpublished observations). Furthermore, the single base mutation occurring in Py mutants selected for growth in F9 mouse embryonal cells generates a carcinoma site for the TGGCA binding protein or nuclear factor I (Borgmeyer *et al.* 1984). However, we could not detect any footprinting on this element with extracts that react with similar sequence elements present in the chicken lysozyme gene or the adenovirus terminal repeats. Gel retardation experiments with different constructions containing the B-C domain as well as the generation of DNase I hypersensitive sites on its boundary suggest that protein binding induces local bending of the DNA fragment (J. Piette *et al.*, in preparation). Such a change in DNA conformation may be related to enhancer function.

The sequence motifs identified in the present study are certainly not unique to the polyoma genome. As already mentioned, some of them are part of other cellular or viral enhancers. Other cellular genes may include similar motifs in their regulatory sequences. The proteins that we have identified here are certainly involved in the regulation of transcription of cellular genes or in the regulation of cellular DNA synthesis. When polyoma virus infects mouse fibroblasts at least three proteins we have identified interact with the enhancer region and activate replication and transcription. Other enhancers, like that of SV40 or the Ig heavy chain or Kappa light chain enhancers, also interact with several proteins (Wildeman *et al.* 1986). Nevertheless, in several cases it was shown that the tandem repetition of one domain of an enhancer can restore full enhancer activity, as was shown for the A and B domains of the polyoma enhancer (Veldman *et al.* 1985, J. A. Hassel, personal communication). How can several tandem repeats of a unique protein DNA complex fulfil a function identical to that of a complex array of DNA-protein interactions? It is possible that all enhancer binding proteins are part of the same class of proteins having distinct sequence specificities located in one polypeptide domain and sharing another functional polypeptide domain which interacts with RNA polymerase, transcription factors or elements of the replication machinery. Alternatively, each of these proteins may interact with a different component of the transcription machinery, their acting being additive. Finally, one could postulate that the binding of different proteins along the enhancer will result in the same structural change around the early promoter or the viral origin of replication.

The action at a distance in an orientation-independent fashion is still difficult to explain. A plausible model involves DNA bending and direct contacts between proteins associated with the enhancer and the transcription initiation site. However, other models cannot be excluded. As suggested, enhancers may serve as entry sites for RNA polymerase or transcription factors. Entry will be followed by a directional sliding of the protein complex formed on the enhancer until it reaches a transcription initiation site. Another possibility is the existence of a glue-like protein that transmits enhancer effects bidirectionally. Finally, although models that involve a topological change induced by the enhancer, or the binding to a specific site on the nuclear matrix, do not agree with some known facts, they cannot be totally excluded yet.

It is clear that a continued effort to characterize and isolate proteins that interact with enhancers and the development of *in vitro* systems will help in understanding the mode of action of these 'magic' elements.

REFERENCES

Banerji, J., Olson, L. & Schaffner W. (1983) A lymphocyte specific cellular enhancer is located downstream of the joining region in immunoglobulin heavy chain genes. *Cell 33*, 729–40.

Borgmeyer, U., Nowock, J. & Sippel, A. E. (1984) The TGGCA binding protein: a eukaryotic nuclear protein recognizing a symmetrical sequence on double stranded linear DNA. *Nucl. Acids Res. 12,* 4295–311.

Cowie, A., Jat, P. & Kamen, R. (1982) Determination of sequences at the capped 5' ends of polyoma virus early region transcripts synthesized *in vivo* and *in vitro* demonstrates an unusual microheterogeneity. *J. Mol. Biol. 152,* 225–55.

Cowie, A. & Kamen, R. (1984) Multiple binding sites for polyomavirus large T antigen within regulatory sequences of polyomavirus DNA. *J. Virol. 52,* 750–60.

Cowie, A., Tyndall, C. & Kamen, R. (1981) Sequences at the capped 5' ends of polyoma virus late region mRNAs: an example of extreme terminal heterogeneity. *Nucl. Acids Res. 9,* 6305–22.

de Villiers, J. & Schaffner, W. (1981) A small segment of polyoma virus DNA enhances the expression of a cloned rabbit β-globin gene over a distance of at least 1400 base pairs. *Nucl. Acids Res. 9,* 6251-64.

de Villiers, J., Schaffner, W., Tyndall, C., Lupton, S. & Kamen, R. (1984) Polyoma virus DNA replication requires an enhancer. *Nature 312,* 242–6.

Edlund, T., Walker, M. D., Barr, P. J. & Rutter, W. J. (1985) Cell specific expression of the rat insulin gene: evidence for role of two distinct 5' flanking elements. *Science 230,* 912–6.

Ephrussi, A., Church, G. M., Tonegawa, S. & Gilbert, W. (1985) B lineage specific interactions of an immunoglobulin enhancer with cellular factors. *Science 227,* 134–40.

Fujimura, F. K. (1986) Nuclear activity from F9 embryonal carcinoma cells binding specifically to the enhancers of wild type polyoma virus and PyEC mutant DNAs. *Nucl. Acids Res. 14,* 2845–61.

Galas, D. J. & Schmitz, A. (1978) DNase footprinting: a simple method for the detection of a protein: DNA binding specificity. *Nucl. Acids Res. 5,* 3157–70.

Hassel, J. A., Mueller, C. R. & Muller, W. J. (1985) The polyoma virus enhancer: multiple sequence elements required for transcription and DNA replication. *Current Comm. in Mol. Biol.,* Eukaryotic transcription, pp. 33–40. Cold Spring Harbor Laboratory, New York.

Hearing, P. & Shenk, T. (1983) The adenovirus type 5 E1A transcriptional control region contains a duplicated enhancer element. *Cell 33,* 695–703.

Herbomel, P., Bourachot, B. & Yaniv, M. (1984) Two distinct enhancers with different cell specificities coexist in the regulatory region of polyoma. *Cell 39,* 653–62.

Herbomel, P., Saragosti, S., Blangy, D. & Yaniv, M. (1981) Fine structure of the origin proximal DNAse I hypersensitive region in wild type and EC mutant polyoma. *Cell 25,* 651–8.

Luthman, H., Nilsson, M. G. & Magnusson, F. (1982) Non-contiguous segments of the polyoma genome required in cis for DNA replication. *J. Mol. Biol. 161,* 533–50.

Miller, J., McLachlan, A. D. & Klug, A. (1985) Repetitive zinc-binding domains in the protein transcription factor III A from Xenopus oocytes. *EMBO J. 4.* 1609–14.

Muller, W. J. & Hassel, J. A. (1986) *Cancer Cells 4,* Cold Spring Harbor Laboratory, New York (in press).

Muller, W. J., Mueller, C. R., Mes, A-M. & Hassel, J. A. (1983) Polyoma virus origin for replication comprises multiple genetic domains. *J. Virol. 47,* 586–99.

Sakonju, S. & Brown, D. D. (1982) Contact points between a positive transcription factor and the Xenopus 5S RNA gene. *Cell 31,* 395–405.

Sassone-Corsi, P., Wildeman, A. & Chambon, P. (1985) A transacting factor is responsible for the SV40 enhancer activity in vitro. *Nature 313,* 458–63.

Schöler, H. R. & Gruss, P. (1984) Specific interaction between enhancer containing molecules and cellular components. *Cell 36,* 403–11.

Siebenlist, U. & Gilbert, W. (1980) Contacts between E. coli RNA polymerase and an early promoter of phage T7. *Proc. Natl. Acad. Sci. 77,* 122–6.

Strauss, F. & Varshavsky, A. (1984) A protein binds to a satellite DNA repeat at three specific sites that would be brought into mutual proximity by DNA folding in the nucleosome. *Cell 37,* 889–901.

Tyndall, C., La Mantia, G., Thacker, C. M., Favaloro, J. & Kamen, R. (1981) A region of the polyoma virus genome between the replication origin and late protein coding sequences is required in cis for both early gene expression and viral DNA replication. *Nucl. Acids Res. 9,* 6231–50.

Veldman, G. M., Lupton, J. & Kamen, R. (1985) Polyoma virus enhancer contains multiple redundant sequence elements that activate both DNA replication and gene expression. *Mol. Cell. Biol. 5,* 649–58.

Weiher, H. & Botchan, M. R. (1984) An enhancer sequence from bovine papillomavirus DNA consists of two essential regions. *Nucl. Acids Res. 12,* 2901–16.

Weiher, H., König, M. & Gruss, P. (1983) Multiple point mutations affecting the SV40 enhancer. *Science 219,* 626–31.

Wildeman, A. G., Zenke, M., Schatz, C., Wintzerith, M., Grundstrom, T., Matthes, H., Takahashi, K. & Chambon, P. (1986) Specific protein binding the SV40 enhancer in vitro. *Mol. Cell Biol. 6.* 2098–105.

Wirak, D. O., Chalifour, L. E., Wassarman, P. M., Muller, W. J., Hassel, J. A. & de Pamphilis, M. L. (1985) Sequence-dependent DNA replication in preimplantation mouse embryos. *Mol. Cell Biol. 5,* 2924–35.

Yaniv, M. & Cereghini, S. (1986) Structure of transcriptionally active chromatin. *CRC Crit. Rev. Biochem. 21,* 1–26.

DISCUSSION

KJELDGAARD: Have you looked at protein binding of DNA probes in Western blots?

YANIV: We have done one or two experiments without much success. You see very nicely histones or other abundant DNA binding proteins with this approach. We and other people (e.g. S. L. McKnight) believe that you need partially purified proteins to use this kind of Western blot approach. With total extracts everything binds.

CELIS: How abundant are these proteins?

YANIV: Astonishingly, these proteins should be either quite abundant if you can detect them in crude extracts or in nuclear wash by band shift or by direct footprinting, or alternatively they have a very high affinity for their targets.

CELIS: What sort of percentage of the total cellular proteins?

YANIV: I don't know what is the estimate for Spl for example, but it is not 10^5 or 10^6 molecules per cell. On the other hand, it is not, say, 20 or 100 molecules per cell. If it was in this range, we would not see them in these assays.

DYNAN: The latest estimate from Tjian's lab is something like 10 000 molecules of Spl per cell. They arrive at this number by quantitating the number of DNA molecules that are occupied in a binding reaction.

YANIV: Yes, it is in this range. We estimated 10 000 to several times ten thousand. It is astonishingly high for regulatory proteins. But still, in the case of a virus you finish with a cell that has 50 000 or 100 000 copies of the viral genome after infection. Not all of them are transcriptionally active, but still if 1% of them are transcriptionally active, there is quite a lot. For example, in the case of SV40 infection our DNase 1 protection studies *in vivo* indicated that all the SV40 molecules contain Sp1 because they are protected against DNase 1 in that region, and there are 50–100 000 DNA molecules (Piette *et al.* 1986 *Cancer Cell 4,* (in press)).

VAN DER EB: Did you test the effect of adenovirus E1A on your enhancer mutant that is active in the EC cells?

YANIV: There is a recent paper from Chambon's lab on this (Hen *et al.* 1986 *Nature 321,* 249–251). Using the same mutant they showed that E1A represses the wild type but does not repress the mutant enhancer.

HANAHAN: In view of the implication that you have tissue-specific factors, can you review what is known about the tissue specificity of polyoma virus in mice? Does this make sense in context of its normal properties?

YANIV: The wild type virus does not grow in embryonal carcinoma cells of the mouse, and when you select for something that grows – that was done by Blangy, by Levine, by Fujimura and Linney – you get modifications in the enhancer region, and one of the modifications that people both in Paris, Stony Brook, and San Diego got is the F9 mutation that changes an A to a G at position 5133. So just one mutation changes the specificity of the virus. These viruses grow quite well in embryonal carcinoma cell and maintain growth in fibroblasts. When Yoshi Ito selected for polyoma that grows in trophoblasts he got a small deletion between the A and B domain of the viral enhancer, as if you have to bring them closer by a few nucleotides. When Paulo Amati selects for polyoma growth in Friend cells or in neuroblastoma, he gets duplications and deletions. So, it is clear that every time you leave the fibroblastic or epithelial cells of the mouse and you select for growth on a specific type of cells you observe changes in this region (see review Amati 1985 *Cell 43,* 561–562). It appears as if different cells have varying concentrations of enhancer binding factors or different partners of a group of factors, and you have to change the distance or change the affinity by mutations in the DNA sequence in order to get action. Alternatively you can delete some elements if the corresponding factor is missing and duplicate others for which the factor is present. It seems that both mechanisms occur with different polyoma variants. In conclusion, the system behaves as if you have to assemble an active complex with several components along the enhancer; if any of these factors is modified or missing you have to adopt the DNA sequence to the new situations.

II. Transcriptional and Translational Control

Introduction

R. C. Gallo

This afternoon's session will focus on the human retroviruses. I would make some brief introductory remarks. In addition to causing fatal diseases involving T-cells, either malignant growth or a depletion of the same cells, the human retroviruses have been of fundamental interest because of several novel features. One of these is the *in vitro* transforming or immortalizing effects of HTLV-I and II. Of course, the virus that causes AIDS does the opposite, viz. depleting T4 cells *in vivo* and killing these cells *in vitro*. A second unusual feature of these viruses is that very related viruses have been discovered in monkeys, in particular the African green monkey, where a virus about 95% identical to the human leukemia virus was discovered several years ago. In addition, a relative of HTLV-III, the virus that causes AIDS, has also been discovered in the same animal. The third and last point I want to mention is that the proviral sequences do not have a common site of integration. Although the tumors are clonal for the provirus, the provirus can be found in different sites, different chromosomal regions in tumors from different patients. If leukemia induction were to involve activation of cell genes for growth by the viral sequences, then these results indicated that a trans-mechanism must be involved. This session will deal with these new mechanisms which human retroviruses employ and the new genes involved in these mechanisms.

Structure and Function of the HTLV-III Genome

Flossie Wong-Staal

STRUCTURE OF THE HTLV-III GENOME

The human T-lymphotropic retroviruses are important not only because they are the first and only retroviruses isolated from humans and are etiologically associated with leukemia and AIDS, but also because they provide new models for virus-cell interactions and gene regulation. HTLV-I and HTLV-II, both associated with leukemias, transform T-cells *in vitro*, while HTLV-III, also called lymphadenopathy virus (LAV) or human immunodeficiency virus (HIV), kills T-4 cells and is associated with the acquired immune deficiency syndrome.

The unusual biological properties of these viruses can in part be explained by their complex genetic structures. All non-defective retroviruses contain 3 structural genes coding for the core antigens (gag), the polymerase (pol), also called reverse transcriptase, and the envelope glycoproteins (env). These genes are bounded by the long terminal repeat sequences (LTR) which house the regulatory elements for transcription and translation. Most retroviruses contain only these genes (see Figure 1, Mo-MuLV as an example). A minority of retroviruses are the so-called acutely transforming viruses, which have acquired a cellular gene by recombination sometime in their evolutionary past and frequently, in so doing, have lost part of their replicative genes in the exchange and become defective (Figure 1, SSV as an example). All the HTLVs contain gene(s) in addition to gag, pol and env that have no counterparts in normal chromosomal DNA. HTLV-I and HTLV-II contain an extra gene located between the env gene and 3' LTR (Seiki *et al.* 1983, Shaw *et al.* 1984a) (Figure 1). The protein product of this gene, which is highly conserved between the 2 viruses, has been shown to mediate transcriptional activation of the viral TLR and genes linked to it (Haselti-

---

Laboratory of Tumor Cell Biology, National Cancer Institute, National Institutes of Health, Bethesda MD, U.S.A.

VIRAL CARCINOGENESIS, Alfred Benzon Symposium 24,
Editors: N. O. Kjeldgaard, J. Forchhammer, Munksgaard, Copenhagen 1987.

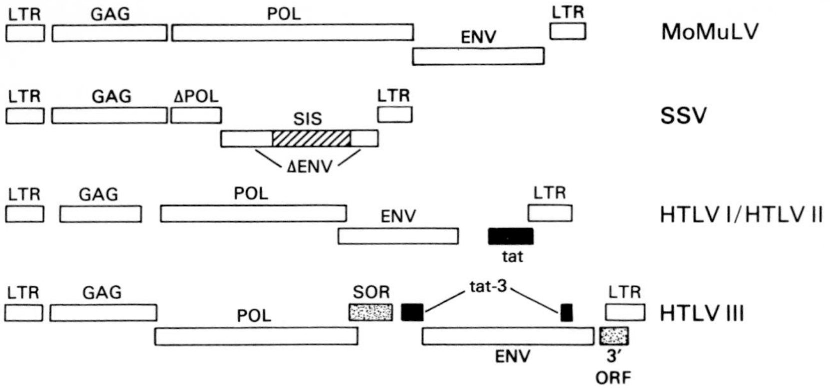

Fig. 1. Genetic structures of retroviruses MoMuLV (Moloney murine leukemia virus) SSV (simian sarcoma virus); GAG, group-specific core antigens; POL, DNA polymerase; ENV, envelope; LTR, long terminal repeat; SIS, the oncogene of SSV; tat, transactivator; SOR, short open reading frame; 3' ORF, 3' open reading frame. The gene trs which overlaps with tat-3 but uses a different reading frame is not shown.

ne *et al.* 1984, Sodroski *et al.* 1984, Felber *et al.* 1985). The target nucleotide sequences for this transactivation have been identified and found to overlap with the enhancer and promotor sequences for transcription (Rosen *et al.* 1985). The presence of this gene, now named tat-1 and tat-2 for transactivator, is critical for virus replication (Chen *et al.* 1985). HTLV-III also encodes a transactivator gene (tat-3) which is specific for its LTR (Sodroski *et al.* 1985a, b, Arya *et al.* 1985, Seigel *et al.* 1986). However, the major effect of this activation is post-transcriptional, as detailed later. The tat gene of HTLV-III is transcribed into a doubly-spliced 2.0 kb mRNA (Arya *et al.* 1985). The relatively small tat coding region of 86 amino acids is split between the 2nd and 3rd exons, with the major and critical functional domain residing in the 2nd exon. Within this domain is a stretch of highly basic residues (8/9 are lysine or arginine) which may be responsible for nucleic acid binding (Arya *et al.* 1985, Sodroski *et al.* 1985b, Seigel *et al.* 1986). In addition to tat-3, the nucleotide sequence of HTLV-III revealed two additional potential coding regions (sor and 3' orf) (Ratner *et al.* 1985, Sanchez-Pescador *et al.* 1985, Wain-Hobson *et al.* 1985) (Figure 1). Although there is now evidence that these are real genes coding for proteins that are immunoreactive with patients' sera (Kan *et al.* 1986, Arya *et al.* 1986, Franchini *et al.* 1986), their functions are as yet unknown. A 7th gene which also regulates virus expression has recently been identified (Sodroski *et al.* 1986, Feinberg *et al.*, 1986). This gene (trs) regulates the levels of spliced and unspliced viral mRNA, and utilizes an alternate reading frame in the same region of tat.

HTLV-III AND CYTOPATHICITY: VIRAL AND CELLULAR FACTORS

HTLV-III infects and kills the T4 cell *in vivo* and *in vitro*. As a first step to dissect the genetic determinants of the cytopathic activity of HTLV-III, we have obtained a cloned virus genome which directs the synthesis of infectious and cytopathic virions (Fisher *et al.* 1985). Normal cord blood T-cells transfected with this clone contained viral DNA and expressed p15 and p24, extracellular reverse transcriptase, and complete virions with the characteristic cylindrical cores as detected by electron microscopy. The percentage of virus-infected cells increased steadily up to 18 d post-transfection, suggesting that infectious virus was being produced. However, after 20 d, there is a dramatic drop in cell viability, rapidly leading to depletion of the infected cells. At 30 d, the culture consisted only of uninfected cells. Furthermore, while more than half the cells were OKT4+ at d 0, greater than 90% of cells are of the OKT8 phenotype at d 30, indicating selective killing of the OKT4+ cells (A. Fisher *et al.*, unpublished). Therefore, the HTLV-III genome alone contains all the essential information for inducing the primary defect in AIDS, i.e., depletion of the T4+ helper cells.

There is also evidence for cellular determinants for the cytopathic response. We have superinfected HTLV-I-transformed cell clones with HTLV-III (De Rossi *et al.* 1986). While most of these cell clones were OKT4+, one cell clone expressed the OKT8 phenotype. HTLV-III replicates efficiently in all these cell clones. However, the OKT4+ cells were susceptible to the cytopathic effect, but the OKT8+ cell clone was completely resistant. This was not due to an adaptation of the virus, since virus produced by the T8 line retained its tropism and cytopathic activity for T4 cells. This result indicates that virus replication alone is not sufficient for cytopathicity and suggests that expression of T4 may be necessary for the cytopathic response. Consistent with this hypothesis is the observation that infection of a variety of T4 cells by DNA transfection had no apparent effect on cell viability. (Feinberg *et al.*, submitted, and our unpublished observations).

ANALYSES OF tat DELETION MUTANS OF HTLV-III

The availability of the biologically active clone, HXB-2, allowed for further dissection of the viral determinants essential for virus replication and cytopathicity. To investigate the role of the tat-III gene in these functions, we constructed two deletion mutants lacking either the major coding region (pΔSal-Sst) or the splice acceptor sequences (pΔSal-R1) of tat-III (Fisher *et al.* 1986). The trans-activating capacity of these deleted genomes was assessed in co-transfection experiments using pCD12CAT as an indicator. These studies revealed no tat activity for pΔSal-Sst and low level activity for pΔSal-RI, suggesting that the latter may express tat inefficiently by utilizing an alternative splice acceptor or from a polycistronic mRNA. These same deleted genomes, as well as the parent pHXB2, were then introduced into H9 cells by protoplast fusion to assay for their replicative capabilities. One week after transfection with pHXB2, 1–15% of H9 cells expressed HTLV-III p17 and extracellular and budding virions were evident. The number of HTLV-III-expressing cells increased to 80–90% within the first 2 wk. In contrast, no virus expression was detectable after transfection with pΔSal-Sst, and only low level expression ($<2\%$ of cells expressing p17) was seen 2 wk after transfection with pΔSal-R1. These results indicate that tat-III is essential for virus replication. Co-transfection of pΔSal-Sst with a functional tat-III cDNA clone (pCV-1) resulted in transient expression of virus which is non-infectious. There are two practical implications of these experiments. First, identification of an additional crucial gene for virus replication provides yet another target for anti-viral therapy. Second, the production of non-infectious virus by complementation of tat deletion mutants may provide new avenues for vaccine development.

VIRUS EXPRESSION IN NON-LYMPHOID CELLS

We also addressed the question of whether the T4-lymphotropism of HTLV-III is confined solely by the nature of the viral receptor, which includes T4 as a necessary component. We introduced the pHXB2 genome into a variety of non-lymphoid cells and found that some but not all would support virus replication. One of these, the African green monkey endothelial cell line (cos-1) was found to express virus upon transfection with a plasmid pHXB2gpt which contains the biologically active HTLV-III genome and the SV40 origin of replication. Since cos-1 cells express the large T antigen of SV40, efficient episomal replication of this plasmid is expected. Culture supernatants of the transfected cultures contained high levels of reverse transcriptase activity and electron microscopy of the transfected cells revealed copious amounts of free and budding virions with morphology characteristic of HTLV-III. The virus thus produced retained infectivity and tropism for T4 lymphoyctes. Northern blot analysis of the transfected cos-1 cells revealed a normal pattern of RNA transcripts which include the 9.7

TABLE I

Parameters of virus expression of wild-type and tat deletion mutants in Cos cells

			Viral Protein		
Clone	Tat-III function	Viral RNA	tat	gag, env	Virus
pHXB2gpt	+++	+++	+	+++	+++
pΔSal-R1	+	+++	−	++	+
pΔSal-Sst	−	+++	−	−	−
pΔSal-Sst+pCV-1	+++	+++	+	++	++

kb genomic RNA, a 4.3 kb spliced mRNA encoding the env gene, as well as a heterogeneous collection of mRNA in the range of 2 kb.

One unusual feature of cells infected by cytopathic retroviruses is the accumulation of unintegrated linear and viral DNA which could derive from continual reverse transcription of viral RNA, episomal replication or failure to establish interference for reinfections. Since cos-1 cells are refractory to infection by HTLV-III virions, reinfection cannot occur with the transfected cos-1 cells. Examination of the state of viral DNA in these cells revealed only that which is physically linked to the episomally replicating plasmid sequences but failed to detect the linear (9.7 kb) or closed circular viral DNA forms. This result suggests that continuous reinfection is necessary for the maintenance of unintegrated DNA in infected T-lymphocytes.

The transfected cos-1 cells also expressed high levels of viral proteins, including several novel species not detectable in infected lymphocytes. These include a 14 kd protein, a 25 kd protein and a 26 kd protein. Both the 14 kd and 26 kd proteins were immunoprecipitated by rabbit sera specific for a 14 kd tat protein expressed in E. coli. The exact constitution and function of the 26 kd tat-related protein or the 25 kd protein are not known.

POST-TRANSCRIPTIONAL ACTIVATION BY tat-III

To investigate the nature of compromise of viral expression in the molecular clones where tat-III expression was either diminished or abrogated, we examined the trans-activating potency, levels of viral RNA transcripts and amounts of viral proteins expressed in cells transfected with the pΔSal-R1 and pΔSal-Sst mutants (Table I). Quantitative measurement of viral mRNA was accomplished by RNA dot blot hybridization and densitometric analysis. Cells transfected by the various HTLV-III clones demonstrate approximately equal levels of viral RNA transcription. Thus, the p⊿Sal-Sst plasmids, which had low or no transactivating capacities, appeared to initiate equal number of transcripts as the wild-type genome. Radioimmunoprecipitation studies revealed no significant difference in the pattern of HTLV-III protein synthesis in cells transfected with pHXB2 or p⊿Sal-R1, with the exception of the 14 kd and 26 kd tat-III-related proteins which were not detectable in the p⊿Sal-R1-transfected cells. This result suggests that a small amount of tat-III protein may be required for viral protein expression. Immunoprecipitation from p⊿Sal-Sst-transfected cells failed to detect any viral protein, suggesting that the major effect of tat-III is post-transcriptional. This conclusion is in agreement with that of Rosen *et al.* (1986), published recently. However, HTLV-III-infected cells do synthesize a transcriptional activator, as demonstrated in an *in vitro* transcription system in which nuclear extracts of infected cells specifically enhanced transcription from an HTLV-III LTR (Okamoto & Wong-Staal, in press). It is not known whether this represents a minor effect of tat-III or involves a different viral or viral induced cellular protein.

DERIVATION OF NON-CYTOPATHIC VARIANTS OF HTLV-III

Two genes, sor and 3' orf, have been identified in HTLV-III by nucleotide sequence analysis and serological studies, but their functions have not been clarified. In a study specifically addressing the possible role of the 3' orf gene in virus replication and cytopathicity, we constructed mutants with deletions around a unique Xho I site in 3' orf of pHXB2. Six variant genomes were generated (Fisher *et al.* 1986b). In the case of plasmids ⊿X-A, ⊿X-B, ⊿X-C and ⊿X-D, deleted segments of 55, 109, 85 and 100 base pairs, respectively, were confined to the 3' orf gene. The clones X10-1 and X9-3 contained more extensive deletions that encompassed the last 14 base pairs of the env gene and the first 182 or 159 base pairs of the 3' orf gene, respectively. All these variants gave rise to cultures expressing HTLV-III gag and env proteins as well as infectious virus particles, suggesting that the 3' orf gene is not necessary for virus replication. The infectiveness of the mutants on H9 cells is comparable to that of the wild-type. The cytopathic potential of these clones was then compared using two target cell types: PHA-stimulated normal cord blood mononuclear cells and the HTLV-I-immortalized cell line ATH8 which has been shown to be highly susceptible to killing by HTLV-III. The results, as summarized in Table II, indicated that, while the ⊿X-A series which contained deletions exclusively in 3' orf maintained full cytopathic activity, X10-1 and X9-3 had no discernible cytopathic activity. Cells infected by X10-1 and X9-3 expressed abundant virus, formed syncytia, exhibited

TABLE II
Infectivity and cytopathicity of 3' deletion mutants

Plasmid	Position of Deletion in 3' orf (a·a)	Virus Production		Cytopathic Effects	
		H9/ATH8	PBL	ATH8	PBL
pHXB2gpt	none	+	+	+	+
ΔX-A	27–45	+	+	+	+
ΔX-B	23–58	+	+	+	+
ΔX-C	25–53	+	+	+	+
ΔX-D	23–55	+	+	+	+
X9-3*	1–49	+	+/−	−	−
X10-1*	1–61	+	+/−	−	−

* Also deleted in 5 amino acids of the carboxyl terminus of envelope.

tat activity and contained large amount of unintegrated DNA. Therefore, although each of these features had been linked to the cytopathicity of HTLV-III, none of them appeared sufficient. Comparison of the non-cytopathic variants with the cytopathic ones suggests that the critical determinant for cytopathicity resides in the env gene. It is notable that the env gene of HTLV-III is much larger than that of most other retroviruses, with a long cytoplasmic domain extending from the transmembrane protein. It is this novel carboxyl-terminal cytoplasmic domain that was tampered with in the noncytopathic mutants. As a consequence of the manipulation, not only were the last 5 amino acids removed but an additional 15 amino acid stretch rich in arginine was added on. This addition could conceivably alter the structural and biochemical properties of this moiety. Clearly, further experiments are required to define the mechanisms of cytopathicity; however, our results bring focus on the env gene as a critical determinant for this biological activity.

GENETIC POLYMORPHISM OF HTLV-III AND THE PROSPECTS FOR A VACCINE

Genomic variation of different HTLV-III isolates was first recognized by restriction endonuclease analysis of a few proviruses (Shaw *et al.* 1984b). As more and more isolates were subjected to restriction enzymes, heteroduplex mapping and nucleotide sequence analyses, the picture that emerged was one of a family of related but distinct viruses evolving through time (Wong-Staal *et al.* 1985, and our unpublished data). The greatest divergence is localized to the extracellular envelope protein (gp120) (Wong-Staal *et al.* 1985, Gonda *et al.*

1985). Since this is expected to be the most antigenic viral protein and the one to contain the epitopes for eliciting neutralizing antibodies in infected individuals, this finding raises concern for the feasibility of developing a broadly crossreactive vaccine for AIDS. Further justification for this concern lies in the finding of increasing parallels between HTLV-III and the ungulate lentiviruses (visna, equine infectious anemia virus or EIAV, for example) (Gonda *et al.* 1985, 1986). Both visna and EIAV are known to undergo changes in their envelope proteins, and this may be one of the strategies that these viruses utilize to escape immunosurveillance of the host (Clements *et al.* 1980, Montelaro *et al.* 1984). To further scrutinize this issue, we have examined the amino acid sequence of the envelope of several divergent HTLV-III isolates, using a computer program that predicts the secondary structure of proteins superimposed with values for hydrophilicity (Starcich *et al.* 1986). Similar analysis of other proteins, including viral envelopes, has shown that continuous antigenic epitopes are often associated with hydrophilic domains containing β-turns. Continuous epitopes are defined as those peptides whose antigenicity is reflected in the secondary structure of the primary amino acid sequence, whereas discontinuous epitopes are antigenic loci formed by the tertiary folds of the native protein. Such analysis showed that the exterior envelope proteins of the 4 retroviruses compared, HTLV-III RF, HTLV-III BH10, LAV and ARV, each contains sites that are likely to be antigenic based on the above-mentioned criteria. Most of these coincide with hypervariable regions which exhibit differences in hydrophilicity, secondary structure and potential glycosylation patterns. Because of this variability, it is likely that the antigenicity of the corresponding envelope proteins of these viruses could differ considerably. This finding also suggests that these changes result, at least in part, from selective pressures present *in vivo* and that they are biologically meaningful.

Interspersed with the variable regions of the exterior envelope protein were highly conserved areas. One of these corresponded to a stretch of 45 amino acid residues immediately adjacent to the processing site of the envelope precursor, was hydrophilic, and contained numerous β-turns. Therefore, this region of the exterior envelope glycoprotein would be expected to be both immunogenic and cross-reactive among isolates. Other conserved, antigenic epitopes (e.g., the discontinuous epitopes) not predicted by this analysis may also be present in the native protein. All these conserved epitopes may possess important biological functions (e.g., for interaction with the cellular receptor), and furthermore serve as targets for recombinant DNA-based vaccines.

SUMMARY AND PERSPECTIVES

The human lymphotropic retroviruses HTLV-I, II and III are the prototypes of a new category of retroviruses linked together not only by their common target cell tropism, but also by novel regulatory mechanisms for virus expression, which involve trans regulation by viral coded proteins. The interplay of these gene products with each other and with cellular factors may modulate latent and lytic phases of infection. Some of these genes appeared to be requisite for virus replication and thus provide additional targets for antiviral therapy. This is particularly pertinent in the case of HTLV-III, in which continual virus synthesis and recruitment of newly-infected cells appear to be important for pathogenesis.

Our studies on the genomic diversity of the HTLV-III genome indicate that, although the major envelope glycoprotein is highly variable overall, there are conserved regions which may contribute antigenic sites which are group-specific. These analyses will be valuable guides for devising recombinant DNA-based vaccines. The derivation of non-cytopathic variants and replication-deficient virions may also facilitate these efforts.

ACKNOWLEDGMENTS

I am grateful to Dr. R. C. Gallo for his collaboration and support in many of these studies, to all my colleagues at the Laboratory of Tumor Cell Biology for their contributions, in particular Drs. A. Fisher, M. Feinberg, A. Aldovini, R. Jarrett, L. Ratner, T. Okamoto for permission to cite unpublished data.

REFERENCES

Arya, S. K., Chan, G., Josephs, S. F. & Wong-Staal, F. (1985) Transactivator gene of Human T-cell leukemia (lymphotropic) virus type II (HTLV-III). *Science 229*, 69–73.

Arya, S. K. & Gallo, R. C. (1986) Three novel genes of human T-lymphotropic retrovirus-III (HTLV-III): Immune reactivity of their products with sera from acquired immune deficiency syndrome patients. *Proc. Natl Acad. Sci. 83*, 2209–13.

Chen, I. S. Y., Slamon, D. J., Rosenblatt, J. D., Shah, N. P., Ouan, S. & Wachman, W. (1985) The x gene is essential for HTLV replication. *Science 229*, 54–7.

Clements, J. E., Pedersen, F. S., Narayan, O. & Haseltine, W. A. (1980) Genomic changes associated with antigenic variation of visna virus during persistent infection. *Proc. Natl. Acad. Sci. USA 77*, 4454–8.

De Rossi, A., Franchini, G., Aldovini, A., Del Mistro, A., Chieco-Bianchi, L., Gallo, R. C. & Wong-Staal, F. (1986) Differential response to the cytophatic effects of HTLV-III superinfection in OKT4 and OKT8 cell clones transformed by HTLV-I. *Proc. Natl. Acad. Sci. USA 83*, 4297–4301.

Feinberg, M. B., Jarrett, R. F., Aldovini, A., Gallo, R. C. & Wong-Staal, F. (1986) HTLV III Expression and production involve complex regulation at the levels of splicing and translation of viral RNA. *Cell 46*, 807–17.

Felber, B. K., Paskalis, H., Kleinman-Ewing, C., Wong-Staal, F. & Pavlakis, G. N. (1985) The pX protein of human T-cell leukemia virus-I is a transcriptional activity of its long terminal repeats. *Science 229*, 675–9.

Fisher, A. G., Feinberg, M. B., Josephs, S. F., Harper, M. D., Marselle, L. M., Reyes, G., Gonda, M. A., Aldovini, A., Debouck, C., Gallo, R. C. & Wong-Staal, F. (1986a) The trans-activator gene of HTLV-III is essential for virus replication. *Nature 320*, 367–71.

Fisher, A. G., Ratner, L., Mitsuya, H., Marselle, L. M., Harper, M. E., Broder, S., Gallo, R. C. & Wong-Staal, F. (1986b) A non-cytopathic variant of HTLV-III generated by altering the 3' region of the virus genome. *Science* (in press).

Fisher, A. M., Collalti, E., Ratner, L., Gallo, R. C. & Wong-Staal, F. (1985) A molecular clone of HTLV-III with biological activity. *Nature 316*, 262–5.

Franchini, G., Robert-Guroff, M., Wong-Staal, F., Ghrayeb, J., Kato, N. & Chang, N. (1986) Expression of the 3'open reading frame of the HTLV-III in bacteria: Demonstration of its immunoreactivity with human sera. *Proc. Natl. Acad. Sci. USA 83*, 5282–5.

Gonda, M. A., Braun, M. J., Clements, J. E., Pyper, J. M., Wong-Staal, F., Gallo, R. C. & Gilden, R. V. (1986) Human T-cell lymphotropic virus type III shares sequence homology with a family of pathogenic lentiviruses. *Proc. Natl. Acad. Sci. USA 83*, 4007–11.

Gonda, M. A., Wong-Staal, F., Gallo, R. C., Clements, J. E., Narayan, O. & Gilden, R. V. (1985) Sequence homology and morphologic similarities of HTLV-III and visna virus, a pathogenic lentivirus. *Science 227*, 173.

Haseltine, W. A., Sodroski, J., Patarca, R., Briggs, D., Perkins, D. & Wong-Staal, F. (1984) Structure of 3' terminal region of type II human T lymphotropic virus: evidence for new coding region. *Science 225*, 419–21.

Kan, N. C., Franchini, G., Wong-Staal, F., DuBois, G. C., Robey, W. G., Lautenberger, J. A. & Papas, T. S. (1986) A novel protein (sor) of HTLV-III expressed in bacteria is immunoreactive with sera from infected individuals. *Science 231*, 1553–4.

Montelaro, R. C., Parekh, B., Orrego, A. & Issel, C. J. (1984) Antigenic variation during persistent infection by equine infections anemia virus, a retrovirus. *J. Biol. Chem. 259*, 10539–44.

Okamoto, T. & Wong-Staal, F. (1986) Demonstration of virus-specific transcriptional activators in cells infected with human T-cell lymphotropic virus type III by *in vitro* cell-free system. *Cell* (in press).

Ratner, L., Haseltine, W., Patarca, R., Livak, K., Starcich, B., Josephs, S., Doran, E. R., Rafalski, J. A., Whitehorn, E. A., Baumeister, K., Ivanoff, L., Petteway, Jr., S. R., Pearson, M. L., Lautenberger, J. A., Papas, T. S., Ghrayeb, J., Chang, N. T., Gallo, R. C. & Wong-Staal, F. (1985) Complete nucletide sequence of the AIDS virus, HTLV-III. *Nature 313*, 277–84.

Rosen, C. A., Sodrowki, J. G., Goh, W. C., Dayton, A. I., Lippe, J. & Haseltine, W. A. (1986) Post-transcriptional regulation accounts for the trans-activation of the human T-lymphotropic virus type III. *Nature 319*, 555–9.

Rosen, C. A., Sodroski, J. G. & Haseltine, W. A. (1985a) Location of cis-acting regulatory sequences in the human T-cell leukemia virus Type I long terminal repeat. *Proc. Natl. Acad. Sci. USA 82*, 6502–6.

Rosen, C. A., Sodroski, J. G. & Haseltine, W. A. (1985b) The location of cis-acting regulatory sequences in the human T-cell lymphotropic virus Type III (HTLV-III/LAV) long terminal repeat. *Cell 41*, 813–23.

Sanchez-Pescador, R., Power, M. D., Barr, P. J., Steinmer, K. S., Stempien, M. M., Brown-Shimer, S. L., Gee, W. W. Renard, A., Randolph, A., Levy, J. A., Dina, D. & Luciw, P. A. (1985) Nucleotide sequence and expression of an AIDS-associated retrovirus (ARV-2). *Science 227*, 484–92.

Seigel, L. J., Ratner, L., Josephs, S. F., Derse, D., Feinberg, M. B., Reyes, B. R., O'Brien, S. J. & Wong-Staal, F. (1986) Transactivation induced by human T-lymphotropic virus Type III (HTLV-III) maps to a viral sequence encoding 58 amino acids and lacks tissue specificity. *Virol. 148*, 226–31.

Seiki, M., Hattori, T., Hirayama, Y. & Yoshida, M. (1983) Human adult T-cell leukemia virus: complete nucleotide sequence of the provirus genome integrated in leukemia cell DNA. *Proc. Natl. Acad. Sci. USA 80*, 3618–22.

Shaw, G., Gonda, M., Flickinger, G., Hahn, B. H., Gallo, R. C. & Wong-Staal, F. (1984a) Genomes of evolutionarily divergent members of the human T-cell leukemia virus family (HTLV-I and HTLV-II) are highly conserved, especially in pX. *Proc. Natl. Acad. Sci. USA, 81*, 4544–8.

Shaw, G. M., Hahn, B. H., Arya, S. K., Groopman, J. E., Gallo, R. C. & Wong-Staal, F. (1984b) Molecular characterization of human T-cell leukemia (lymphotropic) virus type III in the acquired immune deficiency syndrome. *Science 226*, 1165–71.

Sodroski, J., Goh, W. C., Rosen, C., Dayton, A., Terwilliger, E. & Haseltine, W. (1986) A second post-transcriptional *trans*-activator gene required for HTLV-III replication. *Nature 321*, 412–7.

Sodroski, J., Patarca, R., Rosen, C., Wong-Staal, F. & Haseltine, W. A. (1985b) Location of the trans-activating region of the genome of HTLV-III/LAV. *Science 229*, 74–7.

Sodroski, J. G., Rosen, C. A. & Haseltine, W. A. (1984) Transacting transcriptional activation of the log terminal repeat of human T-lymphotropic viruses in infected cells. *Science 225*, 381–3.

Sodroski, J., Rosen, C., Wong-Staal, F. Salahuddin, S. Z., Popovic, M., Arya, S. K., Gallo, R. C. & Haseltine, W. A. (1985a) Trans-acting transcriptional regulation of human T-cell leukemia virus type III long terminal repeat. *Science 227*, 171–3.

Starcich, B., Hahn, B., Shaw, G. M., Modrow, S., Josephs, S. F., Wolf, H., Gallo, R. C. & Wong-Staal, F. (1986) Identification and characterization of conserved and divergent regions in the envelope gene of AIDS viruses. *Cell 45*, 637–48.

Wain-Hobson, S., Sonigo, P., Danos, O., Cole, S. & Alizon, M. (1985) Nucleotide sequence of the AIDS virus. LAV. *Cell 40*, 9–17.

Wong-Staal, F., Shaw, G. M., Hahn, B. H., Salahuddin, S. Z., Popovic, M., Markham, P. D., Redfield, R. & Gallo, R. C. (1985) Genomic diversity of human T-lymphotropic virus type III (HTLV-III). *Science 229*, 759–62.

DISCUSSION

RAPP: I understand that under the conditions where you get cell killing you have small fractions of the cells infected, about 5%. If you think in terms of the gp70, do you know whether there is shedding in that culture?

WONG-STAAL: I think you can find that in the media. There has to be an extracellular excretion.

GALLO: You do find it in the media, that is for sure.

WONG-STAAL: We do find it in the media, but we do not know if it is active shedding or cell lysis. But it is something to look into.

RAPP: I was also thinking in terms of the effect of shed gp70 which could mediate cytotoxicity. Since it would presumably not generate the intracytoplasmic tail into the next cell, that tail structure should not be involved in killing.

WONG-STAAL: Well, it could get into the cell.

SHERR: You have introduced HTLV-III genomes into a variety of cells. Are they cytopathic in all the different cell types or are they strictly cytopathic in T4 cells?

WONG-STAAL: We find that T4 expression may be necessary, and that is supported by the transfection experiments that even though you are able to introduce the genome in all types of cells and have replication in some of them, you never get cytopatic effect unless the cells also express T4.

SHERR: Have you tried to cointroduce T4 into other cells?

WONG-STAAL: We are trying.

RAPP: In those transfected cells, the non-T4 cells, do you get at lot of unintegrated copies there?

WONG-STAAL: You can tranfect *cos*-1 cells that are refractory to HTLV-III infection, and the cells are infected, but you do not get unintegrated DNA, at least not the free proviral DNA, suggesting that the unintegrated DNA requires reinfection. But we do have one cell type which has not T4 that we are able to infect, and in that cell you get a lot of unintegrated DNA.

GALLO: Which would tend to rule out unintegrated DNA as the main direct cause of the killing effect of HTLV-III.

HANAHAN: With the *trs* gene, since this is such a small exon, with all these mutations, is it possible to exclude that this just affects the ability to process RNA – and secondly, can you complement this mutant with the cDNA of *trs*?

WONG-STAAL: Yes, to the second question. You can partly complement by co-transfecting with cDNA that makes both *tat* and *trs*. We do not have clones that express only *trs*, as yet. In terms of the alterations of other things that may be important, we are trying to address that question by using point mutations introduced by site-specific mutagenesis.

GALLO: This gene, *trs*, is the same as the one Haseltine has called *art*, and the same basic observation that it functions as a posttranscriptional regulator was found. However, Dr. Wong-Staal argues that the mechanism has to do with regulation of splicing, and so a more specific name has been suggested.

Replication and Pathogenesis of Human Retroviruses

William A. Haseltine

THE LEUKEMIA VIRUSES

Human retroviruses associated with leukemia, HTLV-I and -II and AIDS (HTLV-III, HIV or LAV) have recently been isolated. These viruses are unusual amongst retroviruses both with respect to their genetic organization and replication.

The human leukemia retroviruses, like the related simian (STLV-I) and bovine (BLV) retroviruses are poorly transmitted agents of long latent period leukemias and lymphomas. The infections are spread sexually from mother to progeny and by exchange of blood via transfusion or intravenously via needles. The human leukemia viruses are also spread sexually from men to women. Purified HTLV-I and -II viruses are very poorly infectious, and natural transmission may occur solely by transfer of infected cells. Co-cultivation of infected cells with uninfected fresh peripheral blood cells leads to outgrowth of immortalized recipient T4+ cells, indistinguishable from cells derived from patients. Examinations of sites of integration within the host genome of naturally arising, virally associated human and bovine leukemias reveals no preferred site of proviral integration. This observation, coupled with the ability of these viruses to immortalize primary cells in culture indicates that the virus itself carries transforming genes. For a recent review see Wong-Staal & Gallo (1985).

The structure of the leukemia retrovirus genomes is unusual in that they contain a region of about 1500 nucleotides, called pX, located between the end of the envelope gene and the 3' long terminal repeat sequence (LTR) (Figure 1). This region shares no close homology to conserved cellular sequences as it does not anneal to most cellular DNA species, including human and other primates, even under non-stringent conditions of hybridization.

The pX region of these viruses has been shown to encode at least 3 proteins

---

Dana-Farber Cancer Institute and Harvard Medical School 44 Binney Street, Boston, MA 02115, U.S.A.

VIRAL CARCINOGENESIS, Alfred Benzon Symposium 24,
Editors: N. O. Kjeldgaard, J. Forchhammer, Munksgaard, Copenhagen 1987.

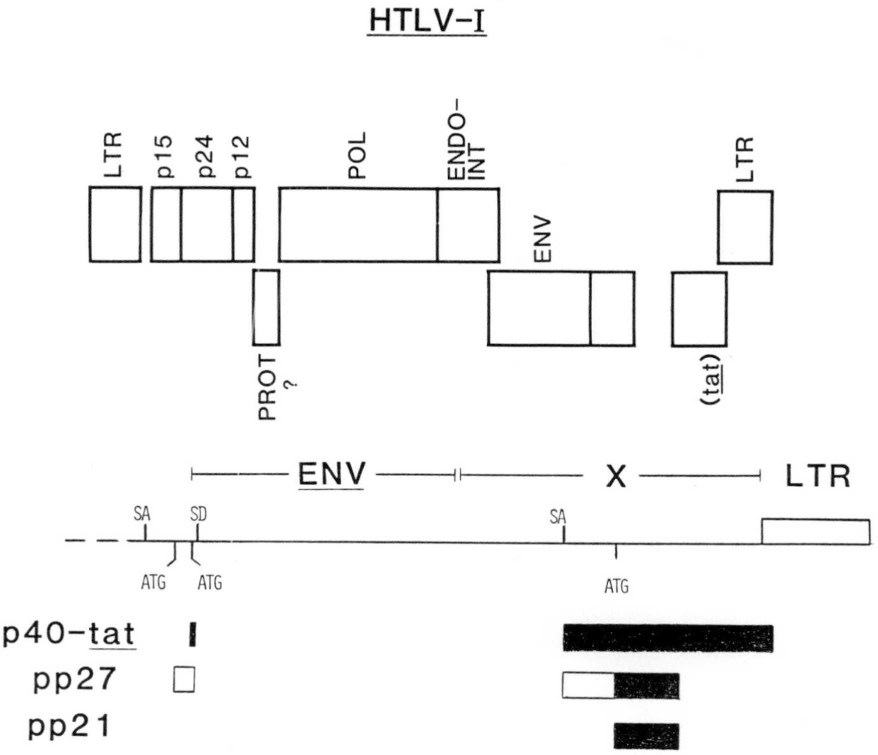

Fig. 1.

(Kiyokawa *et al.* 1985). One of these proteins, that derived from the longest open reading frame within the pX region, encodes a protein termed the transactivator (*tat*) (Sodroski *et al.* 1985). Expression of this protein product in cells leads to greatly increased acceleration of gene expression directed by the viral LTR (Sodroski *et al.* 1984). The sequences within the viral LTRs that respond to the *tat* gene product are termed TAR sequences and are located entirely 3' to the site of RNA initiation (Rosen *et al.* 1985). Levels of steady state RNA directed by the viral LTRs increase in rough proportion to the extent of increase in measureable protein. Nuclear transcription run-on experiments have been said to show increases in the rate of LTR-directed RNA initiation in the presence of the *tat* gene (Pavalakis, personal communication). For these reasons, it is likely that the *tat* gene of the leukemia viruses increases the rate of transcription initiation directed by the LTR.

What role might the *tat* gene of these viruses play in virus replication and cellular transformation? Chen and co-workers (Shimotoyno *et al.* 1984) report that the deletion of the transactivator gene from an infectious HTLV-II provirus destroys the ability of the virus to replicate. The specific defect noted is an inability to produce large amounts of viral RNA. It seems likely that replication of this family of viruses is dependent upon the *tat* gene. However, this hypothesis has not been formally tested for all the viruses as no infectious clones for HTLV-I have been isolated as yet.

The discovery of a transcriptional transactivator gene of the viruses prompted speculation that these gene products might also be involved in cellular transformation via induction of transcription of genes involved in normal growth control (Sodroski *et al.* 1984). In particular, speculation focused on the transcription activity of the T cell growth factor receptor (TCGF receptor, also known as the IL-2 receptor) as cells derived from patients with adult T cell leukemia lymphoma (ATLL) exhibit an abnormally high level of the TCGF receptor.

The ability of the *tat* gene of HTLV-II to induce the TCGF receptor was examined by introduction of this gene in a ZIPneo retrovirus vector into a T cell line that was previously shown to respond to stimulation by phytohemaglutinin (PHA), and phorbol esters (PMA). Examination of stable cell lines that express the tat_{II} product showed modest levels of expression of the TCGF receptor and TCGF mRNA (Greene *et al.* 1986). These experiments support the concept that the *tat* genes are involved not only in the replication of virus but also in deregulation of the growth of T4+ cells. Neither the TCGF receptor nor TCGF expression itself was evident in a B cell line (Raji) that expressed the transactivator gene product.

Recently, it has been shown that the pX region of HTLV-I, II, STLV-I and BLV can encode at least 2 additional proteins from a reading frame alternate to that of the *tat* genes (Kiyokawa *et al.* 1985) (Figure 1). The 2 additional proteins are phosphorylated, and the larger of the two proteins is located in the nucleus. It is likely that these proteins, as well as the *tat* genes, are involved in cellular transformation.

The existing data permit the construction of an interesting hypothesis regarding viral replication in tumorogenesis. These viruses would appear to be capable of prolific replication as they encode an autostimulatory factor. However, prolific replication of such viruses has not been observed either in infected animals or in tissue cultures. Part of the limited replication of these viruses is the apparent requirement for immune activation for activation of viral RNA expression (hence

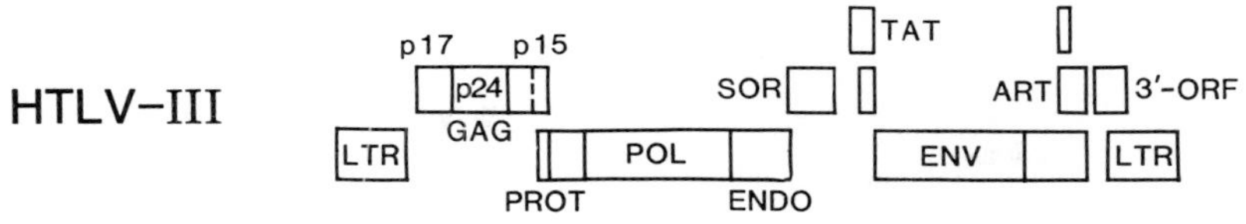

Fig. 2.

the requirement for TCGF, or phytohemaglutinin, for the isolation of these viruses). However, the replication defect present on HTLV-I and -II appears to be even more profound, limiting natural infections to cell-to-cell contacts. If the infected cell itself is the infectious unit, the pX region encoded genes that lead to cell proliferation of the infected cell population may be viewed as unusual replication genes, expanding the number of infectious units in an infected animal. Limited replication and transmission of these viruses may represent the end point of the evolution of a comparatively benign host-parasite relationship. Such limited replication is in sharp contrast to that of other persistent retroviruses in short-lived animals such as chicken and mouse.

THE AIDS VIRUS

The AIDS retrovirus, HTLV-III (HIV or LAV), is distinct from HTLV-I and -II in several respects. The virus is profoundly cytopathic to T4+ cells in culture and is not directly associated with malignant transformation (for a review, see Wong-Staal & Gallo 1986). Epidemiological studies show that the virus is more easily transmitted than is HTLV-I and can be transmitted from person to person by cell-free virus. The genomic organization of the virus is very different from that of HTLV-I. In addition to the *gag, pol,* and *env* genes typical of all retroviruses, the AIDS virus encodes at least 4 additional genes, *sor*, 3' *orf, tat,* and *art* (Sodroski *et al.* 1986b) (Figure 2). The molecular weights of these proteins are 23kD, 27kD, 14kD and 20kD, respectively. All of these additional proteins are antigenic in some, but not all, infected persons and all are made in lytically infected cells (Sodroski *et al.* 1986a; Terwilliger *et al.* 1986, Goh *et al.* 1986a, b).

Mutational analysis of an infectious proviral genome reveals that the 3' *orf* gene is required neither for virus replication nor for cytopathic effect. Indeed, mutants incapable of making the 3' *orf* product replicate somewhat faster than do viruses that do produce the 27kD 3' *orf* protein (Terwilliger *et al.* 1986). Mutations in the *sor* gene retard the growth of the virus considerably but do not eliminate the cytopathic effect of the infection (Sodroski *et al.* 1986a). Both the *tat* and *art* gene products are absolutely required for virus replication (Dayton *et al.* 1985, Sodroski *et al.* 1986b). Mutations in either of these genes eliminate the ability of the proviruses to yield detectable levels of virion structural proteins. However, despite the deficiency in the production of virion structural proteins, mutations in the *art* gene are capable of producing wild type levels of the *tat* gene product.

The mechanism of action of the *tat* and *art* genes is unusual. Mutants defective for either the *tat* or *art* gene are capable of synthesis of viral RNA even though no virion proteins are made (Rosen *et al.* 1986b, Sodroski *et al.* 1986b). Studies using the viral LTR to direct the synthesis of heterologous genes show that the sequences from +1 to +80 (the site of RNA transcription initiation is at +1) are both necessary and sufficient for the response to the *tat* gene provided that they are located at or near the 5' end of the nascent message (Rosen *et al.* 1985, 1986b). Such sequences are called TAR for transacting responsive region. These findings have led to the hypothesis that the *tat* gene product acts by binding to the 5' end of the viral mRNA species and that such binding is required for the synthesis of viral structural proteins (Rosen *et al.* 1985). In a heterologous system where the LTR directs the synthesis of a non-viral gene, this same reaction leads to greatly accelerated protein synthesis as compared to the rate of protein synthesis in either the absence of the *tat* product by the same message or by the same message in the absence of the TAR sequence even in the presence of the *tat* gene product.

The *art* gene appears to be involved in differential expression of discrete viral coding exons. In genetic terms, the *art* gene behaves as if it relieves negative regulation of *gag* and *env* gene expression without dramatically altering the level of *tat* gene expression. As the *gag, tat,* and *env* genes are synthesized from a different mRNA species, these experiments suggest that the *cis*-acting negative regulatory sequences are spliced out of the *tat* gene messenger RNA whereas such sequences remain in the *gag* and *env* mRNA species.

Mutational analyses provide some insight regarding the mechanisms of cytopathicity as well. *In vitro* infections of T4+ cells are characterized by formation of giant multinucleated syncytia and ballooning of individual cells followed by death. Viruses defective for expression of the 3' *orf* gene are fully cytotoxic (Sodroski *et al.* 1986a, Terwilliger *et al.* 1986). Viruses defective for the *sor* gene replicate slowly but are also cytotoxic (Sodroski *et al.* 1986a). Viruses that cannot synthesize the *tat* and *art* gene products are not cytoxic as they do not replicate. The *tat* gene itself is not profoundly cytotoxic to T4+ cell lines, as cell lines that stably express the *tat*III gene product have been produced (Rosen *et al.* 1986a). Viruses deleted for the *tat* gene can replicate in and kill such cell lines. Likewise, complementation of the *art* gene leads to replication in cytopathic effect of *art*-defective viruses. These experiments suggest that the *tat* and *art* genes themselves, while not directly cytotoxic to T4+ cells, are important for expression of other viral genes which are cytotoxic.

As a test of this hypothesis, plasmids were constructed that express only the envelope glycoprotein. Expression of the *env* gene in T4+ cells induces the formation of multinucleated giant cells. Such cells die over a 5- to 10-day period as measured by vital dye exclusion (Sodroski *et al.* 1986c).

Syncytia formation can be blocked by some antibodies to the T4 surface molecule. Expression of the envelope gene on T4- cells does not lead to syncytium formation. However, co-cultivation of T4- cells that express the envelope glycoprotein with the T4+ cell population leads to syncytium formation of the T4+ cells.

These experiments show that a high level of expression of the envelope glycoprotein of the AIDS virus produces a cytotoxic effect in T4+ cells. These and other experiments indicate that binding of the T4 molecule by the envelope and membrane fusion reactions that ensue are cytotoxic to T4+ cells. The binding and fusion reaction is probably the same reaction whereby the virus enters the cell.

The two primary determinants of the cytotoxic reaction appear to be the surface concentration of the T4 molecule and the surface concentration of the envelope glycoprotein. High concentrations of both molecules, as are found upon infection of T4+ cell lines with virus, may lead to a loss of membrane integrity by virtue of autofusion reactions whereby the portions of the membrane bearing the envelope fuse to the portions of the membrane carrying the T4 molecule. If the surface concentration of the T4 molecules is too low, as for example in cells such as monocytes or macrophages that express low but detectable levels of the T4 proteins or in subpopulations of T4+ cell lines that express unusually low concentrations of surface T4 molecules, persistent non-cytopathic infections may be found. If the surface concentration of the envelope gene product is reduced, either as a consequence of mutation in the *env* gene that destabilizes the association of the protein with the membrane or as a consequence of limited viral protein expression, persistent non-cytopathic infections may also occur in T4+ cell lines.

The notion that the principal cytotoxic mechanisms of AIDS virus infection can be attributed to the envelope gene product should be viewed in the context of other retroviruses. As a group, retroviruses are not notably cytopathic. On the contrary they are generally non-cytopathic and establish persistent productive infections. The extreme cytopathic character of the AIDS virus is probably attributable to two features of the virus, an unusually tight binding of the envelope gene product to an abundant surface receptor, and unusually high level of envelope protein synthesis consequent to transactivation of the expression of the envelope gene by the *tat* gene product of the virus.

In this view, the current AIDS epidemic appears to be the result of entry into a mobile urban population of a strain of virus that has two unusual features, high transmissibility and high cellular toxicity. It is likely that these two features are related to the control of virus replication, and to the cytopathic character of the envelope glycoprotein of the virus.

ACKNOWLEDGMENTS

I thank my colleagues, Drs. Joseph Sodroski, Craig Rosen, Wei Chun Goh, Roberto Patarca, Andrew Dayton and Ernest Terwilliger, who did much of the work reported herein and Drs. Robert Gallo and Flossie Wong-Staal who provided many of the starting materials for these experiments.

REFERENCES

Dayton, A. I., Sodroski, J. G., Goh, W. C., Haseltine, W. A. (1986) *Cell 44*, 941.

Goh, W. C., Rosen, C., Sodroski, J., Ho, D. & Haseltine, W. A., (1986a) *J. Virol. 59*, 181.

Goh, W. C., Sodroski, J. G., Rosen, C. A. & Haseltine, W. A. (1986b) *J. Virol* (in press).

Greene, W. C., Leonard, W. J., Wano, J., Svetlik, P. B., Peffer, N. J., Sodroski, J. G., Rosen, C. A., Goh, W. C. & Haseltine, W. A. (1986) *Science. 232*, 877.

Kiyokawa, I, Yoshida, M. 1985. *Proc. Nat. Acad. Sci. USA, 82*, p8359.

Rosen, C. A., Sodroski, J. G., Campbell, K. & Haseltine, W. A. (1986a) *Virology 57*, 379.

Rosen, C. A., Sodroski, J. G., Goh, W. C., Dayton, A. I., Lippke, J. & Haseltine, W. A. (1986b) *Nature 319*, 555.

Rosen, C. A., Sodroski, J. G. & Haseltine, W. A. (1985) *Cell 41*, 813.

Shimotomo, K., Wachsman, W., Takahashi, Y., Golde, O. W., Miwa, M., Sugimura, T. & Chen, I. S. Y. (1984) *Proc. Natl. Acad. Sci. USA 81*, 6657.

Sodroski, J. G., Goh, W. C., Rosen, C. A., Campbell, K. & Haseltine, W. A. (1986c) *Nature 322*, 470.

Sodroski, J. G., Goh, W. C., Rosen, C. A., Dayton, A. I., Terwilliger, E. & Haseltine, W. A. (1986b) *Nature 321*, 412.

Sodroski, J. G., Goh, W. C., Rosen, C. A., Tartar, A., Portetelle, D., Burny, A. & Haseltine, W. A. (1986a) *Science 231*, 1549.

Sodroski, J. G., Rosen, C. A., Goh, W. C. & Haseltine, W. A. (1985) *Science 228*, 1430.

Sodroski, J. G., Rosen, C. A. & Haseltine, W. A. (1984) *Science 225*, 381.

Terwilliger, E., Sodroski, J. G., Rosen, C. A. & Haseltine, W. A. (1986) *J. Virol.* (in press).

Wong-Staal, F. & Gallo, R. C. (1985) *Nature 313*, 395.

DISCUSSION

YANIV: Did you try to reproduce the transactivation of translation in *in vitro* systems taking SP6-made RNA?

HASELTINE: We are trying that, but it has not worked yet. We are helping three or four other groups trying to do that. They are more proficient at protein synthesis than we are. So, I would say we have some of the best groups working on this problem.

ZUR HAUSEN: You mentioned in the first part of your talk that you fell that some growth stimulation takes place during the latency period of HTLV-I infections. What is the actual evidence for that? You quoted as an example EBV. There is really no evidence in EBV that, during the persistent phase of infection, any stimulation of growth in lymphocytes takes place, except for the acute phase of infectious mononucleosis.

HASELTINE: There is considerable evidence that for both for BLV- and HTLV-I-infected persons there is a phase, or can be a phase, which is called persistent lymphocytosis, well-characterized in BLV, and now characterized, in very good detail, by the Japanese. In fact, an entire array of symptoms, ranging from no hematopoietic abnormalities to a polyclonal proliferation called smouldering leukemia, to outright malignant clonal leukemia, are reported in HTLV-I-infected people. There is a whole spectrum of abnormalities infected with BLV. A similar spectrum is observed clearly in sheep and in cattle. So there is, I think, a good deal of evidence to suggest that there are HTLV-I-induced proliferative states short of full malignancy. As is the case for EBV, one sees *in vitro* transformation as well.

ZUR HAUSEN: Yes, but because *in vitro* the stimulation is a kind of secondary event which follows the activation of virus expression.

HASELTINE: You are talking about *in vitro*. Primary cells can be immortalized by HTLV-I and -II. Two independent lines of evidence indicate that HTLV-I and -II possess potent growth-stimulating activities.

WEINBERG: You talked about an autocrine mechanism with HTLV-I, but is there in fact any direct evidence that the growth of the immortalized lymphocytes is in any way dependent on either the display of the receptor or the elaboration of the growth factor that binds to this receptor?

HASELTINE: Let me say, first of all, that it is a uniform property of primary T-cells tumors in people that they have elevated levels of the T-cell receptor. I think that there remains a good deal of work to be done on looking at what the primary cells look like. Once you see cells that are grown out in culture, you are not looking at the primary cell anymore. You are looking at a cell which we know is not the primary cell. Those cells have different arrangements of their T-cell receptor. So, we do not really know yet what the primary cell looks like. What we know is what a highly adapted cell line looks like, and in that case sometimes they express IL-2 as well as the IL-2 receptor, and sometimes they do not.

GALLO: Remember that IL-2 is in the circulation *in vivo*. As the slide suggested, when leukemia develops the cells have IL-2 receptors, but usually do not make IL-2.

HASELTINE: I think that the remarkable thing about the transactivator gene is that it does elicit a very specific response of genes that are normally involved in the proliferation of those cells. It does not induce the response of all genes. There are a number of genes, that I did not describe, which are not induced, and it does so in a lineage-specific manner. I think those are remarkable properties.

GALLO: I want to emphasize that this *in vitro* data mimics the disease probably more than any other system we have for the study of a human cancer.

HASELTINE: It mimics it surprisingly closely.

RIGBY: If I understood the model of HTLV-III replication that was on the slide, it requires a temporal control of the expression of *tat?* So, do you have any evidence of such temporal control, and do you have any idea how it might work?

HASELTINE: Do we have direct evidence? No, we do not. That is why I said that the model is highly speculative. However, there are cases, naturally occurring cases, of virus similar to this, where one sees prolific expression of RNA without concomitant protein synthesis. There are two or three such reports for visna virus in the literature. What surprises me is that the authors do not remark that there is an unusual state of virus expression. They find a lot of viral RNA, but no viral proteins. I think that the argument that I gave you here was that it may be involved in short-term latency, the timing of a virus release. Equally well, the *tat art* cycle could be involved in the maintenance of a latent state.

ROIZMAN: I just want to correct that. Posttranscriptional regulation has been described numerous times since 1975.

HASELTINE: There is a recent, very good report on cytomegalovirus posttranscriptional regulation.

SHERR: I would like to follow up on Robert Gallo's comments and on your model. What is the mechanism for transformation? You have an increase in IL-2 receptors in infected T4 cells, but can you demonstrate normal receptor function; for example, can you demonstrate down regulation?

HASELTINE: Down regulation or the IL-2 receptors is not typical of receptors, and whether or not it is down regulated in the classic sense is debatable. However, anti-IL-2 receptor antibodies existed and IL-2 receptors were measured. In those cells in which the IL-2 receptor is induced, IL-2 binding can be demonstrated, which is about as functional as you can get in a cell line, which are already transformed.

GALLO: Actually, that is one of the things that mimics the tumor; you do not lose the receptor.

SHERR: Is this the so-called low affinity receptor?

HASELTINE: No, it is a high affinity. It is a mixture of high and low affinity receptors. What appears to happen now is that expression of high affinity receptors is conveyed by certain cell lines. You can express the same gene in different cell lines and either get no high affinity receptors or get a small fraction of high affinity receptors. I think the favorite model is that there is an additional lineage-dependent factor which conveys high affinity to a small fraction of the IL-2 receptors.

SHERR: I had the impression that the high affinity receptors could be down regulated?

HASELTINE: They have a very long half-life. It is a very funny kind of down regulation, if you look at it.

LOWY: Two questions, the first relating to HTLV-I disease. I had thought that there were high levels of IL-2 receptors on peripheral blood lymphocytes although many of them did not express any viral RNA.

HASELTINE: There are no IL-2 receptors in normal T cells until they are activated.

WONG-STAAL: Only after activation.

GALLO: If you examine primary leukemic cells immediately the number of IL-2 receptors is not high in most, and virus expression in about 90% of the cases is non-detectable. But if you put them in short-term culture, receptors are found in abundance, and then there is virus expression.

HASELTINE: But Lowy has addressed an issue which is always addressed in any of these discussion, and that is, if you look in the periphery of an infected patient, you do not often see viral gene expression. The question arises, is *tat* a maintenance gene? I would point out that gene expression in these viruses, as has been mentioned several times now, is very tightly linked to the activation of the T cell. One characteristic of all T cells, be they tumor cells or natural T cells, is that there is no mitotic index in the peripheral population. Yet, obviously, the tumor cells are replicating somewhere, because people are dying of leukemia. The fact is, we do not know where they replicate, so we do not know where to look to see the *tat* gene activated. I will bet you that when you see the tumor cell replicate, you will see the transactivator function as well. And so, the apparent paradox has much more to do with natural T-cell biology than with the question of whether this is a maintenance gene. What argues most strongly for maintenance is that every time you see one of these tumors the transactivated gene is present even if other parts of the genome are deleted.

LOWY: I wanted to ask about your northerns blots. Do they have subgenomic mRNA?

HASELTINE: In Flossie Wong-Staal's cDNA blot, there were viral messages and some of them were properly spliced. There are a number of interpretations. One of the interpretations is that there is a splicing abnormality. I can find probably two or three other equally plausible explanations for that. For example, it is very likely that messages, unless they are translated, are not stable, and that larger messages are less stable than smaller messages. That would give you an equally plausible explanation. So, I think that until we actually have the mechanism of action of the *art* gene defined it is premature to name a gene, especially on the basis of a hypothetical mechanism.

GALLO: Did you not call it an anti-repressor?

HASELTINE: Yes, but that is a genetic definition.

GALLO: But have you demonstrated a repressor?

HASELTINE: I would refer you to the early papers of Jacob and Monod where they define what an operator and a repressor is, and their definitions of those genetic functions are very close to the definition of these genetic functions. It is a genetic definition, not a biochemical definition.

LOWY: Is there any evidence that you might need splicing to translate *gag?*

HASELTINE: No, there is not, in fact. I think it is very unlikely that you need splicing to translate *gag*.

WONG-STAAL: I could clarify that. What we show is that in the absence of this gene, it is not that you don't get splicing, but everything is spliced down to 2 kb. So you do not express *gag,* you do not express *env,* because there is no messenger around.

HASELTINE: That is your interpretation. I would take the alternative explanation, that the larger messages are less stable than the smaller.

WONG-STAAL: O.K., but the observation is the same, there is no messenger around in the steady state.

HASELTINE: No, actually I would perhaps quarrel with your definition of "no". If you expose those blots longer, you would see more RNA at a higher molecular weight.

WONG-STAAL: Did you try to immortalize normal T cells with your *tat*-2 retroviral vector, and fail, or did you not do it?

HASELTINE: We have not yet succeeded.

GALLO: It is called tit-for-*tat*.

Activation and Inhibition of *mos* Transformation by Proviral Long Terminal Repeats and c-*mos*-Associated Cellular Sequences

D. G. Blair[1], M. L. McGeady[2], M. K. Oskarsson[2], M. Schmidt[2], F. Propst[2], T. G. Wood[3], A. Seth[2] & G. F. Vande Woude[2]

Moloney murine sarcoma virus (Mo-MuSV) is a replication-defective retrovirus containing an oncogene (v-*mos*) responsible for its ability to phenotypically transform cells in tissue culture and to induce fibrosarcomas *in vivo*. The acute transforming retroviruses have, in general, been useful in demonstrating the role of specific cellular genes in malignant transformation (Bishop 1983), and the viral and cellular homologues of the *mos* oncogene have been particularly useful for studying the molecular elements required for such transformation. These studies provided the first evidence of direct structural identity between a viral oncogene and its cellular homologue (Jones *et al.* 1980, Oskarsson *et al.* 1980). More importantly, they demonstrated that a normal cellular sequence has transforming potential when a proviral transcriptional control element or long terminal repeat (LTR) was coupled to the cellular mouse proto-oncogene sequences (c-mos^{Mu}) (Oskarsson *et al.* 1980, Blair *et al.* 1980, 1981, Wood *et al.* 1983).

The relative ease by which the simple induction of c-*mos* expression in fibroblasts resulted in cell transformation suggested that these sequences could be an extremely potent oncogene *in vivo*, and thus might be stringently regulated during normal cell growth. This seemed consistent with the persistent failure to detect *mos* proto-oncogene expression in normal cells and tissues (Gattoni *et al.* 1982, Goyette *et al.* 1984, Muller *et al.* 1982), and the recent detection of low levels of

---

[1]Laboratory of Molecular Oncology, National Cancer Institute, Frederick, MD 21701-1013, [2]Viral Carcinogenesis Laboratory, Bionetics, Inc., Frederick, MD 21701-1013, and [3]Recombinant DNA Laboratory, Program Resources, Inc., Frederick, MD 21701-1013, U.S.A.

VIRAL CARCINOGENESIS, Alfred Benzon Symposium 24,
Editors: N. O. Kjeldgaard, J. Forchhammer, Munksgaard, Copenhagen 1987.

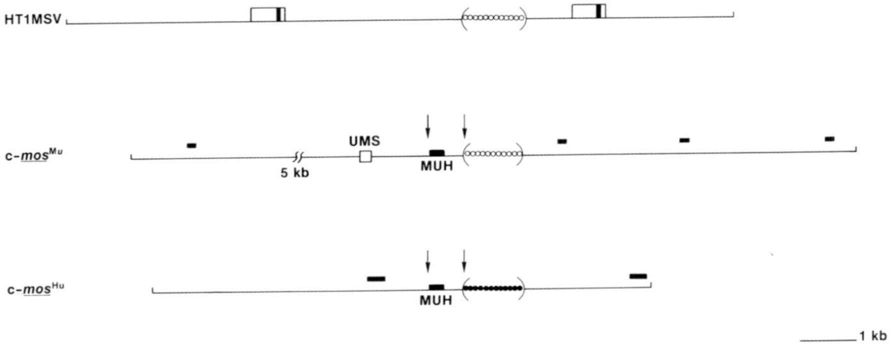

Fig. 1. Structural elements of v-*mos* and the cellular c-*mos*^{Mo} and c-*mos*^{Hu} loci. (○○○○○○) DNA sequences representing the v-*mos*/c-*mos*^{Mo} homology (●●●●●●●) c-*mos*^{Hu} homology; (MUH) the *mos* upstream homology; (UMS) the upstream mouse sequence; (■) DNA sequences representing the mouse B1 repetitive DNA family in c-*mos*^{Mo} and the human *Alu*-repetitive family.

tissue-specific c-*mos* transcripts in normal testes and ovaries (Propst & Vande Woude 1985). We have studied the biological activation of c-*mos* by proviral LTR sequences as a means of identifying transcriptional and other regulatory elements within the normal locus. We have also compared the behavior of the normal *mos* locus of man and chicken to that of mouse, since such comparisons should help to point out significant control elements which have been evolutionarily conserved. These latter studies have also demonstrated the presence and conservation of functional domains within the *mos* protein.

We will discuss here the accumulated evidence, primarily from studies on the mouse and human (c-*mos*^{Hu}) loci, indicating the presence of regulatory sequences which can influence c-*mos* gene expression as detected by biological transformation assays. The presence of similar sequences has also been observed by others in studies of the rat c-*mos* locus (Van der Hoorn *et al.* 1984, 1985). The strong conservation of the c-*mos* open reading frame suggests that the gene must function during the normal growth and development of higher eukaryotes (Van Beveren *et al.* 1981, Van der Hoorn *et al.* 1984, Watson *et al.* 1982).

The c-mos^{Mu} *locus and the function of UMS*

The general structural features of the c-*mos*^{Mu} locus is shown schematically in Figure 1. Two prominent features located upstream from the *mos* coding region are indicated. The MUH region was identified by its 77% homology with a similar sequence located upstream from the major *mos* homologous open reading

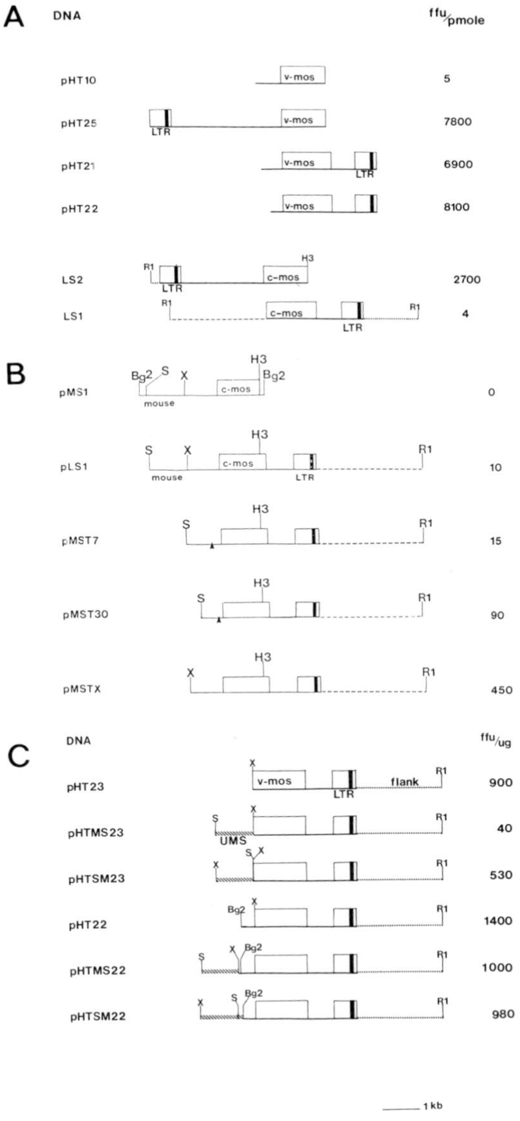

Fig. 2. Biological activity of LTR-activated v-*mos* and c-*mos* constructs. Structure of constructs and activities have been described in detail elsewhere (Blair *et al.* 1981, Oskarsson *et al.* 1980, Wood *et al.* 1983, 1984). Restriction sites shown are: R1-EcoRI; H3-HindIII; Bg2-BglII; S-SacI; X-XbaI. Each plasmid DNA was linearized by digestion with EcoRI prior to transfection into NIH3T3 cells over a range of DNA concentrations (50 ng-1 μg/2.5 $\times$ 10^5 cells). The transforming efficiency in focus-forming units per picomole (A and B) or per microgram of plasmic DNA is calculated from at least four replicate determinations.

frame (Blair *et al.* 1984). Although the degree of sequence conservation is strongly suggestive of some conserved function, its true role or significance remains unknown. The second, more distal region (UMS), was identified as a potentially significant regulatory element following early attempts to activate c-*mos*^{Mu} with LTR elements placed either upstream or downstream of the coding region of the oncogene. Figure 1A and B shows that constructs in which an LTR was placed downstream of *mos* transformed 1000-fold less efficiently than those in which *mos* was activated by an upstream LTR (Oskarsson *et al.* 1980, Wood *et al.* 1984). The element called UMS (for upstream mouse sequence), located approximately 1500 base pairs (bp) upstream to c-*mos*^{Mu}, was subsequently shown to be responsible for this reduction in transforming activity. This inhibitory sequence is an AT-rich region less than 600 bp long. UMS can prevent activation of v-*mos* by a downstream LTR only when it is inserted between sequences which serve as a promotor and the ATG of the viral oncogene (Figure 2C). However, this effect is orientation- and position-dependent. Furthermore, cells transformed by pHTMS22 only contained low levels of *mos* transcripts, while cells transformed by pHT22 exhibited an abundant and complex v-*mos* transcription pattern. These latter transformants had presumably acquired efficient upstream promotors during transfection from either host or carrier DNA sequences, but we reasoned these were probably inefficiently translated or prevented the detection of the true functional v-*mos* transcript since low expression of v-*mos*, 1–10 transcripts per cell, was sufficient to transform NIH3T3 cells (Wood *et al.* 1983). This supported the hypothesis that UMS acts by preventing transcriptional read-through from promoters upstream to UMS.

Direct evidence for this mechanism of UMS action has recently been obtained utilizing constructs in which UMS sequences were used to replace the region of the LTR containing the polyadenylation signal and the region defined as a possible transcriptional termination signal (Honigman & Panet 1983, Honigman *et al.* 1985). In these constructs UMS appears to provide both polyadenylation and transcriptional termination functions (McGeady *et al.* 1986).

The activity of UMS allows us to interpret our results in terms of the structure of the c-*mos*^{Mu} locus. Thus, it appears that there is no promoter capable of functioning in NIH3T3 cells within the 2kb of nucleotide sequences upstream of the c-*mos*^{Mu} proto-oncogene. This had suggested to us that c-*mos*^{Mu} transcription might be activated in trans (Wood *et al.* 1984). Since UMS appears to act as a transcription terminator (McGeady *et al.* 1986), it could be the termination of an as yet undefined gene locus upstream of *mos*. It is also possible that UMS

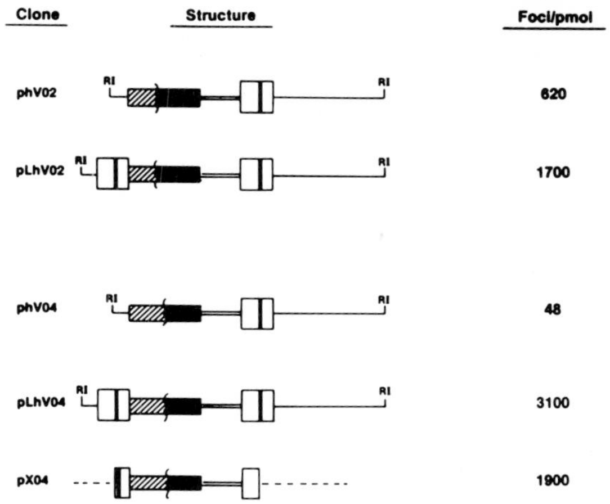

Fig. 3. Evidence for inhibition of *mos* transforming activity by upstream c-*mos*^{Hu}/v-*mos* hybrid constructs shown are described elsewhere (Woodworth *et al.* 1984, Blair *et al.* submitted). In the phV constructs the LTR was inserted 22 bp upstream from the conserved *mos* ATG. pX04 was derived from pLh04 and contains only the promoter sequences of the upstream LTR and the enhancer sequences of the downstream LTR. Values given for activities represent the averages of four assays. EcoR1–R1; – human sequences; ▨ – c-*mos*^{Hu} sequences; ■ – v-*mos* sequences; = MuSV-derived sequences; ◫ – LTR.

functions merely to block transcription into the c-*mos* locus in certain cell types in which, like NIH3T3 cells, low levels of *mos* expression can result in transformation.

Transcripts containing UMS are observed in certain tissues (Propst & Vande Woude 1985) so under these circumstances UMS may be positively regulated to allow, or not interfere with, *mos* expression. Homology to UMS has been reported in rat (Van der Hoorn *et al.* 1985), suggesting that in this case regulation may be similar to that proposed for mouse. However, UMS is not conserved in the c-*mos*^{Hu} locus, indicating that alternative control elements may be possible.

The human mos *locus: Evidence for inhibitory sequences*

The schematic representation of c-*mos*^{Hu} in Figure 1 shows the relative position of the major homologous open reading frame (75% conserved, Watson *et al.* 1982), together with the MUH region defined above. The human homologue also possesses an overlapping open reading frame (OORF) (Watson *et al.* 1982) whose initiation site precedes the conserved *mos* ATG. This feature has only been detected in primate *mos* cellular homologues (Paules and Vera, unpublished data).

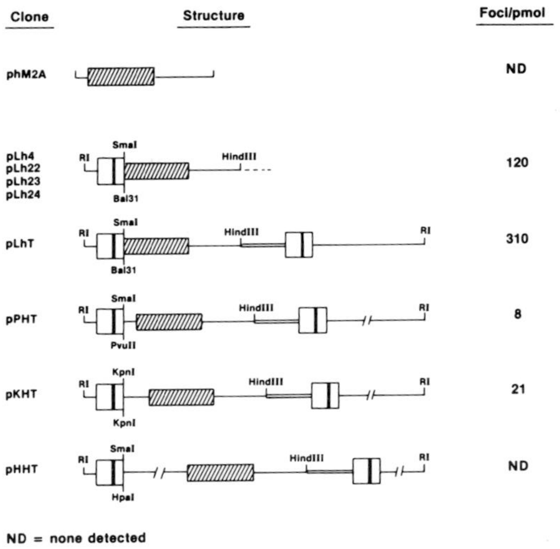

Fig. 4. Structure and biological activity of constructs containing human c-*mos*. The activity represents an average of at least four assays. The structural features of the *mos* constructs are as described in Figure 3.

Early attempts to activate the transforming potential of c-mos^{Hu} with LTR sequences were unsuccessful (Woodworth *et al.* 1984, Blair *et al.* 1984). However, transfection of recombinant DNA constructs containing hybrid human/mouse *mos* revealed a complex pattern of biological activity which defined functional domains of the *mos* gene product and demonstrated that human sequences immediately upstream from the human *mos* ATG reduced the transforming potential (Blair *et al.* submitted). Thus, as shown in Figure 3, the removal of the human sequences present in the hybrid clones phVO2 and phVO4 and the introduction of an LTR 25 bp upstream from the ATG enhanced the transforming potential. The role of the LTR appears to be to provide a promoter, since pXO4, (Figure 3) in which the enhancer is located downstream and the promoter upstream of *mos*, is still significantly more active than phVO4.

Based on these results, we analyzed the transforming activity of c-mos^{Hu} constructs in which the LTR was placed 25 bp upstream from the conserved *mos* ATG, eliminating the OORF (Figure 4; Blair *et al.*, submitted). These constructs (Lh4) all exhibited transforming activity 20- to 50-fold lower than the equivalent c-mos^{Mu} constructs (Figure 4). In addition, the foci were smaller than v-*mos* foci and contained fewer cells with a refractile morphology (Figure 5). Furthermore, when the LTR was positioned further upstream from the c-mos^{Hu} ATG, the

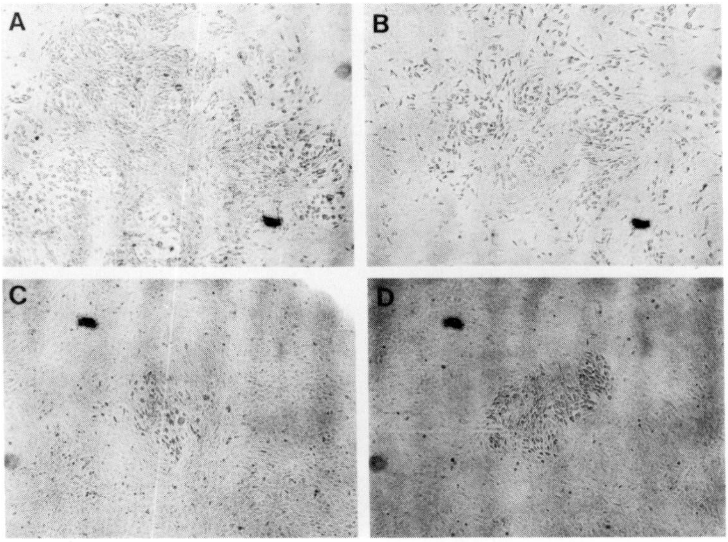

Fig. 5. Morphology of foci induced on NIH3T3 cells by v-*mos* (A and B) or human c-*mos* (C and D) in the presence of 0.125 μM dexamethasone.

activity was reduced or eliminated. It is possible that the elimination of the OORF allows the more efficient expression of the c-mos^{Hu} transforming protein, which appears to be less effective in transforming mouse cells than the mouse c-*mos* product.

Identification of domains in hybrid mouse/human mos *genes which affect activity*

The 50-fold differences between the transforming potential of LTR-activated c-mos^{Hu} and c-mos^{Mu} (Table I) are presumably not due to differences at the transcriptional level, and have been attributed to differences in the protein coding sequence (Blair *et al.* submitted). Analysis of a series of hybrid *mos* constructs for their transforming activity have allowed us to identify three regions, or domains, in the *mos* coding region which can influence the ability of the hybrid gene products to transform NIH3T3 cells. The first domain of c-mos^{Mu} which can increase the transforming activity of hybrid *mos* constructs lies between position 48 and 80 (Figure 6, all positions refer to the numbering of c-mos^{Hu} codons). There are only five amino acid differences between the two *mos* genes in this region, and three of the five (at positions 68, 73 and 80) cluster in a region

TABLE I
mos *transforming efficiencies*

Plasmid	*mos* Species	Structure[a]	FFU/pmole
pTS1	c-*mos*Mu	□---( )	2572
pLh04	c-*mos*Hu	□-( )	54
pPHT	c-*mos*Hu	□---( )	8
pM1CM33	c-*mos*Ch	□-( )	1166
pM1CM36	c-*mos*Ch	□------( )	239

[a] The □ indicates position of LTR sequences relative to the designated *mos* ( )gene.

showing extensive homology between *mos, src* and the bovine cAMP-dependent protein kinase (Barker & Dayhoff 1982). This region is just upstream from a conserved lysine at position 87 which is the proposed ATP binding site for the bovine kinase and *mos* and which is required for v-*mos* transformation (Hannick & Donoghue 1985). Since *mos* protein produced in *E. coli* binds ATP (Seth and Vande Woude 1985), it is possible that differences between human and mouse *mos* within this domain affect ATP binding, and hence influence transforming activity.

The identification of the domain between positions 155 and 172 is based on the observation that when the sequences C-terminal to this domain are not from the homologous species, there is a marked reduction in biological activity. We have hypothesized (Woodworth *et al.* 1984, Blair *et al.* submitted) that this domain, which shows little homology between mouse and man, interacts in some fashion with the C-terminal portion of the *mos* molecule.

However, the region which seems to have the most significant effect on the transforming activity, and the phenotype of the transformed cells, is the C-terminal region (from position 223–346). This is the region which shows significant homology to the *src* kinase domain (Hunter & Cooper 1986) and our results, plus recent evidence that the *mos* product is a serine kinase (Kloetzer *et al.* 1983, Maxwell *et al.* 1985), are consistent with this being an important functional domain of the *mos* gene product.

The biological activity of chicken c-mos

As indicated above, and in Table I, the transforming activity of LTR-activated c-*mos*Hu is relatively low in comparison to c-*mos*Mu, and the phenotype of the transformed cells is also consistent with less efficient transformation (Figure 5).

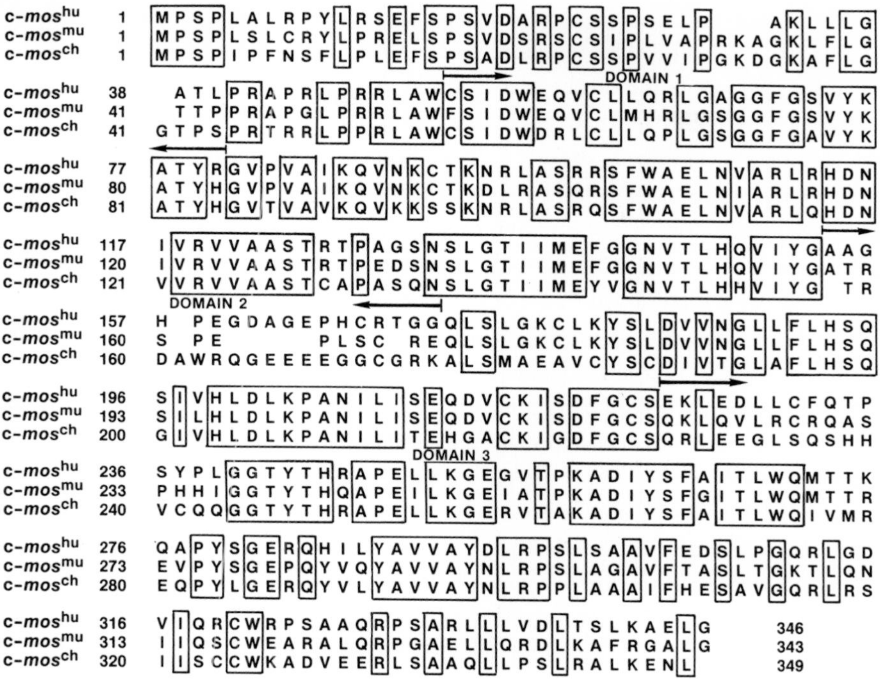

Fig. 6. Comparison of the deduced amino acid sequences of c-mos^{Hu}, c-mos^{Mu} and c-mos^{Ch}. Each domain (I–III) is defined by arrows and is based on data from Blair *et al.* submitted; and Woodworth *et al.* 1984. Sequences are from Van Beveren *et al.* 1981, for c-mos^{Hu}; Watson *et al.* 1982, for c-mos^{Hu}; and Schmidt (in preparation) for c-mos^{Ch}.

We have also recently begun to analyze the structure and function of the chicken homologue of *mos* (c-mos^{Ch}) which, by sequence analysis, is only 62% homologous to c-mos^{Mu} (Schmidt *et al.*, in preparation) (compared to 75% for c-mos^{Hu}). Somewhat unexpectedly, LTR activated c-mos^{Ch} transformed with essentially the same efficiency as LTR activated c-mos^{Mu} (Table I, Schmidt *et al.*, in preparation), and the size and morphology of the foci induced in NIH3T3 cells was also similar. Table I also shows that, when an LTR is placed 900 bp upstream of c-mos^{Ch} (pM1CM36), the transforming activity is 5- to 10-fold lower than when it is immediately adjacent to the c-mos^{Ch} ATG (pM1CM33). This suggests that a negative regulatory element could be located upstream to c-mos^{Ch}. However, no homology with either human or mouse upstream sequences are detectable in this region other than the fact that they are all AT-rich (Schmidt, unpublished data).

As indicated in Figure 6, the conserved regions of the three *mos* genes (mouse, human and chicken) occur in short blocks separated by short stretches of non-homology. If we assume that transformation efficiency differences are due to differences in amino acid sequence, their comparison of the sequences within the three domains described above could suggest specific amino acids which may strongly influence transforming activity. Within the first, or ATP-binding, domain we have noted that the only singificant difference between c-mos^{Hu} and the efficiently transforming c-mos^{Mu} and c-mos^{Ch} is at amino acid 68, *within* the consensus ATP binding site (Barker & Dayhoff 1982). Although similar analyses could be performed on the other two domains, all such interpretations must be viewed as highly speculative in the absence of site-specific mutagenic analyses to confirm some of these assignments. It should also be noted that all of these analyses are based on trans-species transformation analysis. The possibility that human *mos* might be an efficient oncogene in human cells or the appropriate target tissue has not been ruled out.

*c-*mos *expression: Possible roles for inhibitory sequences*

The identification of sequences affecting the LTR-activated expression of c-*mos* loci from several species leads to the question of whether these sequences may play some role in the regulation of c-*mos* expression in normal cells. C-*mos* is expressed at low levels (0.1 to 10 copies/cell) in certain normal tissues, primarily testes, ovaries and whole embryo (Propst & Vande Woude 1985). However, the expression is clearly complex, with transcript sizes varying from 1.3 to 6 kb, depending on the tissue. The variations in size are caused by differences at the 5′ end of the various RNAs (Propst & Vande Woude 1985, Propst, unpublished data), suggesting that different promoters are used in different tissues. The absence of evidence of RNA splicing (Propst, unpublished data), indicates that the c-*mos* open reading frame is the only coding exon in most tissues (i.e., testes and ovaries). Since in gonadal tissue the transforming activity of the *mos* product is suppressed, it is possible that transformation by *mos* is the result of its inappropriate expression in non-target cells.

Our data also suggests that c-mos^{Mu} testes transcripts are initiated just downstream from the conserved MUH region (Propst & Vande Woude 1985, Propst, unpublished data), perhaps lending support to the speculation that MUH could play a role in regulatory expression. On the other hand, UMS sequences are likely to be contained in the large 6 kb transcripts (Propst & Vande Woude 1985), although these transcripts have not been extensively characterized.

The initial analyses of c-*mos* expression suggest that transcription of this gene is regulated in a novel fashion. Similarly, both the structure of *mos*, namely its lack of intervening sequences, and its very high transforming efficiency at low expression levels, make it unique among oncogenes and indeed among the known eukaryotic genes. These factors, coupled with its low expression in normal tissues, suggest that further analysis of the regulation of its expression and its mechanism of action should lead to novel insights into both normal and neoplastic cellular processes.

REFERENCES

Barker, W. C. & Dayhoff, M. O. (1982) Viral *src* gene products are related to the catalytic chain of mammalian cAMP-dependent protein kinase. *Proc. Natl. Acad. Sci. USA 79*, 2836–9.

Blair, D. G., McClements, W. L., Oskarsson, M. K., Fischinger, P. J. & Vande Woude, G. F. (1980) Biological activity of cloned Moloney sarcoma virus DNA: Terminally redundant sequences may enhance transformation efficiency. *Proc. Natl. Acad. Sci. USA 77*, 3504–8.

Blair, D. G., Oskarsson, M., Wood, T. G., McClements, W. L., Fischinger, P. J. & Vande Woude, G. F. (1981) Activation of the transforming potential of a normal cell sequence: A molecular model for oncogenesis. *Science 212*, 941–3.

Blair, D. G., Wood, T. G., Woodworth, A. M., McGeady, M. L., Oskarsson, M. K., Propst, F., Tainsky, M., Cooper, C. S., Watson, R., Baroudy, B. M. & Vande Woude, G. F. (1984) Properties of the mouse and human *mos* oncogene loci. In: *Cancer Cells: Oncogenes and Viral Genes*, Vol. 2, eds. Vande Woude, G. F., Levine, A. J., Topp, W. C. & Watson, J. D., pp. 281–289. Cold Spring Harbor Laboratory, Cold Spring Harbor, New York.

Gattoni, S., Kirschmeier, P., Weinstein, I. B., Escobedo, J. & Dina, D. (1982) Cellular Moloney murine sarcoma (c-*mos* sequences are hypermethylated and transcriptionally silent in normal and transformed rodent cells. *Mol. Cell. Biol. 2,* 42–51.

Goyette, M., Petropoulos, C. J., Shank, P. R. & Fausto, N. (1984) Regulated transcription of c-Ki-*ras* and c-*myc* during compensatory growth of rat liver. *Mol. Cell. Biol. 4*, 1493–8.

Hannik, M. & Donoghue, D. J. (1985) Lysine residue 121 in the proposed ATP-binding site of the v-*mos* protein is required for transformation. *Proc. Natl. Acad. Sci. USA 82*, 7894–8.

Hunter, T. & Cooper, J. A. (1986) Viral oncogenes and tyrosine phosphorylation. In: *The Enzymes: Enzyme Control by Protein Phosphorylation*, eds Boyer, P. D. & Krebs, E. G. Academic Press, New York (in press).

Jones, M., Bosselman, R. A., van der Hoorn, F. A., Berns, A., Fan, H. & Verma, I. M. (1980) Identification and molecular cloning of Moloney mouse sarcoma virus-specific sequences from uninfected mouse cells. *Proc. Natl. Acad. Sci. USA 77*, 2651–5.

Kloetzer, W. S., Maxwell, S. A. & Arlinghaus, R. B. (1983) $p85^{gag\text{-}mos}$ encoded by *ts*110 Moloney murine sarcoma virus has an associated protein kinase activity. *Proc. Natl. Acad. Sci. USA 80*, 412–6.

Maxwell, S. A. & Arlinghaus, R. B. (1985) Serine kinase activity associated with Moloney murine sarcoma virus-124-encoded $p37^{mos}$. *Virology 143*, 321–33.

McGeady, M. L., Wood, T. G., Maizel, J. V. & Vande Woude, G. F. (1986) Sequences upstream to the mouse c-*mos* oncogene may function as a transcription termination signal. *DNA* (in press).

Muller, R., Slamon, D. J., Tremblay, J. M., Cline, M. J. & Verma, I. M. (1982) Differential expression of cellular oncogenes during pre- and post-natal development of the mouse. *Nature 299*, 640–4.

Oskarsson, M., McClements, W. L., Blair, D. G., Maizel, J. V. & Vande Woude, G. F. (1980) Properties of a normal mouse cell DNA sequence (sarc) homologous to the *src* sequence of Moloney sarcoma virus. *Science 207*, 1222–4.

Propst, F. & Vande Woude, G. F. (1985) Expression of c-*mos* proto-oncogene transcripts in mouse tissues. *Nature 315*, 516–8.

Seth, A. & Vande Woude, G. F. (1985) Nucleotide sequence and the biochemical activities of the HT1MSV *mos* gene. *J. Virol. 56*, 144–52.

Van Beveren, C., van Straaten, F., Galleshaw, J. A. & Verma, I. M. (1981) Nucleotide sequence of the genome of a murine sarcoma virus. *Cell 27*, 97–108.

van der Hoorn, F. A. & Firzlaff, J. (1984) Complete c-*mos* rat nucleotide sequence: Presence of conserved domains in c-*mos* proteins. *Nucl. Acids Res. 12*, 2147–56.

van der Hoorn, F. A., Muller, V. & Pizer, L. (1985) Sequences upstream of c-*mos* (rat) that block RNA accumulation in mouse cells do not inhibit *in vitro* transcription. *Mol. Cell. Biol. 5*, 406–9.

Watson, R., Oskarsson, M. & Vande Woude, G. F. (1982) Human DNA sequence homologous to the transforming gene (*mos*) of Moloney murine sarcoma virus. *Proc. Natl. Acad. Sci. USA 79*, 4078–82.

Wood, T. G., McGeady, M. L., Baroudy, B. M., Blair, D. G. & Vande Woude, G. F. (1984) Mouse c-*mos* oncogene activation is prevented by upstream sequences. *Proc. Natl. Acad. Sci. USA 81*, 7817–21.

Wood, T. G., McGeady, M. L., Blair, D. G. & Vande Woude, G. F. (1983) Long terminal repeat enhancement of v-*mos* transforming activity: Identification of essential regions. *J. Virol. 46*, 726–36.

DISCUSSION

RAPP: In your 3′ LTR *mos* transfections, where you did get transformation, where does the transcript start?

BLAIR: With v-*mos* the transcripts start in a variety of places. There are many messages in those cells. Some of them are very large, 8 or 9 kb, and they presumably start far upstream. There are also smaller messages, some of which start in a region just upstream from the v-*mos* coding sequence. If you put UMS upstream of that point, you do get a little bit of expression from this presumably cryptic promoter site, and those cells are transformed. One of the problems is the awful lot of polyA+ RNA in most *mos* transformed cells and very little protein. So, clearly, all of it cannot be functional.

RAPP: Did you test your UMS *mos* construct in, say, testicular tumor cells?

BLAIR: I have not tried that. None of the testicular tumors Fritz Probst tested express c-*mos* normally.

LOWY: Is there any evidence in either humans or avian species for expression of *mos* in gonadal tissues?

BLAIR: Fritz Probst found expression in human testicular tissue at a lower level than what he sees in mouse, which is about 1–10 copies per cell.

SHERR: I think that I am correct in saying that there are examples of tumors with rearranged *mos* genes and also a case of intracisternal A particle insertion in the region of this gene. Do these insertions or rearrangements separate the upstream regulatory sequences from the body of the gene?

BLAIR: The intracisternal A particle sequence insertion is about 40 amino acids downstream of the start of the coding sequence, and that is active as a transforming gene. I am not sure about the human translocations. As I recollect there is no expression of c-*mos* where there has been a translocation involving chromosome 8.

BRUGGE: It seems that part of the cell type specificity of transformation could be due to differences in the human and mouse *mos* sequences that would affect the ability of the gene products to interact with substrates from different species of cells, and that the specificity or the ability to transform might be due to these interacting substrates. To address this, have you tried to transform human cells with the human *mos*?

BLAIR: Yes, we have tried to use both transfection and micro-injection into human cells, and to date we have detected no transformation. Those experiments are really just underway, so it is too early to say.

VENNSTRÖM: Does the human *mos* protein, which you can express in resonable quantities in cells, have the same biochemical properties as the viral *mos* protein in terms of enzyme activity and localization? Is it also a serine kinase with no tyrosine kinase activity?

BLAIR: In terms of the *mos* localization, it appears to be cytoplasmic, *mos* is found at least in acutely infected cells. I think that one of problems is that all localization of *mos* has been done under conditions of very high expression. In stable, normally transformed cells it is a little bit harder to say that it is entirely cytoplasmic. As to the second part of the question, we do not know. It is not as good an ATP binding protein as the mouse *mos,* which is consistent with the fact that it is poorly transforming.

WEINBERG: I am still a bit uncomfortable about the notion that the human and the mouse *mos* proteins have slightly different substrate specificity. There is very little precedence to my knowledge in the whole literature on cellular oncogenes. In fact they seem extremely well conserved. I am wondering whether some of your data may be explainable not, as you imply, by the specific transforming activity of the protein, but rather may be due to factors like stability of various proteins in the mouse cells. It is so difficult to measure protein levels that perhaps much of the phenomenon is simply due to the presence or the absence of sufficient concentrations of protein molecules in the cell to elicit the phenotype you are looking at.

BLAIR: First of all, I think the idea about different substrates is purely speculation at this point.

BRUGGE: I did not mean to say that the human and mouse *mos* proteins would have different substrates. I meant to say that the substrates in the two different cells could have different conformations or different structures, and the *mos* protein would not be able to interact with its substrate in mouse cells.

WEINBERG: But even that violates my sense of absolute conservation of mammalian proto-oncogene proteins and their substrates. What about the issue of protein levels? Could that not explain much of the phenomenon?

BLAIR: In the case of hybrid *mos* genes, recombinant constructs between different portions of viral and human *mos,* I will admit that with non-transforming constructs we have not been able to verify that we get a large and stable expression of the proteins in NIH 3T3 cells. So, in that case, it could be a possible explanation. In the case of pure human *mos,* however, it cannot be the explanation, because there we see abnormally high levels of *mos* protein. The level of *mos* protein in NIH 3T3 cells transformed by human *mos* is as high as it is in cells acutely infected with MoMuSV, most of which die. So, in that case (c-*mos*^{Hu}) there is certainly enough protein. That leaves the possibility that either it does not work well enough in mouse cells, because of some substrate divergence or something else, or that it really intrinsically does not retain the ability to function as a transforming protein. These genes are there not to transform cells, but to provide some other function. At least *a priori* I could conceive of a situation such that mouse c-*mos*, which would cause transformation if it were activated, might change sufficiently in evolving to human c-*mos* that it transforms either mouse or human cells. Maybe that violates the conservation of oncogenes, but it seems to me to be at least theoretically possible. Whether that is the case here, I do not know.

WEINBERG: Could it be that the *mos* protein requires an ancillary regulatory subunit which binds with species specificity?

BLAIR: That is certainly possible. It could be that *mos* is a very lineage-specific functional protein, and that it has become perhaps even more so in human than in the mouse. Whereas mouse c-*mos* will transform and function in fibroblasts, human c-*mos* will not, even in human fibroblasts. I think that if we can look at what happens when we put *mos* expression constructs into human cells and see if we can restore transforming activity, we have an answer to that.

III. Transcriptional Control and Modifications

Some Remarkable Properties of Cell Lines and Transgenic Mice which Received the Gene Encoding the Large T Protein of Polyoma Virus

Nicolas Glaichenhaus[1], Pierre Léopold[1], Evelyne Mougneau[1], Joëlle Vailly[1], Christa Cerni[2], Minoo Rassoulzadegan & François Cuzin[1]

We are currently analyzing the effects on rodent cells of a series of viral and cellular oncogenes, whose prototype is the gene of polyoma virus (*plt*) encoding the large T antigen.

1. STUDIES IN CELL CULTURE

a. *At the cellular level*

A first step in transformation of rat embryo fibroblast cells was shown to be induced by transfer of oncogenes of polyoma virus (*plt*: Rassoulzadegan *et al.* 1982, 1983), of adenoviruses (E1A: van den Elsen *et al.* 1982, Ruley 1983) and of cloned cellular oncogenes (*myc*: Land *et al.* 1983, Mougneau *et al.* 1984, *p53*: Eliyahu *et al.* 1984, Parada *et al.* 1984). The resulting cell lines were characterized not only by their increased growth potential in cell culture ("immortalization"), but also by a characteristic and complex phenotype. Of particular interest is the fact that they are prone to undergo subsequent transformation events, either as a consequence of the introduction of activated *ras* or polyoma virus *pmt* oncogenes, or spontaneously. Stochastic transformation events were observed to occur with high frequencies in the progeny of cells which had received *myc* oncogenes (E. Mougneau, C. Cerni & F. Cuzin, manuscript in preparation). Transformation frequencies were increased after TPA treatment (Connan *et al.* 1985), and also when *myc* expression was boosted by addition of a powerful enhancer. Even

---

[1]INSERM U273, Centre de Biochimie, Université de Nice, France, and [2]Institute for Tumor Biology, University of Vienna, Austria.

VIRAL CARCINOGENESIS, Alfred Benzon Symposium 24,
Editors: N. O. Kjeldgaard, J. Forchhammer, Munksgaard, Copenhagen 1987.

using such vectors, however, both morphological transformation and tumorigenicity were observed in late, but not in early passages of the FR3T3 cell line (Zerlin *et al.* 1986), strongly suggesting that *myc* oncogenes have to cooperate with host functions.

Starting from the assumption that spontaneous transformation of these "high risk" cells corresponds to the activation of cellular oncogenes, we examined whether these cells would show some characteristic sign of genetic instability and, in fact, observed that a large series of cell lines, derived either from primary REF cultures or from the established FR3T3 line after transfer of v-*myc*, c-*myc*, *plt* and E1A oncogenes, consistently exhibited two abnormalities at the chromosomal level: a fraction of the population was undergoing sister chromatid exchanges at very high frequencies and aneuploid cells accumulated in the cultures (Cerni *et al.* 1986).

b. *At the molecular level*

A possible insight into the molecular mechanisms acting at this early step of transformation was provided by the observation that two of the protein products of the immortalizing oncogenes are known to act in the regulation of the expression of both viral and cellular genes (large T proteins of polyomaviruses, E1A proteins of adenoviruses). It is tempting to speculate that a primary effect of these oncogenic determinants is a change in the level of gene products critical for the control of cell growth and of the cell cycle. The expression of either *plt* or *myc* oncogenes was shown to result in a modification in the steady state level of a series of cellular gene products. Differential screening of cDNA libraries yielded 9 clones (pIL plasmids) characteristic of rat cellular mRNAs whose cytoplasmic levels increase upon immortalization and transformation by polyoma virus (Glaichenhaus *et al.* 1985). mRNAs identified by 2 of these probes (pIL1 and pIL7) were also increased in rat embryo fibroblast cells immortalized by adenovirus E1A, by v-*myc* and rearranged c-*myc* genes. Nucleotide sequencing identified pIL7 as a mitochondrial transcript (COII). Transcriptional activation of the mitochondrial genome appears as characteristic of early stages in transformation of REF cells (Glaichenhaus *et al.* 1986, Rigby, this Symposium).

Variations in mRNA levels at different periods of the cell cycle were studied after automated cell sorting. Four pIL mRNAs were found to be expressed mostly in phase G2 in normal FR3T3 cells and their expression shifted to G1 in polyoma-transformed cells. No change in their level of transcription could be evidenced by nuclear run-on experiments, suggesting that the cycle-dependent accumulation of these mRNAs in FR3T3 cells might mostly result of differential degradation when the cells goes through mitosis. Nucleotide sequence analysis indicated that these cell cycle-dependent mRNAs contain a characteristic semi-repetitive sequence.

One intrinsic difficulty in identifying genes by differential screening of cDNA libraries is the lack of *a priori* criteria for estimating the biological significance of a change of the level of a given mRNA. Furthermore, it is not easy to differentiate between the primary regulatory effect(s) and the multiple secondary variations expected from the radical changes occurring in many of the properties of a tumor cell. Whether or not mitochondrial transcription is directly *trans*-activated by the oncogene products, it is not likely to be *the* primary effect of the immortalizing oncogenes. It may nevertheless be of interest because of its obvious physiological significance and because we have very few markers for early stages of the oncogenic process. The changes that we observed in the level of expression of a whole class of mRNAs throughout the cell cycle did not appear to depend on transcriptional changes. It may be premature to conclude from this observation that these changes are not direct effects of large T, *myc* or E1A proteins, because our knowledge of the biochemistry of these complex regulatory proteins is by no mean complete, and because we cannot, therefore, exclude that they might also act at a post-transscriptional level. Assessing the biological significance of the changes in the expression of the pIL genes in transformed cells obviously requires a more extensive knowledge of their protein products.

2. STUDIES IN TRANSGENIC MICE

As we wanted to compare the effects of these oncogenes in cell culture and in the organism, we constructed a series of 13 transgenic mouse strains by microinjecting into fertilized eggs the circular form of plasmid pPyLT1 DNA. pPyLT1 carries a modified form of the genome of polyoma virus, which expresses only one of the viral proteins, the large T antigen, under control of its original promotor (Rassoulzadegan *et al.* 1982). Unlike what had been observed in transgenic mice expressing rearranged *myc* genes (Cory, this Symposium), and much to our surprise, it became evident that mouse strains derived from eggs which had received this DNA maintained purely autonomous genetic elements, which were different from one transgenic strain to another, and different in all cases from pPyLT1 (Rassoulzadegan *et al.* 1986).

Structural analysis of these autonomous genetic elements first revealed that, in several of them, the bacterial vector sequences had undergone only minor rearrangements and thus the murine plasmids could be recovered in *Escherichia coli* by transfection with total genomic DNA. By contrast, in all the plasmids, polyoma sequences were extensively rearranged and in many instances completely lost. Radioactive probes, prepared from the plasmids rescued in bacteria, hybridized with a series of restriction fragments on Southern blots of normal mouse DNA, indicating that they had acquired sequences from the cellular genome. This appeared to result from sequence-specific recombination events, since similar patterns of restriction fragments were obtained with plasmids maintained in different mouse strains.

The autonomous elements were usually present in low copy numbers in somatic tissues (usually 2 per diploid cell genome), with the occasional exception of higher copy numbers in either skin or brain. Southern blot analysis indicated in all cases the presence in sperm DNA of only one copy per haploid cell genome. This low average value does in fact correspond to the presence in each spermatozoon of one plasmid molecule, since, upon mating with normal C57BL/6 mice, the transgenic males in all cases transmitted the autonomous structures with 100% efficiency to their progeny. The same non-mendelian high values were observed when transgenic females were crossed with C57/BL/6. In both sexes, sister copies of the plasmids thus segregated very precisely during meiosis. Efficient segregation depends on the presence of centromeric (CEN) sequences which, so far, could be characterized at the molecular level only in yeast (see Murray 1985 for reviews). The transposed cellular DNA segments are obvious candidates for accomplishing the centromeric function demonstrated by the efficient hereditary transmission and low copy numbers. The complete nucleotide sequence of several of these sequences is being established, in order to identify the transposed cellular sequences, and as a first step towards their further characterization by deletion and *in vitro* mutagenesis. One of them was already found to exhibit similarities with the limited regions which are conserved between yeast centromere sequences (Rassoulzadegan *et al.* 1986).

ACKNOWLEDGMENTS

One of us (PL) was a fellow of the Ligue Nationale Française contre le Cancer, France. We are indebted to L. Carbone, M. J. Gonzalez and F. Tillier for skilled technical help. This work was supported by grants from Centre National de la Recherche Scientifique (ATP 1216) and from Association pour la Recherche sur le Cancer, France. Computer resources used to carry out our studies were provided by the BIONET™ National Computer Resource for Molecular Biology, whose funding is provided by the Biomedical Research Technology Program, Division of Research Resources, National Institutes of Health, Grant 1U41RR--1685-02.

REFERENCES

Cerni, C., Mougneau, E., Zerlin, M., Julius, M., Marcu, K. B. & Cuzin, F. (1986) c-*myc* and functionally related oncogenes induce both high rates of sister chromatid exchange and abnormal karyotypes in rat fibroblasts. *Curr. Top. Microbiol. Immunol.* (in press).

Connan, G., Rassoulzadegan, M. & Cuzin, F. (1985) Focus formation in rat fibroblasts exposed to a tumour promoter after transfer of polyoma *plt* and *myc* oncogenes. *Nature 314,* 277–9.

Elihyahu, D., Raz, A., Gruss, P., Givol, D. & Oren, M. (1984) Participation of p53 cellular tumour antigen in transformation of normal embryonic cells. *Nature 312*, 646–9.

Glaichenhaus, N., Léopold, P. & Cuzin, F. (1986) Increased levels of mitochondrial gene expression in rat fibroblast cells immortalized or transformed by viral and cellular oncogenes. *EMBO J. 5*, 1261–5.

Glaichenhaus, N., Léopold, P., Masiakowski, P., Zerlin, M., Julius, M., Marcu, K. B., Vaigot, P. & Cuzin, F. (1985) Changes in the expression of cellular genes in cells immortalized or transformed by polyoma virus. *Cancer Cells, Cold Spring Harbor Lab. 4*, (in press).

Land, H., Parada, L. F. & Weinberg, R. A. (1983) Tumorigenic conversion of primary embryo fibroblasts requires at least two cooperating oncogenes. *Nature 304*, 596–602.

Mougneau, E., Lemieux, L., Rassoulzadegan, M. & Cuzin, F. (1984) Biological activities of the v-*myc* and of rearranged c-*myc* oncogenes in rat fibroblast cells in culture. *Proc. Natl. Acad. Sci. USA 81*, 5758–62.

Murray, A. W. (1985) Chromosome structure and behaviour. *Trends Biochem. Sci. 10*, 112–5.

Parada, L. F., Land, H., Weinberg, R. A., Wolf, D. & Rotter, V. (1984) Cooperation between gene encoding p53 tumour antigen and *ras* in cellular transformation. *Nature 312*, 649–51.

Rassoulzadegan, M., Cowie, A., Carr, A., Glaichenhaus, N., Kamen, R. & Cuzin, F. (1982) The roles of individual polyoma virus early proteins in oncogenic tranformation. *Nature 300*, 713–8.

Rassoulzadegan, M., Léopold, P., Vailly, J. & Cuzin, F. (1986) Germ line transmission of autonomous genetic elements in transgenic mouse strains. *Cell 46*, 513–9.

Rassoulzadegan, M., Naghashfar, Z., Cowie, A., Carr, A., Grisoni, M., Kamen, R. & Cuzin, F. (1983) Expression of the large T protein of polyoma virus promotes the establishment in culture of "normal" rodent fibroblast cells. *Proc. Natl. Acad. Sci. USA 80*, 4354–8.

Rous, P. & Beard, J. W. (1935) The progression to carcinomas of virus-induced papillomas (Shope). *J. Exp. Med. 62*, 523–48.

Ruley, H. E. (1983) Adenovirus early region 1A enables viral and cellular transforming genes to transform primary cells in culture. *Nature 304*, 602–6.

Vogt, M. & Dulbecco, R. (1963) Steps in the neoplastic transformation of hamster embryo cells by polyoma virus. *Proc. Natl. Acad. Sci. USA 49*, 171–9.

Zerlin, M., Julius, M. A., Cerni, C. & Marcu, K. B. (1986) Biological effects of high level c-*myc* expression in FR3T3 fibroblasts. *Curr. Top. Microbiol. Immunol.* (in press).

DISCUSSION

CORY: I have two questions: First of all, presumably something in the PLT must be catalyzing the acquisition of the mouse sequences even though the polyoma sequences are not subsequently retained. Have you done any work to try to pinpoint which region of the polyoma large T is necessary? And secondly, can you now take the acquired mouse sequences and use them in other constructs for high efficiency transfer?

CUZIN: To the second question I can only answer that the experiments are in progress. We reinjected the plasmids amplified in bacteria into mouse eggs and we obtained the same high efficiency of establishment as with pPyLT1 DNA. What we do not know yet is whether or not they went through a new series of rearrangements.

To answer the first question, we can only assume that large T expression was needed at the beginning. It seems that the complete coding region is needed, not only the amino terminal domain, which is efficient for cellular immortalization, probably suggesting that replication has to occur. Actually, replication of polyoma DNA in early mouse embryos has been evidenced in DePamphilis' laboratory, so that it is reasonable to assume that at the beginning there is some replication of pPyLT1 DNA and that the full size large T protein is necessary at the beginning. We obviously have to assay other constructs and some have already been microinjected, but this is not an easy analysis to do, because in each case we have to make the transgenic mice and wait for the F1 generation.

VENNSTRÖM: In yeast one can also get circular DNA to be inherited and replicate, although they are lost at a rather high frequency. What about your system. How stably are they inherited. How often do your lines, or your animals, actually lose these elements?

CUZIN: We did not go further than the fourth generation of back-crosses, and there are many other things to do, like crosses between lines, etc. But we know that it is stable up to the fourth generation. It is present in low number of copies with some variability in somatic tissues, but in sperm cells there is always one copy per cell. As in the yeast plasmids this stability suggest the presence of centromere sequences.

WILLUMSEN: Does this sequence function in yeast as a centromere?

CUZIN: We do not know. The experiments are being done.

BLAIR: Can you introduce these plasmids into mouse or rat cells in culture, and are they also maintained?

CUZIN: I do not have the answer. What we know is only that we can establish cultures from embryos of the transgenic mice and the plasmids are maintained in primary and secondary cultures. In these cell cultures we found that they are in the Hirt supernatants.

BRUGGE: If you looked at independent mice from different mothers, how many contain a similar sequence?

CUZIN: We have no complete results yet. What we know is that the DNA sequence which shows homologies with the yeast CEN is present in several independent plasmids. It argues with the fact that, in Southern blot on mouse DNA, one pattern with a limited number of bands appears to be common to several plasmids.

BRUGGE: In the analysis F_0 mice, what was the source of the DNA? Was it sperm DNA?

CUZIN: We started with tail DNA, and we could not have more complete data before some of the founders died. In all cases, patterns in somatic cell DNA were complicated, but in sperm we already see a simple pattern.

HANAHAN: Do you have any indications that the events that generated the extrachromosomal copies produced by abnormalities of the chromosomes?

CUZIN: No.

HANAHAN: In the original embryo injections of the polyoma DNA, were there any indications that low litter sizes resulted?

CUZIN: No, the litter size was normal in all these mice except for one mouse which is very peculiar. It is the only one which has maintained a significant piece of polyoma DNA and expresses something from this DNA, possibly a truncated form of large T. This founder was male and upon mating females were repeatedly giving birth to babies which looked normal, but died within the first 1 or 2 days. It is difficult to figure out what happened, in part because the mice are eating their dead or dying babies. We assume that the problem of this mouse has something to do with the residual polyoma expression. All the other mice are normal in terms of litter size and health, and did not develop tumors.

VAN DER EB: Do you have any information on other viral transforming genes?

CUZIN: No.

Activation of Cellular Transcription by Simian Virus 40 Large T-antigen

Barbara I. Skene, Nicholas B. La Thangue, David Murphy & Peter W. J. Rigby[1]

The transformation of established lines of cultured cells requires the action of a single oncogene (reviewed by Marshall & Rigby 1984). However, transformed cells differ from their normal parents in many biochemical and biological properties and it is unlikely that these changes occur without significant reprogramming of cellular gene expression. We have attempted to identify and characterize cellular genes which are regulated as a consequence of viral transformation. We hoped that the availability of molecular clones of such genes would enable us to perform two types of experiments. Firstly, we could construct hybrid genes in which the transformation-regulated gene is controlled by a strong promoter and reintroduce such genes into normal cells in order to determine whether over-expression of a single gene could induce any of the properties characteristic of transformed cells. Secondly, we could analyze the *cis*-acting DNA sequences which control transcription of the regulated gene by DNA-mediated gene transfer and use *in vitro* transcription systems to identify the *trans*-acting protein factors involved in regulation. This latter type of experiment is much more straightforward if the transforming protein is known to be able to affect transcription.

Our work has concentrated on the DNA tumor virus Simian virus 40 (SV40). Its transforming protein, large T-antigen, is a nuclear, sequence-specific DNA binding protein (reviewed by Rigby & Lane 1983). It can regulate the transcription of rDNA by RNA polymerase I, both *in vivo* and *in vitro* (Learned *et al.* 1983, Soprano *et al.* 1983), and, at the time that we began our experiments, there was indirect evidence that it could regulate transcription by RNA polymerase II (Postel & Levine 1976, Williams *et al.* 1977). By using differential cDNA cloning

Cancer Research Campaign, Eukaryotic Molecular Genetics Research Group, Department of Biochemistry, Imperial College of Science and Technology, London, SW7 2AZ, U.K.
[1]Present address: National Institute for Medical Research, The Ridgeway, Mill Hill, London NW7 1AA, U.K.

VIRAL CARCINOGENESIS, Alfred Benzon Symposium 24,
Editors: N. O. Kjeldgaard, J. Forchhammer, Munksgaard, Copenhagen 1987.

TABLE I

cDNA clones corresponding to genes activated in SV40-transformed cells

	Designation			
	Set 1	Set 2	Set 3	Set 4
Prototype	pAG64	pAG59	pAG82	pAG88
Insert length (kb)	1.55	1.8	1.58	0.70
Identity	Class I MHC antigen	Endogenous retrovirus	Cytochrome oxidase subunit I (COI)	Cytochrome oxidase subunit II (COII)

	Designation		
	Set 5	Set 6	Set 7
Prototype	pAG10	pAG23	pAG38
Insert length (kb)	1.2	0.7	0.5
Identity	Related to COII	Unknown	B2 repeat non-Class-I

techniques we have isolated several genes which are expressed at elevated levels in SV40-transformed mouse fibroblasts relative to the normal parental line (Scott *et al.* 1983). We have identified most of these genes and have undertaken a number of studies designed to elucidate the mechanism(s) by which they are regulated by T-antigen.

RESULTS

Subtractive hybridization and differential cDNA cloning were used to isolate 42 cDNA clones which correspond to mRNAs which are present at higher levels in the SV40-transformed cell line SV3T3Cl38 than in the parental Balb/c 3T3 A31 fibroblasts. These clones could be divided into 7 sets on the basis on restriction enzyme mapping and cross-hybridization experiments (Scott *et al.* 1983). These sets are summarized in Table I and most of them have now been identified by determining the nucleotide sequences of the cDNA inserts and comparing them with the data bases (Brickell *et al.* 1983, 1985; P. M. Brickell, D. S. Latchman, M. R. D. Scott and P. W. J. Rigby, unpublished data).

The cDNA clones of Set 1 encode the H-2D^d Class I major histocompatibility complex (MHC) antigen. This gene is expressed at high levels in all of the SV40-transformed cell lines that we have tested and in mouse cells transformed by a variety of other agents, including DNA and RNA tumor viruses and chemical carcinogens (Scott *et al.* 1983). We have initiated studies to identify the *cis*-acting DNA sequences which are responsible for the T-antigen responsiveness of the *H-2D*^{d} gene (Rigby *et al.* 1985). To do this we have constructed a hybrid gene in which the 5′ flanking sequences of *H-2D*^{d} drive the expression of the bacterial chloramphenicol acetyl transferase (CAT) gene. When this gene, called 64-*cat*, is transfected into Balb/c 3T3 cells, only a low level of expression is observed, which is expected because the endogenous gene is inefficiently expressed in these cells. However, if 64-*cat* is co-transfected with a plasmid which encodes SV40 large T-antigen, a marked increase is seen (Rigby *et al.* 1985). This induction is also observed in stable transfectants and we have used primer extension to show that the mRNAs in stable transfectants initiate at the RNA polymerase II cap site of the *H-2D*^{d} gene. By deletion analysis we have located some of the sequences involved in this induction to a region 4Kb upstream of the transcription start site (Rigby *et al.* 1985; B.I.S. and P.W.J.R., unpublished data).

SV40-transformed cells contain high levels of small RNAs transcribed by RNA polymerase III from B2 repetitive elements (Brickell *et al.* 1983, Scott *et al.* 1983, Singh *et al.* 1985). This induction of polymerase III transcription manifests some specificity because the transformed cells do not contain elevated levels of 5S rRNA or of a number of other polymerase III transcripts (Carey *et al.* 1986, Singh *et al.* 1985). We have used *in vitro* transcription systems derived from Hela cells and from a variety of mouse cell lines to show that pure SV40 large T-antigen will stimulate polymerase III transcription of a cloned B2 element (Rigby *et al.* 1985; N.B.L.T. and P.W.J.R., unpublished data).

DISCUSSION

Subtractive cDNA cloning techniques have enabled us to isolate and identify cellular genes which are expressed at elevated levels in SV40-transformed cells (Scott *et al.* 1983). Some of these genes are also activated in cells transformed by other agents (Brickell *et al.* 1983, Scott *et al.* 1983). The over-expression of these genes has not so far been linked to any feature of the transformed phenotype but we have begun to elucidate the mechanisms by which large T-antigen regulates the transcription of these genes.

The experiments described here pave the way for a detailed study of the mechanisms by which SV40 large T-antigen regulates the expression of class I antigen genes. Further analyses should precisely define the *cis*-acting DNA sequences involved and we are attempting to reproduce this regulation of polymerase II transcription by large T-antigen in *in vitro* systems so as to be able to characterize the *trans*-acting protein factors involved.

We do not know the function of the small RNAs transcribed by polymerase III from the B2 repetitive elements. However, we have shown that pure large T-antigen is capable of activating this transcription in *in vitro* systems. Moreover, we and others have shown that the expression of these small RNAs is also regulated by the proliferation state of fibroblasts (Edwards *et al.* 1985, Singh *et al.* 1985, K.-H. Westphal and P.W.J.R., unpublished data) and during the differentiation of embryonal carcinoma cells (Murphy *et al.* 1983). It will be of considerable interest to define the biochemical mechanisms involved in controlling the transcription of these repetitive elements.

Viral transforming proteins have the ability to regulate transcription by all 3 cellular RNA polymerases. It remains likely, although unproven, that such transcriptional regulation is crucial to the mechanisms of transformation and further studies of the type described here should resolve this issue.

ACKNOWLEDGMENTS

B.I.S. and N.B.L.T. were supported by a Research Studentship and a Training Fellowship, respectively, from the Medical Research Council. P.W.J.R. held a Career Development Award from the Cancer Research Campaign which also paid for this work.

REFERENCES

Brickell, P. M., Latchman, D. S., Murphy, D., Willison, K. & Rigby, P. W. J. (1983) Activation of a *Qa/Tla* class I major histocompatibility antigen gene is a general feature of oncogenesis in the mouse. *Nature 306,* 756–60.

Brickell, P. M., Latchman, D. S., Murphy, D., Willison, K. & Rigby, P. W. J. (1985) The class I major histocompatibility antigen gene activated in a line of SV40-transformed mouse cells is *H-2D*^{d}, not *Qa/Tla*. *Nature 316*, 162–3.

Carey, M. F., Singh, K., Botchan, M. & Cozzarelli, N. R. (1986) Induction of specific transcription by RNA polymerase III in transformed cells. *Mol. Cell. Biol. 6*, 3068–76.

Edwards, D. R., Parfett, C. L. J. & Denhardt, D. T. (1985) Transcriptional regulation of two serum-induced RNAs in mouse fibroblasts; equivalence of one species to B2 repetitive elements. *Mol. Cell. Biol. 5*, 3280–8.

Learned, R. M., Smale, S. T., Haltiner, M. M. & Tjian, R. (1983) Regulation of human ribosomal RNA transcription. *Proc. Natl. Acad. Sci., U.S.A. 80*, 3558–62.

Marshall, C. J. & Rigby, P. W. J. (1984) Viral and cellular genes involved in oncogenesis. *Cancer Surveys 3*, 183–214.

Murphy, D., Brickell, P. M., Latchman, D. S., Willison, K. & Rigby, P. W. J. (1983) Transcripts regulated during normal embryonic development and oncogenic transformation share a repetitive element. *Cell 35*, 865–71.

Postel, E. H. & Levine, A. J. (1976) The requirement of Simian virus 40 gene A product for the stimulation of cellular thymidine kinase activity after viral infection. *Virology 73*, 206–15.

Rigby, P. W. J. & Lane, D. P. (1983) Structure and function of the Simian virus 40 large T-antigen. *Adv. Viral Oncol. 3*, 31–57.

Rigby, P. W. J., La Thangue, N. B., Murphy, D. & Skene, B. I. (1985) The regulation of cellular transcription by Simian virus 40 large T-antigen. *Proc. Roy. Soc. Lond., Ser. B 226*, 15–23.

Scott, M. R. D., Westphal, K.-H. & Rigby, P. W. J. (1983) Activation of mouse genes in transformed cells. *Cell 34*, 557–67.

Singh, K., Carey, M., Saragosti, S. & Botchan, M. (1985) Expression of enhanced levels of small RNA polymerase III transcripts encoded by the *B2* repeats in Simian virus 40-transformed mouse cells. *Nature 314*, 553–6.

Soprano, K. J., Galanti, N., Jonak, G. J., McKercher, S., Pipas, J. M., Peden, K. W. C. & Baserga, R. (1983) Mutational analysis of Simian virus 40 T-antigen: stimulation of cellular DNA synthesis and activation of rRNA genes by mutants with deletions in the T-antigen gene. *Mol. Cell. Biol. 3*, 214–9.

Williams, J. G., Hoffman, R. & Penman, S. (1977) The extensive homology between mRNA sequences of normal and SV40 transformed human fibroblasts. *Cell 11*, 901–7.

DISCUSSION

ZUR HAUSEN: Some of the changes which you observe in the transformed cells are in the expression of cellular genes; do you also find those early after infection in not yet transformed cells?

RIGBY: I have not looked systematically at everything. What I do know is that if you infect 3T3 cells with SV40 at high multiplicity and look about 2½ days later, which is the only way we have done the experiment, then you cannot see the endogenous retroviral RNA, but we can already by that time see increased levels of the MHC gene message. I have not looked through some of these other things, and I have not done it at any more times or under any different circumstances. But I am pretty sure that the class I gene does respond to acute infection.

CUZIN: One comment and one question: The comment is that we also see an increase in mitochondrial expression in polyoma transformants (Glaichenhaus *et al.* 1986, *EMBO J.* 5, 1261). It appeared to us as a much more general phenomenon that we also could see in *ras*-transformed cells, in papilloma-transformed cells, and in fact in all the transformed and tumor cells that we studied, including the cells at the immortalization stage. My question is: have you looked at either pol III transcripts or MHC transcripts in polyoma transformants or in transformants other than SV40?

RIGBY: Let me respond firstly to the comment and say that we would agree with you. We have also used those probes to survey mouse cells transformed by a wide variety of different agents, both oncogenes that encode nuclear proteins and *src*, etc. etc., and they all have highly elevated levels of these mitochondrial transcripts. We also see something slightly different which is probably worth pointing out; in the transformed cell you see an increase in the abundance of the mature messenger RNA. But we also see accumulation in the transformed cells of high molecular weight RNAs. We presume that these are incompletely processed mitochondrial precursors, but we do not know that for sure. So, we would agree with you on that. But the questions is then for the other genes, for the class I gene we have looked at and in polyoma-transformed cells there are high levels, we did that originally, and then Luigi Lania in Naples has done it much more systematically and shown that there are high levels of class I MHC transcripts in a wide variety of polyoma transformed rat cells (Majello *et al.* 1985 *Nature 314*, 457). We looked at mouse cells transformed by *src*, by *abl*, and by chemical carcinogens, and they certainly all have high levels of the class I gene. The small pol III transcripts, yes you find high levels of those in a wide variety of tumor cells. They are not universal, but they are very common, and Georgiev has also published quite a lot on that point. It is pretty widespread, but not universal.

VAN DER EB: The last results that you presented on the F9 cells, were they stable or transient expression?

RIGBY: They were all transient.

VAN DER EB: Did you look at transcription initiation by run-on?

RIGBY: No, I have not yet done run-ons to look at class I. They are in the works. It is perhaps worth just noting that Grosveld did some run-ons in Ad12-infected cells, and could show that in that experiment there was an increase in transcription initiation.

VAN DER EB: We have done that also.

WEINBERG: One experiment I found very interesting was the *in vitro* transcription system that you described. It seemed to me that the general phenomenon that you began your talk with could be explained either by the ability of the large T antigen to indirectly enhance transcription by, for example, moderating second messenger levels in the cells, or alternatively, that T antigen somehow directs the transcriptional apparatus. Those are, I would say, the two major ways of describing this mechanistically. You showed us an experiment in which the pol III transcription *in vitro* is enormously enhanced by the T antigen. That seems to be a very provocative experiment, but you did not seem to follow it up in a great number of detailed ways to rule out a variety of artefacts that might arise during the reconstruction of that system.

RIGBY: You are absolutely right that it needs following up in a number of ways. I would guess that it works in a way that is independent of the classical DNA binding ability of T antigen. I say that for a number of reasons that Bill Dynan probably can comment on better than I. Certainly in the experiments from Tjian's lab looking at the effect of T on RNA polymerase-I, they were unable to see T antigen binding to anything within the ribosomal RNA promoter (Learned *et al.* 1983 *Proc. Natl. Acad. Sci. USA 80*, 3558). My interpretation of the data that have been in the literature about the ability of T antigen to activate the late promoter is that they indicate that the sequences involved in that activation are certainly different from the classical T antigen binding sites on the DNA. There could be another binding reaction that we have not seen, but it appears they are different from what we all know about

WEINBERG: Maybe it is not binding to the DNA directly.

RIGBY: Right. I am going to tell you what I think it is. We have not seen binding so far. What I think it may well be derives from the observation that it is vital for this effect of T antigen that the template is delivered to the *in vitro* system as a supercoil. With linear templates there is no response. I will put that together with a couple of other things in an argument which is pure speculation. There are data, which I think are convincing, which show that large T antigen is an ATP-dependent helicase (Stahl *et al.* 1986 *EMBO J. 5*, 1939). Worcel's group has argued that the 5S ribosomal RNA pol III transcription factor, TFIIIA, induces a sequence-specific gyration (Kmiec & Worcel 1985 *Cell 41*, 945). I think that the way T works is that the helicase activity is changing the conformation of template to something that is more accessible to the transcription apparatus. But that is pure speculation, I have no evidence for it.

WEINBERG: This suggests that there is rather little specificity on the part of the T antigen.

RIGBY: That is not true, because there is specificity. Yoy see this increase of pol III transcription when the template is a B2. One thing that I forgot to point out on the slide is that there was a control which was the pol III promotor from an adeno VA gene. That does not respond *in vitro*. Botchan has looked *in vivo* and shown that in a cell line in which you see this really dramatic change in the pol III transcripts of the B2s there is no change in the level of 5S. So, it is not a pan-pol III promiscuous effect. There is some specificity in there. But I grant you that if my helicase explanation is true then I have got some other questions to answer in terms of how the specificity is put on.

WONG-STAAL: In the case of the adenovirus E1A protein where you see activation and other people see repression, are the target sequences the same?

RIGBY: I will make two comments, one of clarification. I think that the situation with regard to the effect of E1A on class I genes is that that one person sees one thing and another person sees another in the same experiment. I think we all pretty much agree. It is just that if you do different kinds of experiments, you see different responses. That is really important to stress. I do not know where the target sequences for E1A are. That is something we are doing now, sorting through the deletions that we made for T in the E1A cotransfection assay to see if they operate on the same or different sequences, but the data are not in yet.

DYNAN: Could you show again the location of the target sequences in the class I MHC. Do you think the mechanism of activation is the same for SV40 late gene activation?

RIGBY: I have no idea what the mechanism is. The target is within a HindIII fragment which is about $4\frac{1}{2}$ kb upstream of the cap site. Such far upstream regulatory regions are interesting but in no way unique; albumin and alpha fetoprotein, for example, have their regulators even further upstream of the cap site. So, that is where it is. I stress it is purely a deletion analysis. I have not put that fragment onto another promoter. In terms of how it works and whether it is the same as late gene activation, I have no data and I have no idea. What we are trying to do now, which would obviously help us address those issues, is to see if we can do this *in vitro* as well.

HANAHAN: Have you or anyone tried to take a cell and alter its growth rate, say by the amount of serum, to see if some of these effects are due to rapid proliferation as opposed to specific transformation?

RIGBY: Yes, we and Mike Botchan have tried to address that issue, which is always a real worry. When you do the experiments in the first place you try as hard as you know how to have all the cells growing at the same rate every time you take the RNA out. We have then done some direct experiments to ask what is the relationship, is it really the fact that you are transformed or is it the fact that you are growing faster, however careful you are in doing the experiment? The answer, so far as we can tell, is that certainly the class I gene is not regulated by proliferation. But equally it is clear that if you do a classical serum stimulation experiment in BALB/3T3s and probe with the B2, then those small RNAs come up to a high level as a result of serum stimulation. So they are proliferation-regulated, and that actually squares with the fact that two groups, those of Dan Nathans and David Denhardt, have done differential cDNA cloning exercises, where the differential that is put on is between quiescent and serum-stimulated cells, and, as you would except from what I just told you, they both cloned these B2s. So, they are clearly proliferation-regulated. As far as I can tell the class I gene and the retrovirus are not proliferation-regulated.

VAN DER EB: Do you know anything about class I expression in other haplotypes than the mouse d haplotype?

RIGBY: I know a little about k from looking at C3H cells, and again everything we looked at had high levels of RNA that hybridized with pAD64. But I have no idea in the k haplotype which gene is responsible.

The Role of Papillomaviruses in Human Cancer

Harald zur Hausen

INTRODUCTION

During the past 10 years the plurality of human papillomavirus types has been well established (see review zur Hausen & Schneider 1986). At present about 40 distinct types of human papillomaviruses (HPV) have been analyzed. It is likely that this number will increase in the future.

Papillomaviruses reveal a very characteristic infectious cycle: they appear to require the availability of cells still capable of dividing for successful infection. The uptake of viral DNA by such cells, usually exposed to the surface in microlesions of epidermis or mucosa or at specific sites (e.g. the transition zone of the cervix) results in its episomal persistence. The persisting DNA stimulates enhanced proliferation. Its own independent replication seems to be blocked by host cell factors. During subsequent steps in differentiation and keratinization this block obviously is released, resulting in viral DNA replication, synthesis of structural proteins and the maturation of viral particles. Thus, the keratinized superficial layer of a wart containing infectious particles indicates an underlying proliferating layer of cells haboring viral DNA which stimulates cell proliferation.

The dependence of particle maturation on specific stages of cell differentiation is probably the main reason for the present inability to propagate papillomaviruses in tissue culture. The analysis of their biological functions is further restricted by a remarkable host specificity with a pronounced preference of individal types for specific types of tissue. Thus, human papillomaviruses do not produce recognizable changes in animal hosts.

The DNA of several papillomavirus types has been sequenced. In contrast to polyoma-type viruses, transcripts are read from one strand only, revealing the

---

Deutsches Krebsforshungszentrum, Im Neuenheimer Feld 280, D-6900 Heidelberg, Fed. Rep. of Germany

VIRAL CARCINOGENESIS, Alfred Benzon Symposium 24,
Editors: N. O. Kjeldgaard, J. Forchhammer, Munksgaard, Copenhagen 1987.

same polarity. Several open reading frames (ORF) exist, 2 of them (L1 and L2) appear to code for structural proteins, a varying number of additional ORFs (E1 to E8) seem to contain information responsible for early functions (see review Pfister 1984). E6 and E5 of bovine papillomaviruses apparently code for transforming functions (Schiller *et al.* 1984, Yang *et al.* 1985), E2 reveals transactivating activity for the E6, E7 region (Spalholz *et al.* 1985) and E1, and possibly also E7 somehow regulate the episomal state of persisting papillomavirus DNA (Lusky & Botchan 1984). Very little is known of the proteins coded for by early ORFs. Their expression in bacterial vector systems is presently being explored.

GENITAL PAPILLOMAVIRUS INFECTIONS

Twelve types of papillomaviruses have been isolated from the human genital tract (zur Hausen 1986). The majority of these types has only rarely been found in genital tumors. The most prevalent types clearly are HPV 6, 11, 16 and to a lesser extent 18.

HPV 6 and 11 are closely related, 82% of their nucleotides are identical (Dartmann *et al.* 1986). These viruses are found in typical genital warts (condylomata acuminata). In these tumors HPV 6 is found in about 60%, HPV 11 in about 30%. Invasively growing, non-metastasizing giant condylomata have been described, frequently labelled as Buschke-Löwenstein tumors. HPV 6 DNA was found in almost all of such tumors so far analyzed, the only exceptional one contained HPV 11 (Boshart & zur Hausen 1986).

The histology of genital warts is characterized by the exophytic papillary growth and the typical appearance of koilocytotic cells. The latter represent the sites of viral DNA replication, protein synthesis and particle maturation, thus being an expression of cytopathogenic changes induced by events leading to the synthesis of infectious virus.

Although condylomatous changes are also noted upon infections of the vaginal wall, most notably HPV 11 infections of cervical sites reveal a different pattern (see review zur Hausen & Schneider 1986). Kolposcopically they are observed as flat dysplastic lesions which histologically show features of mild dysplasias (cervical intraepithelial neoplasia, CIN-1), usually with extensive koilocytosis.

HPV 6 and 11 infections are extremely rare at non-genital epidermal sites. They occur, however, at low frequency at oral sites or within the respiratory tract (de Villiers *et al.* 1986). The most frequent non-genital predilection site is infection of the vocal cords, resulting in laryngeal papillomatosis (Gissmann *et al.* 1982, Mounts *et al.* 1982). This represents a serious clinical condition, sometimes spreading into the bronchial tract. Laryngeal papillomatosis occurs at higher frequency as a perinatal infection, most likely occurring during delivery due to maternal genital warts. HPV 11 is found in this condition more often than HPV 6.

HPV 6- or 11-containing condylomatous proliferations are occasionally found in the buccal mucosa, the lips, or located directly on the tongue. The histology shows less koilocytotic changes, probably reflecting a substantially reduced virus production at these sites.

Recently a technique has been developed which permits the direct study of the causative role of paillomavirus infection in the induction of dysplastic and papillomatous proliferations (Kreider *et al.* 1985). Inoculation of normal cervical tissue beneath the renal capsule of nude mice results in the development of cysts outlined at their inner surface by cervical epithelium. These cysts persist for several months. Infection of the cells with HPV 11 prior to inoculation results in dysplastic proliferations revealing extensive koilocytosis and, within the koilocytes, virus-specific antigens and particles.

In addition, tissue culture studies indicate the transforming potential of HPV 16 infections. Yasumoto *et al.* (1986) were able to malignantly transform NIH 3T3 cells after HPV 16 DNA transfection. Genomic DNA obtained from an HPV 16-positive cervical cancer biopsy also transformed NIH 3T3 cells (Tsunokawa *et al.* 1986). The transformed cells contained HPV 16 DNA which was transcribed. More recently, Dürst *et al.* (1986) showed that transfection with HPV 16 DNA leads to immortalization of human foreskin epithelial cells. The immortalized cells revealed an aneuploid karyotype, contained integrated HPV 16 DNA and transcripts from the early region. The cells were non-tumorigenic in nude mice.

In HPV 16 and 18 infections, the macroscopic and microscopic pattern of the induced lesions differs markedly from that described for HPV 6 and 11. Most available data result from HPV 16-containing proliferations. This virus type is by far the most frequent HPV found in Bowenoid papulosis or genital Bowen's disease, which represent rather discrete and inconspicuous looking white or reddish plaques found at vulvar, penile and perianal sites (Ikenberg *et al.* 1984). Their histology, however, reveals marked atypia and usually all characteristics of a carcinoma *in situ*. Spontaneous regression of these tumors does occur, although the majority of these lesions seem to persist for long periods of time, in many instances for years, and possibly for decades.

HPV 16 infections of the cervix are associated with marked atypia, changes characteristic for CIN II or CIN III and carcinoma *in situ* (Crum *et al.* 1985).

Usually, little or no koilocytosis is noted in these proliferations, although in some biopsies an extensive koilocytosis may originate from simultaneous infections with additional HPV types.

Simultaneous infections with either HPV 6/11 or HPV 16 and 18 appear to be relatively frequent. A recent survey of smears analyzed from more than 10 000 women indicated the presence of both groups of viruses in about 12% of this population (de Villiers *et al.* unpublished data). More than 5% of these women revealed evidence of infections by more than one of these agents.

PAPILLOMAVIRUSES IN ANOGENITAL CANCER

The original isolation of HPV 16 and HPV 18 DNA were both obtained from cervical cancer samples (Dürst *et al.* 1983, Boshart *et al.* 1984). Thus, it was of interest to analyze other tumor samples for the presence of HPV DNA. Up to now a large number (< 200) of cervical cancer biopsies and a limited number of penile and vulvar cancers have been analyzed (Dürst *et al.* 1983, Boshart *et al.* 1984, Scholl *et al.* 1985). HPV 16 DNA is found in close to 50% of biopsies tested from various parts of the world. Individual results differ in the range between 30 and 80%. HPV 18 DNA is less frequently found. In our own studies the percentage of positive biopsies (including penile and vulvar cancer) comes close to 20%. In additional tumors, other types of papillomaviruses have been detected. In a few samples, HPV 11 DNA (Gissmann *et al.* 1983); in others, more recently, HPV 31 (Lorincz & Temple 1985), HPV 33 (Beaudenon *et al.* 1986) and HPV 35 (Lorincz and Temple, personal communication) have been found. Approximately 80% of all of these biopsies contain specific types of HPV DNA. In the majority of the remaining tumors hybridization under conditions of low stringency discloses the presence of HPV-related, yet unidentified sequences, most likely due to the presence of new types of papillomaviruses. Thus, we encounter the regular presence of DNA from specific HPV types in cervical, vulvar and penile cancer.

Perianal and anal cancer have also been found to frequently contain HPV 16 DNA (Beckmann *et al.* 1985).

Metastatic tissue derived from cervical cancer usually contains HPV DNA in quantities similar to the primary tumor tissue.

A number of cell lines have been derived from cervical cancer, including HeLa cells. Southern blot analysis of these cells revealed the presence of HPV DNA in the majority of lines thus far tested (Boshart *et al.* 1984, Schwarz *et al.* 1985, Yee *et al.* 1985, Pater & Pater 1985). Interestingly, the otherwise rare HPV 18 DNA is

TABLE I

HPV DNA in cancer of the oral mucosa and the respiratory tract

HPV-type	Site of cancer	Authors
HPV – 11	Buccal mucosa	Löning *et al.* (1985)
HPV – 16	Buccal mucosa	Löning *et al.* (1985)
HPV – 2	Tongue	de Villiers *et al.* (1985)
HPV – 16	Tongue	de Villiers *et al.* (1985)
HPV – 16	Tongue	de Villiers *et al.* (1985)
HPV – 16	Larynx	Scheurlen *et al.* (1985)
HPV – 30	Larynx	Kahn *et al.* (1985)
HPV – 16	Lung	Stremlau *et al.* (1986)

present in most lines analyzed up to now, including HeLa cells. The copy number of HPV genomes in these cells differs considerably from about 1 genome copy in C4-1 cells (Schwarz *et al.* 1985) to up to 600 or more copies in the Caski line.

The availability of HPV-positive cell lines permitted a convenient testing for the state of the persisting HPV DNA. At least the vast majority of HPV DNA in these lines, but also HPV DNA in fresh biopsy samples from cervical cancer, contains integrated sequences. The integrated viral DNA is frequently amplified, usually involving the flanking host cell DNA. Some primary tumors contain, in addition to the integrated sequences, episomal viral DNA.

It is interesting to note that HPV 16-positive precursor lesions (cervical dysplasias and Bowenoid papulosis) seem to contain only episomal DNA (Dürst *et al.* 1985). It therefore appears that malignant conversion is associated with a shift in the state of persisting viral DNA.

The persistence and chromosomal localization of viral DNA can be visualized by *in situ* hybridizations, most easily in the Caski line with a large number of viral genome copies. Several integration sites are evident from these studies, revealing high copy numbers for individual sites (A. Mincheva, L. Gissmann and H. zur Hausen, submitted).

The integration pattern reveals some specifity (Schwarz *et al.* 1985). In most primary tumors, metastases – and in all positive cell lines tested so far, at least some of the viral DNA molecules – integrate within the E1–E2 open reading frames. This obviously disrupts an intragenomic regulation which has recently been reported by Spalholz and her associates (1985) for bovine papillomavirus genomes.

The integrated viral DNA is transcribed in cell lines exclusively involving the E6–E7 open reading frames (Schwarz *et al.* 1985). In addition, fusion transcripts between E6–E7 and adjacent host cell sequences are formed. Their biological significance is at present unknown. Transcriptions from the same regions also appear to be a regular feature of primary tumors, although in some of them the transcriptional pattern appears to be more complicated.

Recent data suggest a negative control of papillomavirus transcription in proliferating normal cells which appears to be switched-off in the process of cell differentiation (reviewed in zur Hausen 1986). According to a model proposed recently (zur Hausen 1986), cervical cancer results from a failing cellular control of persisting viral DNA. Thus, cancer development would require the persistence of viral DNA and functional modification of a homozygous set of cellular genes. This could readily explain basic features of HPV-linked carcinogenesis in man such as the long interval between primary infection and tumor appearance, the small number of infected individuals in whom cancer develops, and the monoclonality of the tumors.

PAPILLOMAVIRUS DNA IN NON-GENITAL CANCER

Specific types of papillomaviruses, most notably HPV 5, have been detected in squamous cell carcinomas of patients with a rare condition, epidermodysplasia verruciformis (Orth *et al.* 1980).

Our own group became interested in the presence of papillomavirus DNA in carcinomas of the human respiratory tract and the oral mucosa. Since papillomavirus infections seem to interact synergistically with chemical or physical carcinogens in carcinoma development (zur Hausen 1977, 1982), human tumors clearly linked etiologically to chemical factors (smoking) like laryngeal and lung cancers were the primary target for this investigation. The analysis of more than 100 individual cancer biopsies indeed led to the identification of HPV-positive carcinomas which are listed in Table I. It is interesting to note that 5 of these tumors contained HPV-16 sequence, demonstrating that this type of infection also occurs at extragenital sites, leading again to a remarkable association with malignant growth.

Some rather prelimary data point to the existence of additional, yet not fully characterized HPV types in other laryngeal and lung carcinomas. The availability of some HPV-containing carcinomas derived from this region provides an encouraging baseline for the concept that carcinomas of the oral mucosa and the respiratory tract may originate from an interaction of cells persistently infected by specific HPV types and environmental chemical factors. This, in addition, would open a new pathway for strategies in preventing very common types of human cancers.

REFERENCES

Beaudenon, S., Kremsdorf, D., Croissant, O., Jablonska, S., Wain-Hobson, S. & Orth, G. (1986) A novel type of human papillomavirus associated with genital neoplasias. *Nature 321,* 246–249.

Beckmann, A. M., Daling, J. R. & McDougall, J. K. (1985) Human papillomavirus DNA in anogenital carcinomas. *J. Cell Biochem. Suppl. 9c,* 68.

Boshart, M., Gissmann, L. Ikenberg, H. Kleinheinz, A., Scheurlen, W. & zur Hausen, H. (1984) A new type of papillomavirus DNA, its presence in genital cancer biopsies and in cell lines derived from cervical cancer. *EMBO J. 3,* 1151–1157.

Boshart, M. & zur Hausen, H. (1986) Human papillomaviruses (HPV) in Buschke-Löwenstein tumors: physical state of the DNA and identification of a tandem duplication in the non-coding region of a HPV-6 subtype. *J. Virol. 58,* 963–966.

Crum, C. P., Mitao, M. Levine, R. U. & Silverstein, S. (1985) Cervical papillomaviruses segregate within morphologically distinct precancerous lesions. *J. Virol. 54,* 675–681.

Dartmann, K., Schwarz, E., Gissmann, L. & zur Hausen, H. (1986) The nucleotide sequence and genome organization of human papillomavirus type 11. *Virology 151,* 124–130.

de Villiers, E.-M., Neumann, C., Le, J.-Y., Weidauer, H. & zur Hausen, H. (1986) Infection of the oral mucosa with defined types of human papillomaviruses. *Med. Microbiol. Immunol. 174,* 287–294.

de Villiers, E.-M., Weidauer, H., Otto, H. & zur Hausen, H. (1985) Papillomavirus DNA in human tongue carcinomas. *Int. J. Cancer 36,* 575–579.

Dürst, M., Dzarlieva-Pertrusevzka, R. T., Boukamp, P., Fusenig, N. & Gissmann, L. (1986) Transfection of primary human foreskin keratinocytes with human papillomavirus (HPV) type 16 DNA leads to escape from senescence. (Submitted for publication).

Dürst, M., Gissmann, L., Ikenberg, H. & zur Hausen, H. (1983) A papillomavirus DNA from a cervical carcinoma and its prevalence in cancer biopsy samples from different geographic regions. *Proc. Natl. Acad. Sci. U.S.A. 80,* 3812–3815.

Dürst, M., Kleinheinz, A., Hotz, M. & Gissmann, L. (1985) The physical state of human papillomavirus type 16 DNA in benign and malignant tumors. *J. Gen. Virol. 66,* 1515–1522.

Gissmann, L., Diehl, V., Schultz-Coulon, H.-J. & zur Hausen, H. (1982) Molecular cloning and characterization of human papillomavirus DNA derived from a laryngeal papilloma. *J. Virol. 44,* 393–400.

Gissmann, L., Wolnik, H., Ikenberg, H., Koldovsky, U., Schnürch, H. G. & zur Hausen, H. (1983) Human papillomavirus type 6 and 11 sequences in genital and laryngeal papillomas and in some cervical cancers. *Proc. Natl. Acad. Sci. U.S.A. 80,* 560–563.

Ikenberg, H., Gissmann, L., Gross, G., Grussendorf, E.-I. & zur Hausen, H. (1983) Human papillomavirus type 16-related DNA in genital Bowen's disease and in Bowenoid papulosis. *Int. J. Cancer 32,* 563–565.

Kahn, T., Schwarz, E. & zur Hausen, H. (1986) Molecular cloning and characterization of the DNA of a new human papillomavirus (HPV 30) from a laryngeal carcinoma. *Int. J. Cancer 37,* 61–65.

Kreider, J. W., Howett, M. K., Wolfe, S. A., Bartlett, G. L., Zaino, R. J., Sedlacek, T. V. & Mortel, R. (1985) Morphological transformation in vivo of human uterine cervix with papillomavirus from condylomata acuminata. *Nature 317,* 639–641.

Löning, T., Ikenberg, H., Becker, J., Gissmann, L., Hoepfer, I. & zur Hausen, H. (1985) Analysis of oral papillomas, leukoplakias and invasive carcinomas for human papillomavirus related DNA. *J. Invest. Dermatol. 84,* 417–420.

Lorincz, A. T., Lancaster, W. D. & Temple, G. F. (1985) Detection and characterization of a new type of human papillomavirus. *J. Cell Biochem., Suppl. 9c,* 75.

Lusky, M. & Botchan, M. R. (1984) Characterization of the bovine papillomavirus plasmid maintenance sequences. *Cell 36,* 391–401.

Mincheva, A., Gissmann, L. & zur Hausen, H. (1986) Chromosomal integration sites of human papillomavirus DNA in three cervical cancer cell lines mapped by in situ hybridization. (Submitted for publication).

Mounts, P., Shah, K. V. & Kashima, H. (1982) Viral etiology of juvenile and adult onset squamous papilloma of the larynx. *Proc. Natl. Acad. Sci. U.S.A. 79,* 5425–5429.

Orth, G., Favre, M., Breitburd, F., Croissant, O., Jablonska, S., Obalek, M., Jarzabek-Chrozelska, M. & Rzesa, G. (1980) Epidermodysplasia verruciformis: A model for the role of papillomaviruses in human cancer. In: *Viruses in Naturally Occurring Cancers,* eds. Essex, M., Todaro, G. & zur Hausen, H., pp. 259–282. Cold Spring Harbor Lab. Press, Cold Spring Habor, N.Y.

Pater, M. M. & Pater, A. (1985) Human papillomavirus types 16 and 18 sequences in carcinoma cell lines of the cervix. *Virology 145,* 313–318.

Pfister, H. (1984) Biology and biochemistry of papillomaviruses. *Rev. Physiol. Biochem. Pharmacol. 99,* 111–181.

Scheurlen, W., Stremlau, A., Gissmann, L. Höhn, D., Zehner, H.-P. & zur Hausen, H. 81986) Rearranged HPV 16 molecules in an anal carcinoma and in a laryngeal carcinoma. *Int. J. Cancer* (in press).

Schiller, J. T., Vass, W. C. & Lowry, D. R. (1984) Identification of a second transforming region in bovine papillomavirus DNA. *Proc. Natl. Acad. Sci. U.S.A. 82,* 7880–7884.

Scholl, S. M., Pillers, E. M., Robinson, R. E. & Farrell, P. J. (1985) Prevalence of human papillomavirus type 16 DNA in cervical carcinoma samples in East Anglia. *Int. J. Cancer 35,* 215–218.

Schwarz, E., Freese, U. K., Gissmann, L, Mayer,W., Roggenbuck, B., Stremlau, A. & zur Hausen, H. (1985) Structure and transcription of human papillomavirus sequences in cervical carcinoma cells. *Nature 314,* 111–114.

Spalholz, B. A., Yang, Y.-C. & Howley, P. M. (1985) Transactivation of a bovine papillomavirus transcriptional regulatory element by the E2 gene product. *Cell 42,* 183–191.

Stremlau, A., Gissmann, L., Ikenberg, H., Stark, M., Bannasch, P. & zur Hausen, H. (1985) Human papillomavirus type 16-related DNA in an anaplastic carcinoma of the lung. *Cancer 55,* 737–740.

Tsunokawa, Y., Takebe, N., Kasamatsu, T., Terada, M. & Sugimura, T. (1986) Transforming activity of human papillomavirus type 16 DNA sequences in a cervical cancer. *Proc. Natl. Acad. Sci. U.S.A. 83,* 2200–2203.

Yang, Y. C., Okayama, H. & Howley, P. M. (1985) Bovine papillomavirus contains multiple transforming genes. *Proc. Natl. Acad. Sci. U.S.A. 82,* 1030–1034.

Yasumoto, S., Burkhardt, A. L., Doniger, J. & DiPaolo, J. A. (1986) Human papillomavirus type 16 DNA-induced malignant transformation of NIH 3T3 cells. *J. Virol. 57,* 572–577.

Yee, C., Krishnan-Howlett, I., Baker, C. C., Schlegel, R. & Howley, P. M. (1985) Presence and expression of human papillomavirus sequences in human cervical carcinoma cell lines. *Am. J. Pathol. 119,* 361–366.

zur Hausen, H. (1977) Human papillomaviruses and their possible role in squamous cell carcinomas. *Curr. Top. Microbiol. Immunol. 78,* 1–30.

zur Hausen, H. (1982) Human genital cancer: synergism between a two virus infections or between a virus infection and initiating events. *Lancet 2,* 1370–1372.

zur Hausen, H. (1986) Evidence for an association between human papillomaviruses and neoplasia. *Int. Med. 7,* 66–79.

zur Hausen, H. (1986) Intracellular surveillance of persisting viral infections. Human genital cancer results from deficient cellular control of papillomavirus gene expression. *Lancet 2,* 489–491.

zur Hausen, H. & Schneider, A. (1986) The role of papillomaviruses in human anogenital cancer. In: *The papillomaviruses,* eds. Howley, P. M. & Salzmann, N. P. (in press).

DISCUSSION

NORRILD: When you say that there might be interaction between the virus and cellular genes, are there then in the HeLa cells any indications of e.g. possible rearrangement?

ZUR HAUSEN: Yes, there are some rearrangements of the papillomavirus DNA, not particularly in the region that is supposed to be the transforming region, E6, E7, of the papillomavirus DNA but there are deletions and there are also other rearrangements in other regions of the papillomavirus DNA. In addition, in other lines like the 756 line there are no indications for a rearrangement of the papillomavirus DNA, except for the fact that again the integration site affects the E1–E2 open reading frames of the HPV genome. There are clearly a number of cellular rearrangements in HeLa cells in view of the long maintenance of these cells, but it is interesting to point out that we tried to obtain the early passages of HeLa cells and could compare the pattern of HPV integration, and secondly the pattern of HPV transcription of these early isolates which were continuously kept in tissue culture. The earliest we could obtain was from the American Type Culture Collection, which was supposed to be passage 119, and I think they were frozen down around the end of the 1950s. There was virtually no difference either in the structure of the viral DNA or in the transcription pattern between these early passage and the late passages. A number of HeLa contaminants, like KB cells, HEp2 cells, contain the same type of integrational pattern and transcriptional pattern, and all seem to show some stability of at least the modified genomes as they are present in HeLa cells right now.

CUZIN: What is known about expression of cellular oncogenes in cervix cancers especially at late stages? And what is the present situation regarding a possible role for herpes viruses in cervix cancer?

ZUR HAUSEN: As far as the expression of cellular oncogenes is concerned most of the work has been done by Riou in Paris. He reported that a number of cervical cell lines, but also primary tumors contain amplified *myc* sequences and noted an increased expression in some of these lines. We have done a few studies along these lines ourselves, and certainly confirmed his observations. There are clearly some lines which do not show an amplification of *myc*. As far as other oncogenes are concerned, there is no evidence available yet either for amplification or enhanced expression. Concerning the role of herpes virus it is much more difficult to deal with, because herpes simplex viruses, as many of you know, have been implicated mainly on the basis of sero-epidemiological studies to play a role in human cervical cancer. We are very much interested in some properties of herpes viruses showing that the infection shares properties with chemical and physical carcinogens in that they lead to mutational events within the host cell DNA, and they also are very active in amplifying cellular sequences under those conditions, including selective DNA amplification. These properties are clearly shared with chemical and physical carcinogens. It appears that other herpes viruses, for instance murine cytomegalovirus, pseudo-rabies virus share this type of property. It is possible that there exists some kind of interaction of herpes virus infections with papillomavirus infections, probably cytomegalovirus may be a very good candidate, in view of its less lytic type of infections, by modifying host cells sequences according the model which I outlined before, but the direct experimental evidence is still missing.

WEINBERG: The experiments of Stanbridge with somatic cell hybrids seem to have great conceptual importance, because this work has been often interpreted to indicate that there are specific cellular genes in the normal cell which are responsible for repressing malignancy in general, thereby limiting the growth of tumor cells. This has been extrapolated to the case of retinoblastoma, but if I understood what you were telling us, this hypothesized cellular anti-tumor gene is not one which is necessarily directly involved in regulating malignancy, but rather appears to be only a cellular gene that specifically limits the growth or expression of papilloma virus. Therefore this whole concept that Stanbridge has developed may all be a consequence of this very specific interaction of the viral genome with a cellular gene rather than a generalizable phenomenon which could be applied to all kinds of tumors interacting with anti-tumor genes.

ZUR HAUSEN: That is what we suspect, indeed, that it is a rather specific phenomenon. We suspect that there exists a kind of intracellular surveillance mechanism which normally controls these viruses which probably co-evolved during evolution with man and they are with us already with early stages of our evolution. Actually, it is probably an immunological must to assume that some kind of control exists of persisting viral DNA, if those viruses are indeed transforming viruses and therefore potentially lethal agents. It is rather logical that during evolution a mechanism should have developed which is able to control these types of infection. It seems to be an ancestral defense mechanism, which preceeds the immunological control of those types of infections, and only a failing host cell control of these types of infections which, according to this model, would lead to cancer.

WEINBERG: But there is a control experiment which you might consider doing which, it seems to me, could make the case solid. Put the cryptic papilloma virus you cloned out of HeLa cells back into the HeLa cells under a heterologous promoter, one that is no longer repressible by the normal chromosome, and then show that the result in hybrid cells formed with these tranfected HeLa cells is still tumorigenic.

ZUR HAUSEN: That is exactly what we are doing right now.

LOWY: One of these striking features of these tumors, as you point out, is the loss of functional E2 protein, and yet there seems to be abundant transcription in HeLa cells. What about the possibility that HeLa cells contain some kind of E2-like factor that is able to stimulate transcription and normal cells somehow suppress that E2-like factor.

ZUR HAUSEN: It is quite possible that there is a more indirect mechanism than pointed out here on my model, but basically I think that it does not change the features of the model. You are quite right, I think that is a real possibility.

BLAIR: I just want to comment on the hybrids. In collaboration with Harold Klinger, Bryan O'Hara in my lab has looked at a similar set of tumorigenic segregants for expression of 23 oncogenes, and we found that there was a correlation between the tumorigenicity and the overexpression of both *fos* and *ets*2, and a reduction of *myb* expression. I wondered if you had looked for expression of any other oncogenes or any other sequences?

ZUR HAUSEN: No, we did not. We looked for the expression of *myc* and it was clearly there, but otherwise we did not look.

HANAHAN: Have you looked at other cell lines containing independent insertions of HPV to see whether you can reproduce this with a different integration site?

ZUR HAUSEN: We are presently trying it, but we do not have any data. These are really laborious and time-consuming experiments. It usually takes about 3 months before one of these experiments is finished. Unfortunately, no, it has not been done.

KJELDGAARD: You suggest a constant integration pattern and still you have integration into several chromosomes. Does that mean that you have a constant integration pattern in all cell lines isolated?

ZUR HAUSEN: Probably I did not clarify it sufficiently, but we have a kind of specificity of the integrational pattern on the site of the viral genome. The viral genome is opened, in the E1 or E2 operating frames, but there exists no specificity as far as the host cell genome is concerned. The integrational sites appear to be random, at least according to the *in situ* hybridization data, but also to the analysis of the flanking sequences.

WONG-STAAL: Are either the tumors or the immortalized cells clonal or polyclonal?

ZUR HAUSEN: Both are clonal. The majority of tumors in which you can identify the clonality by the integration pattern are clearly clonal. In some of the tumors it is difficult to judge it, due to the fact that too much episomal DNA is present, but all cell lines are clearly clonal.

CORY: Is there a precancerous stage, or is that not clear?

ZUR HAUSEN: I did not mention it here. I am grateful that you ask the question. The viral DNA in the precancer lesions appears to exist mainly in the episomal form, there are very few precancer lesions up to now identified which contain in addition some integrated DNA, but the majority of them clearly contains episomal DNA quite in contrast to the malignant tumors.

Cis-Acting Elements and *Trans*-Acting Factors Involved in the Expression and Regulation of the *a* Genes of Herpes Simplex Virus Type 1

Bernard Roizman & Thomas M. Kristie

Herpes simplex virus 1 (HSV-1) genes form three major groups, α, β, and γ, whose expression is coordinately regulated and sequentially ordered in a cascade fashion (Honess & Roizman 1974, 1975, Roizman *et al.* 1974, Roizman & Batterson 1985). The 5 α genes (α 4, 0, 22, 27, and 47) are the first viral genes expressed after infection and do not require any *de novo* protein synthesis for their transcription (Honess & Roizman 1984, 1985, Mackem & Roizman 1981). Several of the gene products comprising this group have been shown to be involved in the regulation of viral gene expression. Cells infected with some temperature sensitive (*ts*) mutants in the α4 gene produce only α proteins at the nonpermissive temperature, and the transition from α to β to γ protein synthesis does not ensue (Dixon & Schaffer 1980, Watson & Clements 1980, Morse *et al.* 1978, DeLuca *et al.* 1984). An interesting property of the α4 protein is that it also appears to regulate the synthesis of α proteins. Shift up of some *ts* mutants in the α4 protein to the nonpermissive temperature during the later β and γ gene expression stage results in the reinitiation of the synthesis of α proteins (Watson & Clements 1980, Preston 1979).

ts mutations in the α27 gene and deletion mutants in the α22 gene also affect the synthesis of proteins produced later in infection (Sacks *et al.* 1985, Sears *et al.* 1985). The α0 protein has been reported to enhance the expression of viral genes in transient expression systems, but its function in infected cells is not known (Gelman & Silverstein 1985, O'Hare & Hayward 1985a, 1985b, Mavrom-

---

Marjorie B. Kovler Viral Oncology Laboratories, University of Chicago, 910 East 58th Street, Chicago, Illinois 60637, U.S.A.

VIRAL CARCINOGENESIS, Alfred Benzon Symposium 24,
Editors: N. O. Kjeldgaard, J. Forchhammer, Munksgaard, Copenhagen 1987.

ara-Nazos *et al.* 1986a). Little is known of the function of the α47 gene product since deletion of this gene has no appreciable effect on viral gene expression (Mavromara-Nazos *et al.* 1986b).

The β genes, segregated into β_1 and β_2 based on the temporal pattern of their expression, require α proteins for their transcription (Honess & Roizman 1974, 1975, Dixon & Schaffer 1980, Watson & Clements 1980, Pereira *et al.* 1977). β proteins are primarily involved in viral DNA metabolism. The γ genes have also been separated into two groups, γ_1 and γ_2, based on the stringency of the requirement for DNA synthesis for their expression (Jones & Roizman 1979, Holland *et al.* 1980). These proteins appear to be predominantly structural components of the HSV-1 virion.

The initial focus of the studies described here centered on the mechanism by which the transcriptional machinery of the infected cell differentiated among α, β, and γ genes and specifically transcribed only the α genes immediately after infection. These studies led to the identification of a series of the *cis*-acting elements and *trans*-acting factors which are involved in the expression and viral-directed regulation of the α genes.

α gene expression is induced by a structural component of the virion

Early studies, designed to identify the *cis*-acting elements involved in α gene expression and regulation, utilized chimeric gene constructs consisting of the 5' nontranscribed and leader sequences of α genes fused to the leader and structural sequences of the TK gene, normally a β gene product (Post *et al.* 1981, Mackem & Roizman 1982a, 1982b, 1982c). These constructs, designated as α-TK chimeras, were recombined into the viral genome and were also used to convert TK^- cells to the TK^+ phenotype. In the environment of the viral genome, the α-TK chimeric genes were expressed as α genes inasmuch as they were transcribed in the absence of *de novo* protein synthesis. The significant findings emerged from analyses of α-TK gene expression in cells converted to the TK^+ phenotype. Specifically, (i) the α-TK genes were inducible by superinfection with TK^- virus, (ii) *ts* mutants in the α4 protein efficiently induced the chimeric gene at the nonpermissive temperature, and (iii), the transcription of the α-TK gene was induced by infection of cells in the presence of an inhibitory concentration of cycloheximide. The hypothesis that the α-TK gene was induced by a structural protein of the virus was reinforced by the observation that the expression of the resident α-TK gene was induced with high efficiency by HSV-1*ts*B7 at the nonpermissive temperature (Batterson & Roizman 1983). Previous studies have shown that this mutant is able to infect cells and the capsid is translocated normally to the nuclear pore. However, at the nonpermissive temperature, the capsid retains its integrity and does not release the viral DNA (Batterson *et al.* 1983).

As viral genes are not transcribed at the nonpermissive temperature, the *α* gene *trans*-activating factor (*α*-TIF) had to be introduced into the cell during infection as either a component of the tegument structure surrounding the capsid or in the viral lipid envelope. Early evidence that *α*-TIF was a structural component of the virion, and not a product of the host genome induced by infection, was based on the observations that: i) induction of *α*-TK chimeras correlated with infectivity; and ii) infection with other herpesviruses (*e.g.* cytomegalovirus, bovine mammilitis virus, pseudorabies virus) or with adenovirus did not induce the expression of the *α*-TK gene resident in the transformed cell lines (Batterson & Roizman 1983).

The gene specifying *α*-TIF was mapped by cotransfection of cloned HSV-1 DNA fragments with an *α*-TK chimeric gene (Campbell *et al.* 1984). The map location of the gene corresponds to that of the tegument protein designated as infected cell protein No. 25 (ICP25); a protein that is present in 500 to 1000 copies per virion (Heine *et al.* 1974). The nucleotide sequence of the gene predicts a protein of 59000 translated molecular weight, but the predicted amino acid sequence does not have features that might suggest how *α* gene induction is effected (Pellett *et al.* 1985).

A specific sequence confers inducibility upon α genes by α-TIF

The studies described in the preceeding section centered on the use of the *α*-TK chimeric gene as the indicator of *α* gene expression. The studies described below centered on identification of the *α* gene *cis*-acting elements that determined *α* gene regulation. Justification for the continued use of the *α*-TK chimeric gene for these studies rested on the observation that the ability of the *α* gene 5' nontranscribed and leader sequences to confer *α* regulation was not affected by the nature of the gene to which they were juxtaposed. The *α* promoter-regulatory domains could confer *α* regulation upon the chick oviduct ovalbumin gene, the hepatitis B S gene, and a variety of viral genes (Herz & Roizman 1983, Shih *et al.* 1984). This key observation then led to systematic analyses of the *α* gene promoter-regulatory domains.

The initial studies on the sequences 5' to the *α*4 gene differentiated between the promoter domain (−110 to +33), that enabled efficient expression of the recipient gene, and the regulatory domain (−110 to −4500), that conferred the

```
DESIGNATION                              NUCLEOTIDE SEQUENCE

α Gene Donor Fragments

                                 A+T Homolog
SEQUENCE 3          5'  CCGTGC ATGCTAATGATATTCTT TGGGGG  3'
                        GGCACG TACGATTACTATAAGAA ACCCCC

                                                  A+T Homolog
SEQUENCE 2          5'  CGGAAGCGGAACGGTGTATGTGAT ATGCTAATTAAATACAT GCCACGT  3'
                        GCCTTCGCCTTGCCACATACACTA TACGATTAATTTATGTA CGGTGCA

                                    <<<<<<<< ********                        >>>>>>>>
SEQUENCE 1          5'  CGGATGGCGGG GCCGGGG TTCGACCAACGGGCCGCGGCCACGGG CCCCCGGC GTGCCG  3'
                        GCCTACCCGCCC CGGCCCCC AAGCTGGTTGCCCGGCGCCGGTGCCC GGGGGCCG CACGGC
                            ------  <<<<<<<<<<<< ←←←←←←←    →→→→→→→ >>>>>>>>>>>>

Recipient Promoter Domains

                   -80********           1st Distal                          Proximal                     +1
    β        5'  GAATTCGAACACGCAGATG CAGTCGGGGCGGCG CGGTCCGAGGTCCAC TTCGCATATTAAGGTG ACGCGTGTGGCCTCGAACACC  3'
Promoter         CTTAAGCTTGTGCGTCTAC GTCAGCCCCGCCGC GCCAGGCTCCAGGTG AAGCGTATAATTCCAC TGCGCACACCGGAGCTTGTGG
                                            ------

                 -108        ------              ------ <<<<<<<<<<<                         >>>>>>>>>>>
    α4       5'  GAATTCCGTTC GGGGCGGGCCCGCCTGGGGGGCGG GGGGCCGGCGG CCTCCGCTGCTCCTCCTTC CCGCCGGCCCC
Promoter         CTTAAGGCAAG CCCCGCCCGGGCGGACCCCCCGCC CCCCGGCCGCC GGAGGCGACGAGGAGGAAG GGCGGCCGGGG

                          Tata                   +1
                 TGGGA CTATATGA GCCCGAGGACGCCCCGATCG  3'
                 ACCCT GATATACT CGGGCTCCTGCGGGGCTAGC
```

Fig. 1. Nucleotide sequences of cloned *α* donor fragments and the recipient *α* and *β* gene promoter domains (Mackem & Roizman 1982a, McKnight 1980). In Sequences 2 and 3, "A+T homolog" refers to the 17-nucleotide sequence reiterated within the *α* gene regulatory domains (Mackem & Roizman 1982a). In Sequence 1, < < </> > > identifies the G+C rich inverted repeats present in numerous copies within the regulatory domains of *α* genes, while ←←←/→→→ represent sequences reiterated in a direct manner. The remaining CCAAT box sequences (*), first distal signal, and proximal signal are identified within the *β* gene promoter (McKnight & Kingsbury 1982). Nucleotide sequences homologous to regions within the *β* TK CCAAT box region (*) and the first distal signal (−) are indicated in Sequence 1. The inverted repeats < < </> > > as well as elements analogous to the *β* distal signal GC box are identified in the *α*4 promoter +1 is the first nucleotide to be transcribed.

ability to be induced as an *α* gene by the infecting virus when fused to a promoter-indicator gene (Mackem & Roizman 1982c). Progressive deletions from the 5' end of the *α* gene sequences indicated that the elements responsive to the *α*-TIF were reiterated within the *α* gene regulatory domains. Sequence analysis of the promoter-regulatory domains of *α* genes 4, 0 and 27 revealed that these genes contained numerous G+C rich repeats and from one (*α*27) to several (*α*0) homologs of an A+T rich element (Mackem & Roizman 1982a). Confirmation of the hypothesis that these elements play a role in the expression and regulation of *α* genes emerged from studies utilizing small cloned DNA fragments representative of these reiterated elements (Kristie & Roizman 1984). Specifically, sequences derived from *α* genes 4, 0 and 27 (Figure 1), were fused to a chimeric gene consisting of the *α*4 gene promoter linked to the leader and coding sequences of the TK gene. To control for possible complementing elements contained within the *α*4 promoter domain, these sequences were also fused to the *β* promoter-TK gene which consisted of the natural *β*-TK sequences downstream from nucleotide −80, relative to the transcription initiation site of the TK gene. The chimeric TK genes were tested for basal level expression and viral-directed regulation by infection of cells transformed to the TK^+ phenotype; cells co-transfected with pSV2neo and selected for resistance to G418 antibiotic; and cells transiently expressing the TK chimera.

These experiments demonstrated that the *α*4 regulatory domain (−330 to −110) contained all of the elements necessary to confer high basal level expression and induction by *α*-TIF upon both the *α* and *β* promoter-TK genes. Sequence 1 (Figure 1), a 59bp element derived from the *α*4 regulatory domain, conferred high basal level expression in an orientation semi-dependent manner. In addition, this G+C rich element complemented the *β* class regulation of the truncated *β*-TK promoter, but did not confer response to the *α*-TIF. As will be discussed later, Sequence 1 contains an SP1 factor binding site (Jones & Tjian 1985) and a binding site for the major HSV-1 regulatory protein, *α*4. Sequence 2, a 47bp element from *α*27 gene and Sequence 3, a 29bp element from *α*0 gene each contain a homolog of the A+T element. Both sequences conferred *α* gene regulation upon the *α*4-TK chimeric gene in an orientation-independent manner. However, only sequence 2 contained all the components necessary to convert the *β*-TK gene to an *α*-regulated gene. Consistent with these findings, transient expression systems also indicated a requirement for the presence of the A+T rich homolog for induction of *α*-TK expression by DNA fragments encoding the *α*-TIF gene (McKnight *et al.* 1986).

α4 protein binds to α promoter-regulatory domains

In an attempt to determine whether *α*-TIF binds to promoter-regulatory domains of *α* genes, we employed DNA-protein retardation assays to map and identify the viral proteins specifically binding to specific viral DNA sequences (Kristie & Roizman 1986a, 1986b). These studies detected the binding of numerous host cell proteins but failed to produce evidence that viral gene products bind to DNA fragments containing the A+T homolog which is required for induction of chimeric genes by *α*-TIF. Of particular interest, however, was the observation that promoter-regulatory domains of HSV-1 *a* and γ_2 genes contain binding sites for the HSV-1 *a*4 protein.

In denaturing gels, this protein forms 3 bands ranging from 160000 to 172000 apparent molecular weight (Morse *et al.* 1978). In its native form, the protein is phosphorylated and exists as a homodimer (Wilcox *et al.* 1980, Metzler & Wilcox 1985). It has been reported to bind DNA (Wilcox *et al.* 1980, Freeman & Powell 1982). The evidence that this protein binds to specific viral DNA sequences may be summarized as follows: Figure 2 illustrates the use of the gel retardation assay (Fried & Crothers 1981, Garner & Revzin 1981) to detect infected cell proteins bound to the *a*0-promoter DNA (−110 to +72) in the presence of excess competitor nucleic acids. The identification of the HSV-1-specific protein in the major infected cell protein-*a*0 P DNA complex was approached using protein extracts of cells infected with viral *ts* mutants defective in the transition from *a* to β or β to γ_2 protein synthesis. Since the complex was detected in all infected cell extracts, but was not found in extracts of cells infected and maintained in the presence of inhibitory concentrations of actinomycin D, the proteins in the complex were likely to belong to the *a* regulatory group.

The identification of the *a*4 protein as a component of the infected cell protein-*a*0 P DNA complex utilized monoclonal antibodies against several of the *a* gene products. Figure 3 illustrates that only the *a*4 monoclonal antibody retarded the electrophoretic mobility of the *a*0 P DNA-protein complex when added either before or after the binding reaction. Similar studies demonstrated that the *a*4 protein was present in complexes containing its own promoter and regulatory domains as well as in those formed by the γ_2 promoter-regulatory region and the promoter-regulatory domains of the *a*27 gene. Although the *a*0 and *a*27 proteins have been shown to play a role in the regulation of β and γ gene expression, these proteins were not detected in any of the complexes with the monoclonal antibodies tested.

As the focus of these studies centered on the factors affecting the expression of *a* genes, the binding site of the *a*4-protein complex was mapped within the *a*0-promoter domain DNA and the *a*4 regulatory domain DNA. As shown in Figure 4, exonuclease III was used to delineate the boundaries of the binding site of the *a*4 protein complex within the *a*0-promoter DNA. To control for the specific interaction of the *a*4 protein in the crude nuclear extracts used, the reactions were done in the presence or absence of monoclonal antibody H950. This antibody, directed against the *a*4 protein, significantly inhibited the formation of the *a*4 protein-*a*0 DNA complex. The DNA fragments specifically protected by the

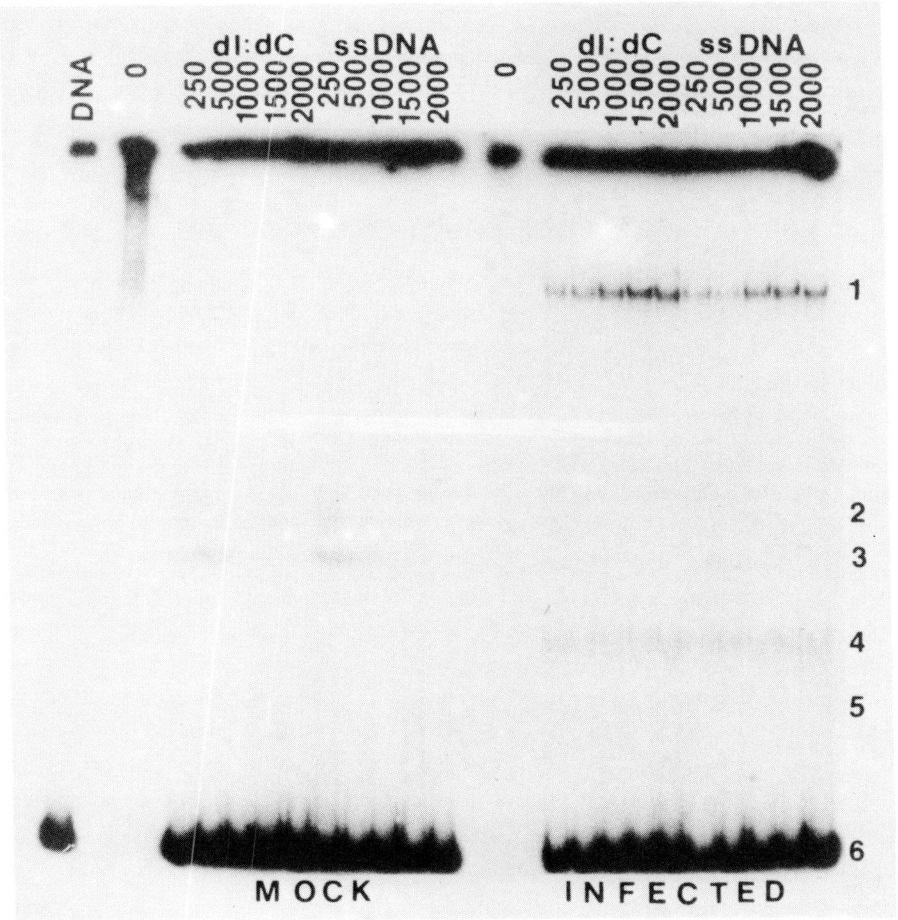

Fig. 2. Autoradiographic images of labeled α0 P DNA complexed with protein in cell lysates in the presence of increasing amounts of competitor nucleic acids. All protein-DNA binding assays were done as described (Kristie & Roizman 1986a, 1986b). For this experiment, the increasing quantity of competitors, poly(dI)·poly(dC), dI:dC, or salmon sperm DNA, ss DNA, were added prior to the addition of extracts of cells harvested 12 hours after mock-infection (MOCK) or HSV-1 (F) infection (INFECTED). The quantity of competitor, in ng, is shown at the top of each lane. The lane marked DNA contained the α0 Promoter probe DNA only. Lanes marked 0 contain the probe DNA and cell lysate in the absence of competitor nucleic acids. The DNA-protein complexes are numbered 1–5 while the unbound DNA migrates at position 6.

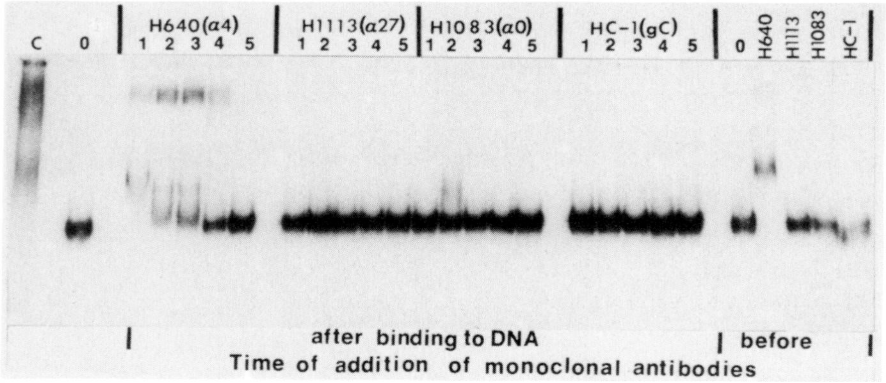

Fig. 3. Autoradiographic images of α0 P DNA-infected cell protein complex in the presence of murine monoclonal antibodies against α4 (H640), α27 (H1113), α0 (H1083) and γ_2gC (HC-1). The monoclonal antibodies were either added after protein-DNA binding or were preincubated with the protein extract for 0.5 hour prior to the incubation of the extract with labeled DNA. Lane C, DNA-protein complexes formed in the absence of poly(dI)·poly(dC); lane 0, no monoclonal antibody added. Lanes 1–5 differ with respect to the amount of monoclonal antibody added to the reaction as follows: 1, 500 ng; 2, 250 ng, 3, 100 ng; 4, 10 ng; 5, 1 ng. Preincubations were done with 500 ng antibody/1000 ng protein extract.

α4 protein from digestion with exonuclease III located the α4 protein binding site to −46 to −71, relative to the α0 transcription initiation site. The protected sequence element contains a portion of the dyad constituting the α0 promoter CCAAT box.

The binding site for the α4 protein within the α4 regulatory domain was localized using small cloned subfragments of this region. These studies mapped the binding site within the 59bp fragment (−135 to −194) designated Sequence 1 in Figure 1. Small deletions made in this 59bp fragment indicated that critical elements required for the α4 protein binding were contained within the promoter proximal region of Sequence 1. The SP1 binding site also contained in Sequence 1 was not required for the binding of the α4 protein.

DISCUSSION

Figure 5 illustrates the locations of the *cis*-acting elements mapped *in vitro* using the α4 gene as a model of the regulation of HSV-1 α genes. The promoter-regulatory domains of these genes consist of a complex mixture of elements which respond to host transcription components and viral *trans*-acting factors which

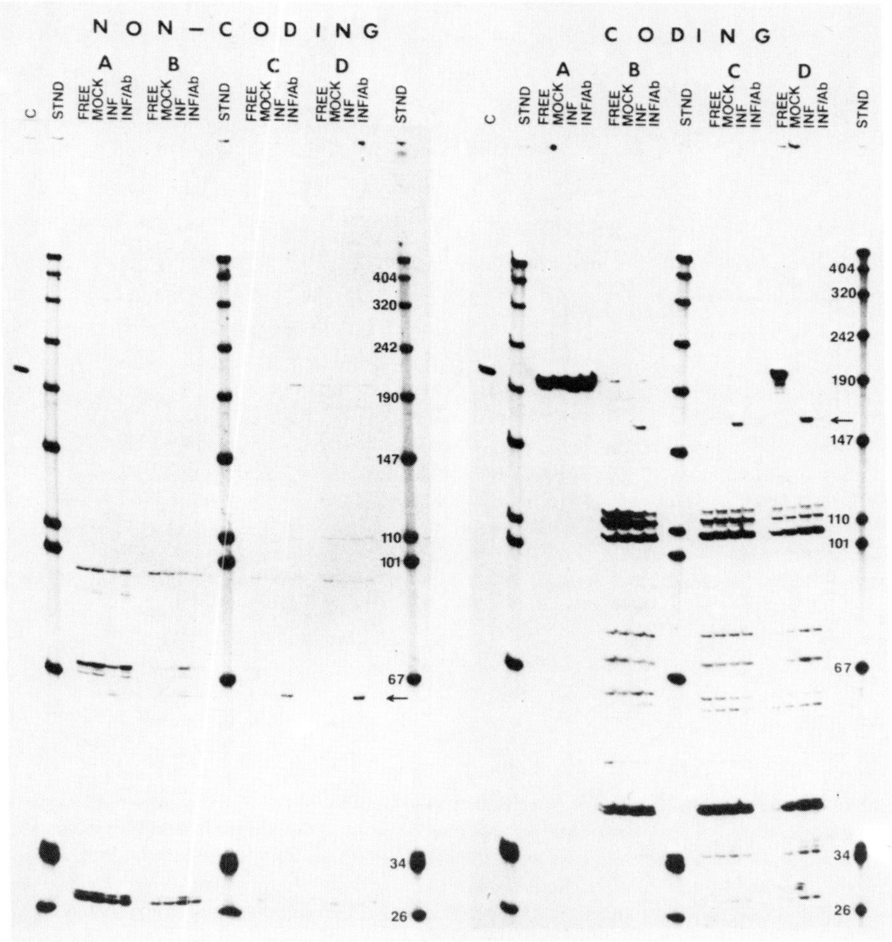

Fig. 4. Autoradiogram of electrophoretically separated exonuclease III digests of 182 DNA in the absence or presence of available α4 protein complex. C, non-digested 182 DNA; FREE, exonuclease III digest after incubation without protein extract; MOCK, exonuclease III digest after incubation with mock-infected cell extract; INF, exonuclease III digest after incubation with infected cell extract; INF/Ab, exonuclease III digest after incubation with infected cell extract that had been preincubated for 0.5 hour with monoclonal antibody H950. The left panel represents digests of 182 DNA labeled at the 5' end of the noncoding strand. The exonuclease III digestions were: lane A, 100 units of enzyme for 15 minutes; lane B, 100 units of enzyme for 30 minutes; lane C, 100 units of enzyme for 45 minutes; lane D, 200 units of enzyme for 30 minutes. The right panel shows digests of 182 DNA labeled at the 5' end of the coding strand. The exonuclease III digestions were for 15 minutes at the following enzyme concentrations: lane A, 1 unit; lane B, 25 units; lane C, 50 units; lane D, 100 units. pUC9 digested with HpaII restriction enzyme and end-labeled with γ^{32}P ATP (STND) served as size markers. The arrows indicate the position of bands representing DNA protected by the α4 protein complex.

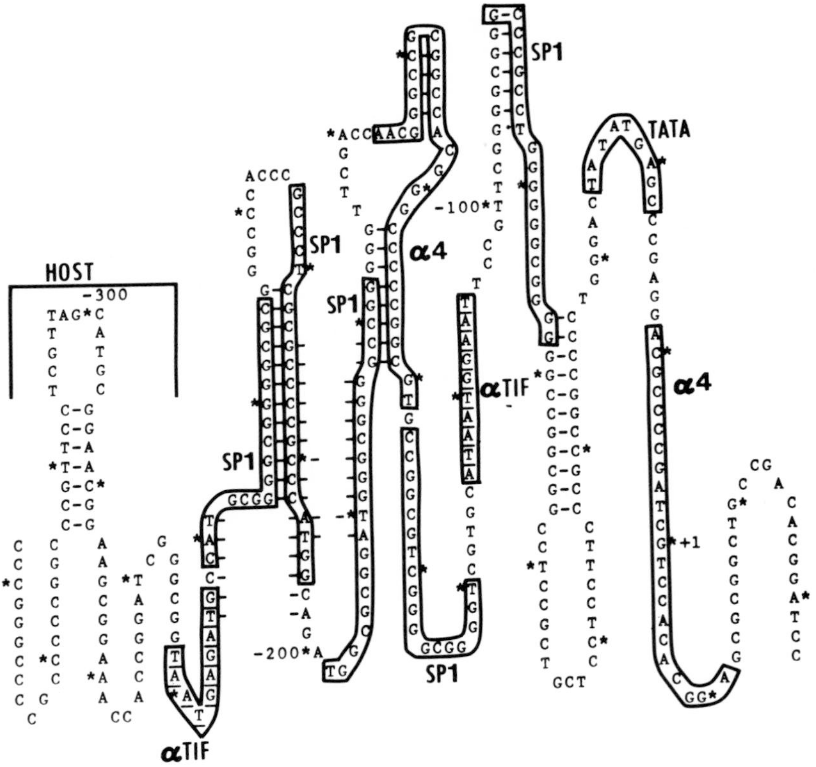

Fig. 5. Summary of the *cis*-acting elements and *trans*-acting factors involved in *α* gene expression and regulation. Shown is a representation of the nucleotide sequence of the *α*4 promoter-regulatory domains. *α*TIF, sequences which enable response to the *α*-specific virion regulatory factor; SP1, location of the *in vitro* binding sites of the SP1 factor (Jones & Tjian 1985); *α*4, location of the mapped *α*4 protein binding sites in the *α*4 gene regulatory domain (Kristie & Roizman 1986b) and the promoter domain (Kristie & Roizman 1986a, Mark Mueller personal communication); HOST, sequences which bind host protein factors. All numbers refer to the site of transcription initiation.

positively and negatively influence the expression of these genes in the infected cell. Of the *cis*-acting elements, the evidence strongly supports the hypothesis that the reiterated A+T rich homolog present in the *α* gene domains is a component necessary for the positive response to the viral *α*-TIF. This element shares homology with a portion of the SV40 72bp enhancer as well as with other mammalian cell regulatory elements (Roizman *et al.* 1984). Although necessary, it is unclear that the A+T element is sufficient to confer the specific *α* regulation upon a non-*α* promoter domain in an efficient manner.

Although many of the *cis*-acting elements which define the positive regulation of the *a* genes are known, the mechanism by which these elements respond to the viral *α*-TIF is unclear. The *α*-TIF protein has not been shown to interact with the A+T elements or to bind any viral DNA directly. The evidence that fragments containing the A+T rich homologs, required for induction by *α*-TIF, bind host proteins suggests the possibility that *α*-TIF may function by acting on a host transcriptional factor.

The second group of defined *cis* elements are the numerous SP1 binding sites. As the SP1 binding sites are a feature shared by both *α* and *β* gene promoter-regulatory domains (Jones & Tjian 1985), they are not sufficient to account for the discrimination between these two gene classes. However, they probably contribute to the ultimate level of transcription of an activated promoter.

The *α*4 binding site constitutes the third group of *cis*-acting elements. In the case of the 59bp sequence from the regulatory domain of the *α* genes (Sequence 1 in Figure 1), the *α*4 protein binding site probably accounts for the strong *β* regulation conferred by this element upon the −80 *β* promoter-TK gene. The function of the *α*4 binding site in its native location in the *α*4 regulatory domain is less certain. Although *α* gene expression appears to be reduced by DNA fragments containing *α*4 genes in transient expression systems, the function of the *α*4 binding site in this particular location is not clear.

The location of the *α*4 protein binding sites and the evidence that the *α*4 protein regulates the expression of *α* genes offers a better insight into possible mechanisms by which the *α*4 protein may regulate gene expression. The protein interacts with the promoter and regulatory domains of *α*, *β*, and *γ* genes. The location of the binding site in the *α*0 promoter domain (−46 to −71) and the *α*4 regulatory domain (Sequence 1) exhibits the interesting feature of proximity to binding sites for host transcriptional components such as the SP1 and the CCAAT factors. As the protein has both positive (*β* and *γ* genes) and negative (*α* genes) regulatory functions, the context of the binding site may play the most critical role in determining the net effect of the *α*4 protein binding upon gene expression.

REFERENCES

Batterson, W., Furlong, D. & Roizman, B. (1983) Molecular genetics of herpes simplex virus. VII. Further characterization of a *ts* mutant defective in release of viral DNA and in other stages of viral reproductive cycle. *J. Virol. 45,* 397–407.

Batterson, W. & Roizman, B. (1983) Characterization of the herpes simplex virion-associated factor responsible for the induction of *α* genes. *J. Virol. 46,* 371–7.

Campbell, M. E. M., Palfreyman, J. W. & Preston, C. M. (1984) Identification of herpes simplex virus

DNA sequences which encode a trans-acting polypeptide responsible for stimulation of immediate early transcription. *J. Mol. Biol. 180,* 1–19.

DeLuca, N. A., Courtney, M. A. & Schaffer, P. A. (1984) Temperature sensitive mutants in herpes simplex virus type 1 ICP4 permissive for early gene expression. *J. Virol. 52,* 767–76.

Dixon, R. A. F. & Schaffer, P. A. (1980) Fine-structure mapping and functional analysis of temperature-sensitive mutants in the gene encoding the herpes simplex virus type 1 immediate early protein VP175. *J. Virol. 36,* 189–203.

Freeman, M. J. & Powell, K. L. (1982) DNA-binding properties of a herpes simplex virus immediate early protein. *J. Virol. 44,* 1084–7.

Fried, M. & Crothers, D. (1981) Equilibria and kinetics of lac repressor-operator interactions by polyacrylamide gel electrophoresis. *Nucl. Acids Res. 9,* 6505–25.

Garner, M. M. & Revzin, A. (1981) A gel electrophoresis method for quantifying the binding of proteins to specific DNA regions. *Nucleic Acids Res 9,* 3047–60.

Gelman, I. H. & Silverstein, S. (1985) Identification of immediate early genes from herpes simplex virus that transactivate the virus thymidine kinase gene. *Proc. Natl. Acad. Sci. USA 82,* 5265–9.

Heine, J. W., Honess, R. W., Cassai, E. & Roizman, B. (1974) Proteins specified by herpes simplex virus. XII. The virion polypeptides of type 1 strains. *J. Virol. 14,* 640–51.

Herz, C. & Roizman, B. (1983) The *a* promoter regulator-ovalbumin chimeric gene resident in human cells is regulated like the authentic *a*4 gene after infection with herpes simplex virus 1 mutants in *a*4 gene. *Cell 33,* 145–51.

Holland, L. E., Anderson, K. P., Shipman Jr., C. & Wagner, E. K. (1980) Viral DNA synthesis is required for the efficient expression of specific herpes simplex virus type 1 mRNA species. *Virology 101,* 10–24.

Honess, R. W. & Roizman, B. (1974) Regulation of herpesvirus macromolecular synthesis. I. Cascade regulation of the synthesis of three groups of viral proteins. *J. Virol. 14,* 8–19.

Honess, R. W. & Roizman, B. (1975) Regulation of herpesvirus macromolecular synthesis requires functional viral polypeptides. *Proc. Natl. Acad. Sci. USA 72,* 1276–80.

Jones, K. A. & Tjian, R. (1985) Sp1 binds to promoter sequences and activates herpes simplex virus 'immediate-early' gene transcription in vitro. *Nature 317,* 179–85.

Jones, P. C. & Roizman, B. (1979) Regulation of herpesvirus macromolecular synthesis. VIII. The transcription program consists of three phases during which both the extent of transcription and accumulation of RNA in the cytoplasm is regulated. *J. Virol. 31,* 299–314.

Kristie, T. M. & Roizman, B. (1984) Separation of sequences defining basal expression from those conferring *a* gene regulation within the regulatory domains of herpes simplex virus 1 *a* genes. *Proc. Natl. Acad. Sci. USA 81,* 4065–9.

Kristie, T. M. & Roizman, B. (1986) *a*4, the major regulatory protein of herpes simplex virus 1, is stably and specifically associated with promoter-regulatory domains of *a* genes and of selected other viral genes. *Proc. Natl. Acad. Sci. USA 83,* 3218–22.

Kristie, T. M. & Roizman, B. (1986) DNA-binding site of major regulatory protein *a*4 specifically associated with promoter-regulatory domains of *a* genes of herpes simplex virus type 1. *Proc. Natl. Acad. Sci. USA 83,* 4000–7.

Mackem, S. & Roizman, B. (1981) Regulation of herpesvirus macromolecular synthesis: Temporal order of transcription of *a* genes in not dependent on the stringency of inhibition of protein synthesis. *J. Virol. 40,* 319–22.

Mackem, S. & Roizman, B. (1982a) Structural features of the herpes simplex virus *a* genes 4, 0 and 27 promoter-regulatory sequences which confer *a* regulation on chimeric thymidine kinase genes. *J. Virol. 44,* 939–49.

Mackem, S. & Roizman, B. (1982b) Regulation of *a* genes of herpes simplex virus: The *a*27 promoter-thymidine kinase chimeric is positively regulated in converted L cells. *J. Virol. 43,* 1015–23.

Mackem, S. & Roizman, B. (1982c) Differentiation between *a* promoter and regulatory domains and sequence of a movable *a* regulator. *Proc. Natl. Acad. Sci. USA 79,* 4917–21.

Mavromara-Nazos, P., Ackermann, M. & Roizman, B. (1986) Construction and properties of a viable herpes simplex virus 1 recombinant lacking the coding sequences of the *a*47 gene. *J. Virol.* (in press).

Mavromara-Nazos, P., Silver, S., Hubenthal-Voss, J., McKnight, J. L. C. & Roizman, B. (1986) Regulation of herpes simplex virus 1 genes: *a* sequence requirements for transient induction of indicator genes regulated by β or late γ_2 promoters. *Virology 149,* 152–64.

McKnight, J. L. C., Kristie, T. M., Silver, S., Pellett, P. E., Mavromara-Nazos, P., Campadelli-Fiume, G., Arsenakis, M. & Roizman, B. (1986) Regulation of herpes simplex virus gene expression: The effect of genomic enviroments and its implications for model systems. In: *Cancer Cells 4, Control of gene expression and replication,* eds. Botchan, M., Grodzicker, T., & Sharp, P. Cold Spring Harbor Laboratories, Cold Spring Harbor, New York.

McKnight, S. L. (1980) The nucleotide sequence and transcript map of the herpes simplex thymidine kinase gene. *Nucl. Acids Res. 24,* 5949–64.

McKnight, S. L. & Kingsbury, R. (1982) Transcriptional control signals of a eukaryotic protein-coding gene. *Science 217,* 316–24.

Metzler, D. W. & Wilcox, K. (1985) Isolation of herpes simplex virus regulatory protein ICP4 as a homodimeric complex. *J. Virol. 55,* 329–37.

Morse, L. S., Pereira, L., Roizman, B. & Schaffer, P. (1978) Anatomy of HSV DNA. XI. Mapping of viral genes by analysis of polypeptides and functions specified by HSV-1 × HSV-2 recombinants. *J. Virol. 26,* 389–410.

O'Hare, P. & Hayward, G. S. (1985a) Three trans-acting regulatory proteins of herpes simplex virus modulate immediate-early early gene expression in a pathway involving positive and negative feedback regulation. *J. Virol. 56,* 723–33.

O'Hare, P. & Hayward, G. S. (1985b) Evidence for a direct role for both the 175,000 and 110,000 molecular weight immediate-early proteins of herpes simplex virus in the transactivation of delayed-early promoters. *J. Virol. 53,* 751–60.

Pellett, P., McKnight, J. L. C., Jenkins, F. & Roizman, B. (1985) Nucleotide sequence and predicted amino acid sequence of a protein encoded in a small herpes simplex virus DNA fragment capable of *trans*-inducing *a* genes. *Proc. Natl. Acad. Sci. USA 82,* 5870–4.

Pereira, L., Wolff, M., Fenwick, M. & Roizman, B. (1977) Regulation of herpesvirus synthesis. V. Properties of *a* polypeptides specified by HSV-1 and HSV-2. *Virology 77,* 733–49.

Post, L. E., Mackem, S. & Roizman, B. (1981) The regulation of *a* genes of herpes simplex virus: expression of chimeric genes produced by fusion of thymidine kinase with *a* gene promoters. *Cell 24,* 555–65.

Preston, C. M. (1979) Control of herpes simplex virus type 1 mRNA synthesis in cells infected with wildtype virus or the temperature-sensitive mutant tsK. *J. Virol. 29,* 275–84.

Roizman, B. & Batterson, W. (1985) The replication of herpesviruses In: *Virology,* ed. Fields, B., pp. 493–526. Raven Press, New York.

Roizman, B., Kristie, T. M., Batterson, W., & Mackem, S. (1984) The regulation of *a* genes of herpes simplex virus 1. In: *Proceedings of the P & S Biomedical Sciences Symposium on Transfer and Expression of Eukaryotic Genes,* eds. Vogel, H. & Ginsberg, H. S., pp 227–38. Academic Press, New York.

Roizman, B., Kozak, M., Honess, R. W. & Hayward, G. S. (1974) Regulation of herpesvirus macromolecular synthesis: Evidence for multilevel regulation of herpes simplex 1 RNA and protein synthesis. *Cold Spring Harbor Symp. Quant. Biol. 39,* 687–701.

Sacks, W. R., Greene, C. C., Aschman, D. P. & Schaffer, P. A. (1985) Herpes simplex virus type 1 ICP27 is an essential regulatory protein. *J. Virol. 55,* 796–805.

Sears, A. E., Halliburton, I. W., Meignier, B., Silver, S. & Roizman, B. (1985) Herpes simplex virus mutant deleted in the $\alpha 22$ gene: growth and gene expression in permissive and restrictive cells, and establishment of latency in mice. *J. Virol. 55,* 338–46.

Shih, M. F., Arsenakis, M., Tiollais, P. & Roizman, B. (1984) Expression of hepatitis B virus S gene by herpes simplex virus 1 vectors carrying α and β regulated gene chimeras. *Proc. Natl. Acad. Sci. USA 81,* 5867–70.

Watson, R. J. & Clements, J. B. (1980) A herpes simplex virus type 1 function continuously required for early and late virus RNA synthesis. *Nature 285,* 329–30.

Wilcox, K. W., Kohn, A., Sklyanskaya, E. & Roixman, B. (1980) Herpes simplex virus phosphoproteins. I. Phosphate cycles on and off some viral polypeptides and can alter their affinity for DNA. *J. Virol. 33,* 167–82.

DISCUSSION

LOWY: In your mutants that are unable to replicate, either in primary or in established cells, which phenotype is dominant if you do cell fusions?

ROIZMAN: We have not done that.

NORRILD: Concerning the sequence necessary for induction of alpha genes by alpha-TIF, is it necessary for several sequences to be present in the promoter, or is one sufficient?

Roizman: One is sufficient but may not be optimal. If you have more that one sequence as, for example, if you put the entire 320 bp sequence, you get a higher level of induction than if you add just one element.

NORRILD: Is the higher activity when multiple sequences are present due to recognition of all sequences, or due to the possibility that one, perhaps the one upstream, is able to activate transcription to a higher level than the ones downstream?

ROIZMAN: The *cis*-acting sites required for induction of alpha genes by alpha-TIF are not equivalent. Position effect and interactions with proteins binding to adjacent sites, e.g. SP1 sites, may be responsible for the effect we see.

NORRILD: Concerning the protein that binds to the *cis*-acting site necessary for induction of alpha genes by alpha-TIF, is anything known about the protein *per se?*

ROIZMAN: We have identified at least 3 different DNA protein complexes. It seems that there may be at least 3 proteins that bind to the *cis*-acting site. We are in the process of purifying them.

RIGBY: Do you know if the alpha-TIF will have an effect on any cellular promoters?

ROIZMAN: We have not tested directly whether alpha-TIF induces host proteins. However, the *cis*-acting site has sequence homology with sequences in promoter domains of SV40, heat shock proteins and other cellular genes. The experiment remains to be done.

YANIV: What is the degree of homology between the signal for alpha-4 in the alpha genes where it depresses expression and in the beta, gamma genes where it induces?

ROIZMAN: The binding sites are similar but not identical. Insufficient numbers of binding sites have been sequenced to determine whether there are two or only one consensus sequence.

YANIV: Did someone succeed doing this stimulation *in vitro?*

ROIZMAN: Yes, it was done recently by a former associate of mine, Dr. Kent Wilcox, and he stimulated transcription in an *in vitro* system with a partially purified protein.

YANIV: When you delete a viral function and show that it does not affect the ability of the virus to multiply, are you sure that this function is not redundant in the viral gene and in the sense that the genome contains 2 genes of similar function?

ROIZMAN: That would have to be functional rather than sequence homology. Redundancy is possible; it is conceivable that the virus carries similar functions by 2 totally dissimilar genes. I doubt it, however. I think there are other plausible explanations.

The Influence of Adenovirus Transformation on Cellular Gene Expression

*A. J. van der Eb, R. T. M. J. Vaessen, H. Th. M. Timmers, R. Offringa, A. G. Jochemsen, J. L. Bos & E. J. J. van Zoelen**

Human adenoviruses have been shown to be very suitable models for studies on the mechanism of carcinogenesis. Major advantages of adenoviruses are that both oncogenic and non-oncogenic types are found within the same group of closely related viruses and that complete oncogenic transformation requires the cooperation of 3 to 4 viral gene products rather than a single multifunctional protein, as in the case of SV40. This latter property should considerably facilitate the molecular genetic analysis of transformation.

The division into oncogenic and non-oncogenic adenoviruses is based on the ability of the viruses to induce tumors in hamsters, when the animals are infected within a few days after birth. According to this criterion, a non-oncogenic adenovirus (Ad), e.g. Ad5, does not induce tumors, whereas an oncogenic virus, such as Ad12, induces tumors with high frequency. However, all adenoviruses are capable of transforming cells *in vitro*. The *in vitro*-transformed cells exhibit the oncogenic properties of the viruses by which they are transformed, i.e. cells transformed by oncogenic Ad12 are oncogenic in newborn syngeneic animals, and cells transformed by non-oncogenic Ad5 are not. Interestingly, cells transformed by Ad5 are weakly oncogenic in T-cell-deficient nude mice, which indicates that the failure to induce tumors in immunocompetent animals is due, at least in part, to the immune defence of the host.

The transforming and oncogenic potential of adenoviruses is localized within the left terminal 4000 bp, which coincides with early region 1 (E1), one of the early regions expressed in the early phase of lytic infection. Region E1 is divided

---

Department of Medical Biochemistry, Sylvius Laboratories, P. O. Box 9503, 2300 RA Leiden, and *Hubrecht laboratory, International Embryological Institute, 3584 CT Utrecht, The Netherlands.

VIRAL CARCINOGENESIS, Alfred Benzon Symposium 24,
Editors: N. O. Kjeldgaard, J. Forchhammer, Munksgaard, Copenhagen 1987.

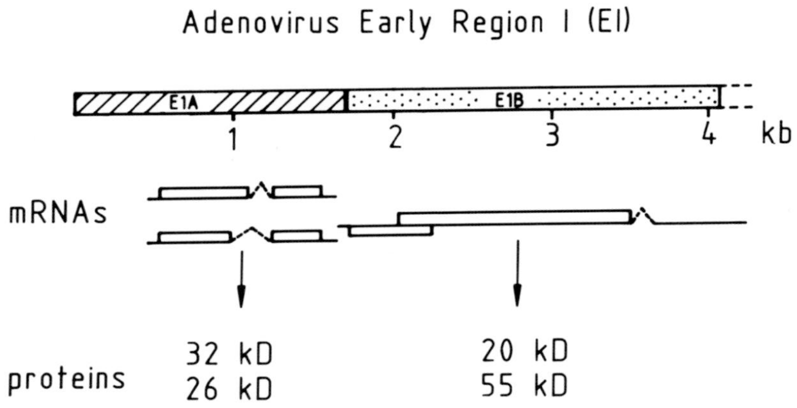

Fig. 1. Organization of the transforming region of human adenoviruses. The open boxes on the mRNAs represent the segments translated into protein. The calculated molecular weights of the protein are indicated.

into subregions E1A and E1B, which each code, in transformed cells, for 2 proteins (Figure 1). The 2 E1A-encoded proteins are identical to each other, except that the larger protein contains an internal stretch of amino acids which is lacking in the smaller. The 2 E1B-encoded proteins are not related to each other (reviewed in Bernards & Van der Eb 1985, Branton *et al.* 1985).

RESULTS AND DISCUSSION

Oncogenicity in immunocompetent animals is determined by the nature of the E1A region

To investigate which viral gene product(s) is responsible for the oncogenicity of transformed cells in immunocompetent animals, Bernards and coworkers constructed hybrid transforming regions, consisting of region E1A of Ad12 and E1B of Ad5, and vice versa. BRK cells transformed by the hybrid E1 constructs were tested for their oncogenicity in nude mice and immunocompetent rats, and the data were compared with those obtained with cells transformed by the intact Ad5 and Ad12 E1 regions. The results showed that the ability to induce tumors in nude mice is a function of the E1B region (oncogenicity is high when E1B is derived from Ad12, and low when it is derived from Ad5; Bernards *et al.* 1982). Oncogenicity in immunocompetent animals, however, is determined in addition by the origin of the E1A region, i.e., cells are only oncogenic when the E1A region is from Ad12. In fact, cells expressing Ad12 E1A have the same degree of oncogenicity in nude mice as in immunocompetent rats, suggesting that these cells are capable of evading the T-cell immune defence (Bernards *et al.* 1983).

Studies by Schrier *et al.* (1983) have provided an explanation for this phenomenon. They noticed that cells transformed by Ad12 E1 or Ad12 E1A+Ad5 E1B lack a number of proteins that are present in untransformed and Ad5 E1-transformed cells. One of these proteins was subsequently identified as the heavy chain of the class I Major Histocompatibility Complex (MHC) antigens. Class I MHC antigens play a crucial role in the T-cell-mediated recognition and destruction of cells expressing foreign antigens (e.g. viral proteins). A foreign antigen can only be recognized if the cell also expresses its own class I MHC antigens, and absence of the latter proteins consequently leads to insensitivity to the cytotoxic T-cell defence. This may explain, at least in part, why cells expressing Ad12 E1A, but not Ad5 E1A, can form tumors in immunocompetent hosts.

Since the inhibition of expression of class I MHC antigens by Ad12 region E1A is an example of an interaction of a viral gene product with the expression of a cellular gene, we have investigated this phenomenon in more detail.

Mechanism of inhibition by Ad12 E1A of class I MHC gene expression

The results of our previous studies on the mechanism of suppression of class I antigens can be summarized as follows:

1. Inhibition of class I gene expression not only occurs at the protein level but also at the level of cytoplasmic mRNA (Schrier *et al.* 1983).

2. The phenomenon seems to be characteristic for cells transformed by highly oncogenic adenoviruses only, and is not found in cells transformed by weakly oncogenic or non-oncogenic adenoviruses (our unpublished observations).

3. Suppression is not restricted to transformed rat cells but also occurs in transformed human, hamster and mouse cells. However, reduction of class I mRNA only occurs when primary cultures are transformed by Ad12, and not when the Ad12 E1 region is introduced into established cell lines (Vaessen *et al.* 1986).

4. If cells are transformed simultaneously with Ad5 and Ad12 E1A, class I expression is normal, indicating that Ad5 prevents Ad12 from suppressing the activity of class I genes (Schrier *et al.* 1983). If a class I MHC-negative Ad12-transformed cell is supertransfected with Ad5 E1, class I expression is restored (Vaessen *et al.* 1986). This shows that Ad5 E1A is dominant over Ad12 E1A in the modulation of class I gene expression and suggests that the E1A products

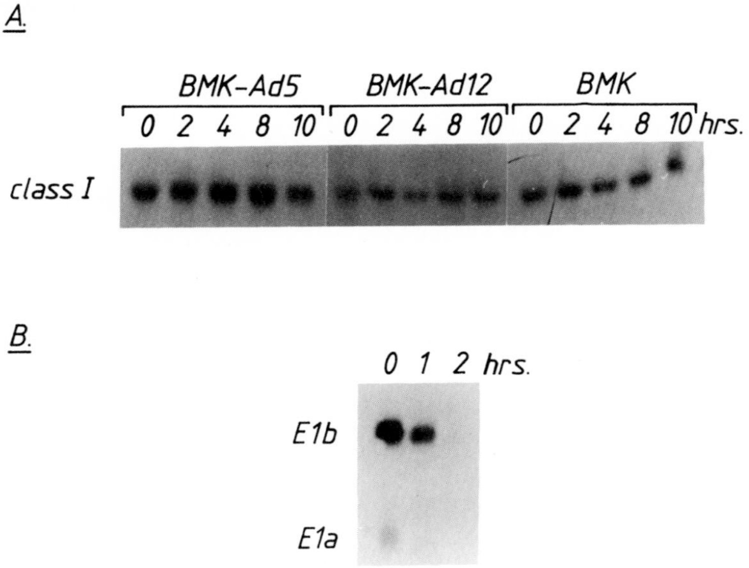

Fig. 2. Determination of stability of class I MHC RNA by actinomycin D treatment of untransformed, Ad5- and Ad12-transformed BMK cells. Actinomycin D (5 μg/ml; Calbiochem) was added to the medium of the cell cultures for the indicated periods. *A*. Cytoplasmic RNA of 2×10^6 BMK and Ad5-BMK cells, and 6×10^6 Ad12-BMK cells was isolated at each time point and subjected to Northern blot analysis as described previously (Schrier *et al.* 1983), using the homologous nick-translated 820-bp *Eco*RI-*Pvu*II cDNA fragment isolated from pB4 (Lalanne *et al.* 1983) as probe. *B*. Northern analysis of cytoplasmic RNA from 2×10^6 Ad12-BMK cells using a cloned Ad12 EcoRI C fragment as probe.

can compete for the same target, Ad5 E1A having a stronger affinity than Ad12 E1A.

5. Both the Ad5 and Ad12 E1A activities are functions of the 13S mRNA product, more specifically of the part encoded by the first exon (Van der Eb *et al.* 1984, Jochemsen *et al.* 1984).

6. The class I Major Histocompatibility Complex consists of 3 loci, designated H-2K, –D and –L in the mouse (HLA-A, –B and –C in humans). In the case of Ad12-transformed mouse cells, all 3 class I loci are suppressed to about the same extent, both at the protein and cytoplasmic mRNA level (Vaessen *et al.* 1986).

7. The inhibition of expression of class I MHC genes is caused by an active switching-off process by Ad12 E1A, and does not result from selective transformation of cells that happen to express low levels of class I antigens (Vaessen *et al.* 1986).

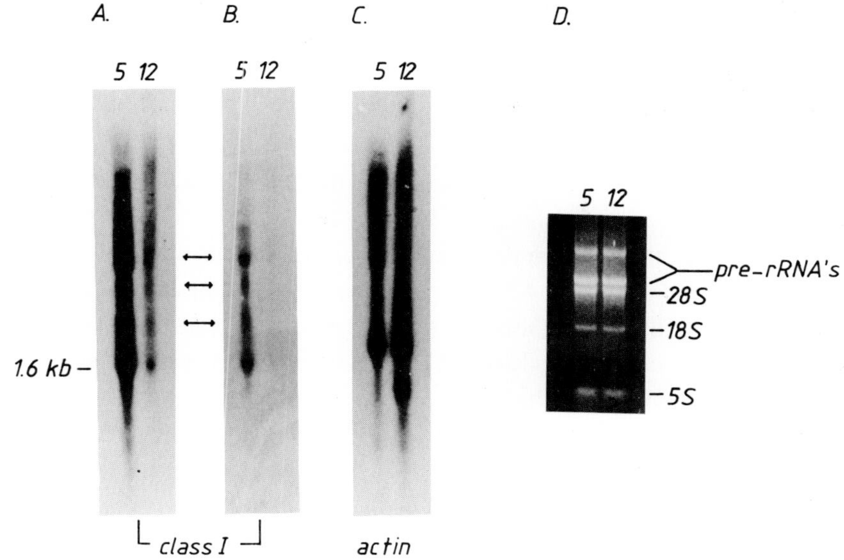

Fig. 3. Northern analysis of class I MHC transcripts in nuclear RNA extracted from Ad5- and Ad12-transformed BMK cells. Nuclei from Ad5- and Ad12-transformed BMK cells were prepared as described in the legend to Figure 4. After washing in 1 × TMKD-10% glycerol, nuclei were mixed vigorously in 3M LiCl/6M Urea. After precipitation overnight at 4°C, the RNA was pelleted by centrifugation (10000 rpm in Sorvall HB-4 rotor at 4°C). The pellet was washed once with LiCl/Urea, once with 70% ethanol, and dissolved in TES (10 mM Tris-Cl (pH 7.8), 5 mM EDTA, 1% SDS). Proteinase K (Boehringer) was added to 100 μg/ml final concentration; digestion was for 1 h at 37°C. The solution was extracted once with phenol/chloroform/isoamyl alcohol and once with chloroform/isoamyl alcohol, and the RNA was precipitated with 2 volumes of ethanol at −20°C. *A.* Equivalent amounts of the nuclear RNA samples were subjected to Northern blot analysis, using the class I cDNA fragment described in the legend to Figure 2 as probe. *B.* Shorter exposure of the filter shown in panel A. *C.* The same filter shown in A and B was stripped in H_2O (30 min, 80°C), and hybridized with a nick-translated mouse γ-actin cDNA fragment. *D.* Equivalent amounts of the nuclear RNA samples, electrophoresed in a sterile ethidium-bromide-containing agarose-TBE gel, photographed under UV light.

A subsequent question that arose was: at which level is the expression of class I MHC genes inhibited? In an attempt to answer this question, the following experiments were carried out.

Firstly, we have tested whether the low levels of class I mRNA in Ad12-transformed cells can be explained by a decreased stability of cytoplasmic mRNA. Cells were exposed to Actinomycin D for various lengths of time and the amount of class I MHC mRNA was then determined in the cytoplasm by Northern blotting analysis. It was found that the residual class I mRNA found in the cytoplasm of Ad12-transformed cells has the same stability as the mRNA in Ad5-transformed and untransformed cells (Figure 2). Secondly, we have analyzed the levels of nuclear class I RNA by Northern blotting. These experiments showed that Ad12-transformed mouse cells also contain highly reduced levels of class I MHC-specific transcripts in the nucleus (Figure 3). Thirdly, we have investigated the transcriptional activity of the class I MHC genes by nuclear "run-on" analysis. It was found that the transcriptional activity of class I genes was the same in untransformed and Ad5- and Ad12-transformed cells of mouse, rat and human origin (Figure 4). Based on these results, we tentatively conclude that the reduction of class I mRNA levels in Ad12-transformed cells is not caused by a suppression of transcription but must be the result of a post-transcriptional process in the nucleus. It cannot yet be excluded that the reduced class I RNA level in nuclei of Ad12-transformed cells is caused by premature termination (resulting in unstable (pre)-mRNAs). Furthermore, since we used a double-stranded class I cDNA clone as probe in the "run-on" assay, we cannot exclude that the positive hybridization signals are caused by anti-sense RNA transcribed from class I loci in the opposite direction. Experiments designed to answer these questions are in progress.

To identify sequences in or around the class I MHC genes that are sensitive to the suppressing effect of Ad12 E1A products, we have investigated whether class I MHC genes, that are introduced by transfection, are also differentially expressed in Ad5- and Ad12-transformed cells. Preliminary results have indicated that a H-2K^b gene, introduced by transfection into transformed mouse cells of the d haplotype, is efficiently expressed in Ad5-transformed cells, but only weakly in Ad12-transformed cells. This observation opens possibilities for identifying sequences that are involved in the Ad12 E1A-mediated suppression of class I gene activity.

Expression of the fibronectin gene is inhibited at the transcriptional level

In most transformed cells the expression of fibronectin is reduced. In both Ad5- and Ad12-transformed cells the reduction is at least 5-fold relative to that in untransformed cells, as measured by Northern blotting of cytoplasmic mRNA. We have recently observed that a much stronger reduction occurs in BRK cells transformed by the combination of Ad5 E1A and T24 *ras,* but not in cells transformed by Ad12 E1A+T24 *ras*. The Ad5 E1A+T24 *ras*-transformed cells show a highly transformed phenotype: they are extremely rounded, attach poorly

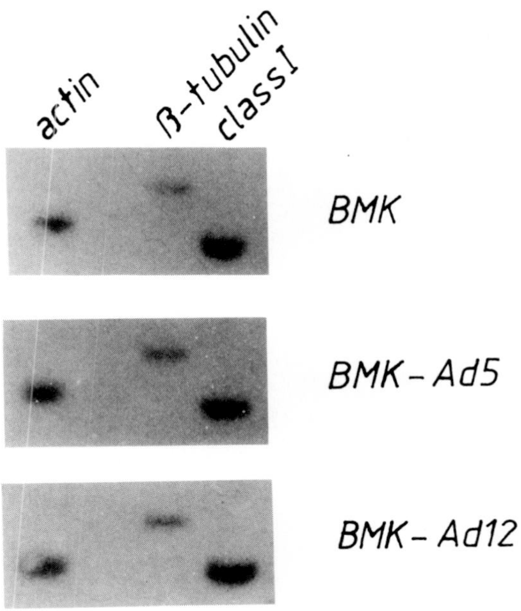

Fig. 4. In vitro run-on transcription in isolated nuclei (adapted from Friedman *et al.* 1984). Run-on RNA from untransformed Baby Mouse Kidney (BMK) cells and Ad5- and Ad12-transformed BMK cells was hybridized against filters containing the following cDNAs: mouse γ-actin (1.1 kb), chicken β-tubulin (1.7-kb *Hind*III fragment isolated from pT2 (Valenzuela *et al.* 1981)) and mouse class I MHC (820-bp *Eco*RI-*Pvu*II fragment isolated from pB4 (Lalanne *et al.* 1983)). After the cell cultures were trypsinized, all handlings were carried out at 4°C. Cells were washed with PBS and nuclei were prepared by Dounce homogenization (5 to 10 strokes, tight pestle) of the cell suspensions in a buffer containing $1 \times$TMKD (10 mM Tris, pH 7.8, 5 mM $MgCl_2$, 150 mM KCl and 5 mM DTT) and 0.1% NP-40. Nuclei were pelleted by centrifugation (3 min $600 \times g$) and washed in a buffer containing $1 \times$TMKD, 10% glycerol and 1 mM $MnCl_2$. 10^7 nuclei were gently resuspended in 75 μl reaction buffer ($1 \times$TMKD, 10% glycerol) and kept on ice. 30 μl of $5 \times$ reaction buffer were added to 1.2 mCi (120 μl) α ^{32}P-UTP (Amersham) and mixed. 25 μl of this UTP solution (200 μCi) were added to the nuclei suspensions. Cold UTP was added to a final concentration of 4 μM, the other NTP's and $MnCl_2$ to final concentrations of 250 μM and 1 mM, respectively. The nuclei were incubated under regular whirling at 30°C for 30 min. Nuclei were pelleted for 15 s in an Eppendorf centrifuge at room temperature and suspended in 300 μl DNAse buffer (20 mM HEPES (pH 7.5), 5 mM $MgCl_2$ and 1 mM $CaCl_2$). RQ1 DNAse (Promega Biotec) was added to a final concentration of 20 μg/ml and incubated at 37°C for 5–10 min. Extraction of RNA from the nuclei was according to McKnight & Palmiter (1979). Equivalent amounts of labeled RNA from the various cell types were hybdridized to Gene Screen Plus filters (NEN) containing different cDNA inserts, in a buffer containing 40% formamide, 20 mM PIPES (pH 7.5), 1% SDS, 0.5 M NaCl and 200 μg/ml E. coli tRNA. Hybdridization was for 48–72 h at 42°C. Washing of the filters was in $2 \times$SSC/0.1% SDS up to 60°C, and included an RNAse A digestion (10 μg/ml) for 30 min at 37°C.

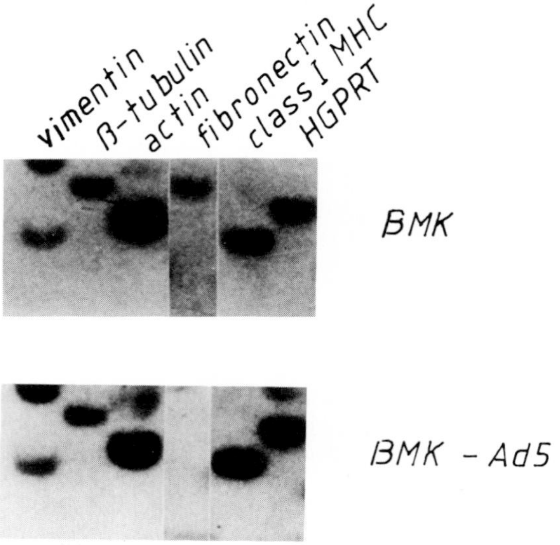

Fig. 5. Run-on RNA from untransformed Baby Mouse Kidney (BMK) cells and Ad-transformed BMK cells was hybridized against filters containing cDNA specific for hamster vimentin (Quax-Jeuken *et al.* 1983), chicken *β*-tubulin (Valenzuela *et al.* 1981), mouse *γ*-actin, human fibronectin (Kornblihtt *et al.* 1985), mouse class I MHC (Lalanne *et al.* 1983) and mouse hypoxanthin guanine phosphoribosyl transferase (HGPRT, Konecki *et al.* 1982). The run-on assay was performed as described in Figure 4.

to tissue culture dishes and are highly oncogenic. In contrast, cells transformed by Ad12 E1A+T24 *ras* are adherent and only weakly oncogenic (Jochemsen *et al.* 1986). Interestingly, Ad5 E1A+*ras*-transformed cells do attach to culture dishes which have been coated with fibronectin. This suggests that the absence of fibronectin gene expression is responsible for the transformed phenotype of the cells in tissue culture. To determine the mechanism by which fibronectin mRNA levels are modulated in Ad-transformed cells, nuclear "run-on" analyses were carried out. These experiments indicated that the transcriptional activity of the fibronectin gene is strongly reduced in Ad5-transformed BMK cells as compared to untransformed BMK cells (Figure 5). Therefore, inhibition of fibronectin expression seems to occur at the level of transcription. This conclusion is supported by *in vivo* RNA labeling experiments with ^{32}P-orthophosphate and hybridization of labeled RNA to Southern blots containing fibronectin cDNA. No fibronectin RNA could be detected, either in the nucleus or in the cytoplasm of Ad-transformed cells. In contrast, fibronectin RNA could easily be detected in untransformed cells (data not shown).

These results show that adenovirus transformation can influence the expression of cellular genes, both at the level of transcription (fibronectin) and at a post-transcriptional level (class I). (Reduction of fibronectin expression is probably not a direct effect of the activity of adenovirus transforming genes but a general result of the transformed state of the cell).

Regulation of cell proliferation in adenovirus-transformed cells

Transformation can be defined as a state in which the regulatory mechanisms of cell proliferation are disturbed to a certain extent. If this definition also applies to adenovirus-transformed cells, one would expect to observe abnormal responses to growth factors, or altered expression of genes that are involved in cell proliferation.

To examine the growth factor responsiveness of our transformed cell lines, we have tested untransformed and Ad5 E1- or Ad12 E1-transformed BRK cells for their ability to incorporate ^{3}H-thymidine into cellular DNA in a "proliferative response" assay and a "mitogenic response" assay. In the proliferative response assay, subconfluent cell cultures are incubated in growth factor-free medium (medium with 10% thiol-serum; Van Zoelen *et al.* 1985), supplemented with the growth factors to be tested, for 48 h. The fraction of cells in S-phase is then measured by pulse-labeling the cells for 16 h with ^{3}H thymidine.

In the mitogenic response assay, cells are allowed to reach a confluent state, after which serum-free medium is added for 48 h. Subsequently, the growth factors of interest are added for 16 h, and the cells are then pulse-labeled for 6 h with ^{3}H-thymidine. This procedure measures the mitogenic activity of the added growth factor. The results of these assays show that Ad5- and Ad12-transformed cells differ in a number of properties (Figure 6). Ad5 cells are relatively independent of growth factors in both assays as they still incorporate appreciable amounts of ^{3}H thymidine in their absence. Furthermore, the cells are stimulated only 2-fold by serum or insulin and respond only weakly to EGF. Nevertheless, Ad5-cells do contain free EGF receptors (not shown), suggesting that they do not produce TGFα, which would occupy the free EGF receptors. However, they may produce other growth factors which do not bind to the EFG receptor. Ad12-cells, on the other hand, are strongly stimulated by serum or insulin, but not by EGF. Moreover, Ad12 cells do not contain free EGF receptors, or only small amounts of it (not shown), which may indicate that these cells produce a growth factor that binds to their own EGF receptor (e.g., TGFα). Studies with cells transformed by Ad5 E1A+Ad12 E1B, or the reverse combination, have shown

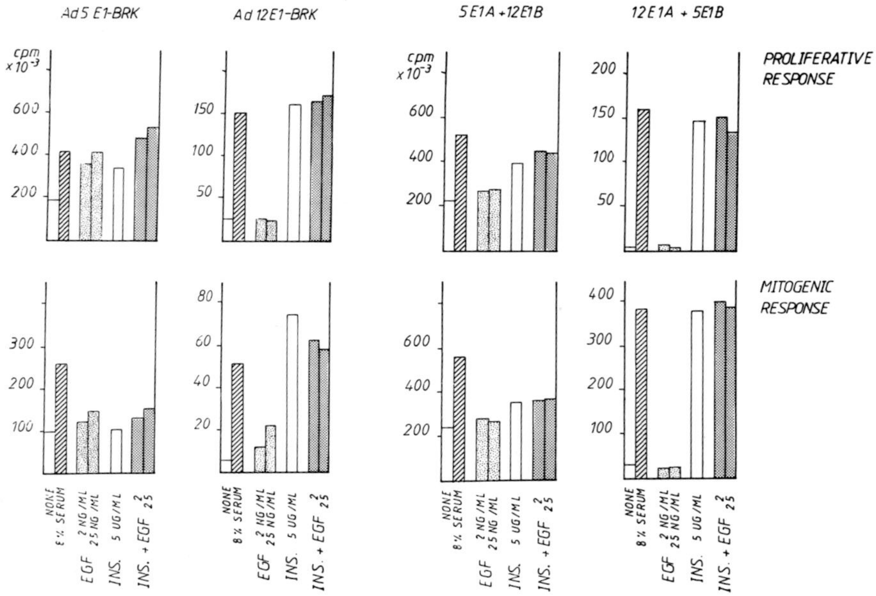

Fig. 6. Proliferative and mitogenic stimulation of Ad-transformed BRK cells. For a proliferative response 25–75 × 10^3 cells were seeded per 16 mm well (Costar, Cambridge MA, USA) in Basal Medium (BM) consisting of a 1:1 mixture of DMEM (Dulbecco's MEM) and Ham's F12 medium containing 10 μg/ml transferrin and 10% thiol-fetal calf serum (Van Zoelen *et al.* 1985), and supplemented with growth factors as indicated below the diagrams. After 48 h cells were exposed to 0.5 μCi ^{3}H-thymidine (Amersham) for 16 h. Incorporation was determined as described by Van Zoelen *et al.* (1985) and expressed in cpm. Standard deviations were omitted for clarity. For a mitogenic response, cells were seeded in DMEM containing 7.5% newborn calf serum in 16 mm wells and were allowed to reach confluency. The medium was then removed and cells were incubated in serum-free DMEM. After 48 h cells were stimulated with the indicated growth factors for 16 h. After this period cells were exposed to 0.5 μCi ^{3}H-thymidine for 6 h. Incorporation was determined as above.

that the growth factor-responsiveness clearly is a function of region E1A (Figure 6). Taken together, these results indicate that Ad5 E1A-expressing cells are largely independent of growth factors for cell proliferation, whereas Ad12 E1A-expressing cells are dependent on the presence of insulin and possibly PDGF (present in serum), but not EGF.

In a second approach, we have started a study on the expression of genes that play a role in cell proliferation. A gene that has been examined in some detail is the JE gene isolated by Stiles and coworkers (Cochran *et al.* 1983). JE is not expressed as cytoplasmic RNA in quiescent 3T3 cells but expression is induced

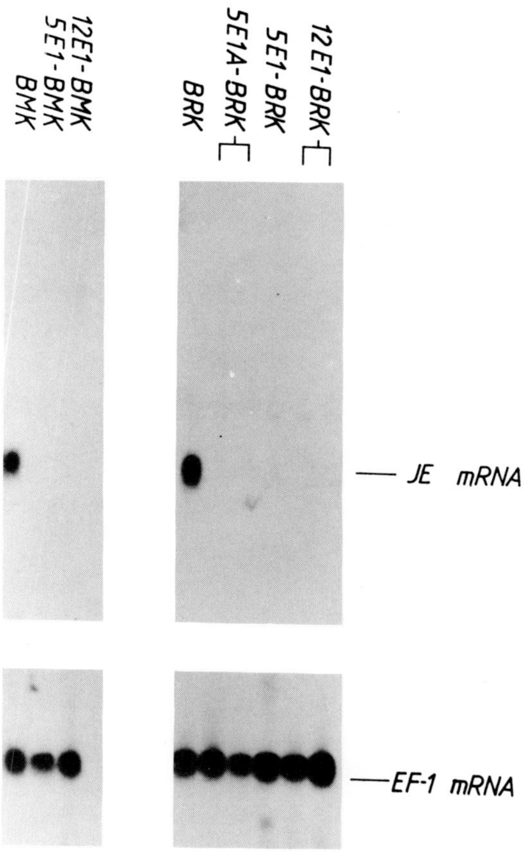

Fig. 7. Northern analysis of JE transcripts in cytoplasmic RNA from untransformed, Ad5-transformed, Ad12-transformed BMK or BRK cells. Each lane contains about 20 μg of cytoplasmic RNA. After Northern blotting the filter was hybridized to a nick-translated JE probe (the 750 bp PstI-fragment of pJE (Cochran *et al.* 1983)) (upper panels). As a control for the amount of RNA loaded, the probe was melted off and the filter was rehybridized to the 1450 bp PstI-fragment of phEF-1α (Brands *et al.* 1986), a cDNA clone of the human elongation factor-1α gene (lower panels). Hybridization stringency was $2 \times$ SSC/50°C.

as a response to addition of serum or PDGF. Figure 7 shows that JE mRNA is present in considerable amounts in the cytoplasm of untransformed growing cultures of BMK and BRK cells but, surprisingly, is absent in the cytoplasm of the corresponding Ad5- or Ad12-transformed cultures. Similar results were obtained with untransformed and transformed NRK cells, although the trans-

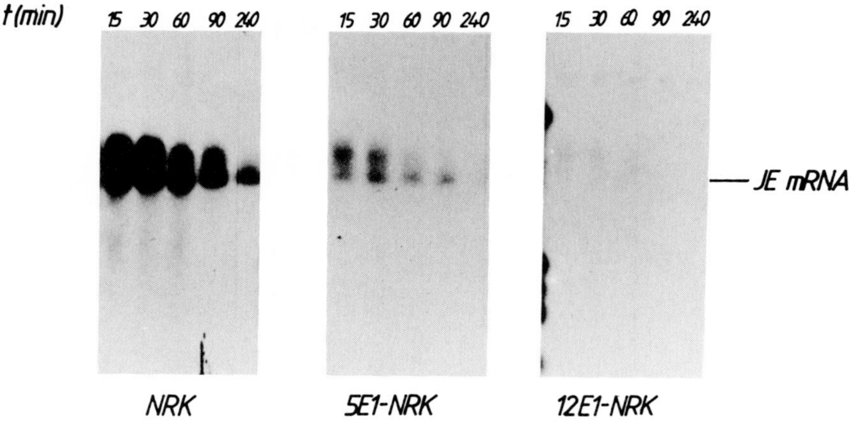

Fig. 8. Determination of the stability of JE RNA by Actinomycin D treatment of untransformed, Ad5-transformed and Ad12-transformed NRK cells. Actinomycin D was added to the medium of subconfluent cell cultures for the indicated periods, after which cytoplasmic RNA was isolated. Northern blot analysis was as described in Figure 7. Hybridization stringency was $2 \times$ SSC/50°C. The amount of RNA loaded per lane was checked by rehybridization with the EF-1α probe (not shown). Since there is a time lag before Actinomycin D can exert its inhibitory activity, the 15-min incubation was taken as the reference point. The half-life of JE mRNA in the cytoplasm is estimated to be between 30 and 60 min in all 3 cell types.

formed NRK cells still contained a small amount of JE mRNA (NRK is an established rat cell line). To test whether the low levels of JE RNA are due to a decreased stability, the half-life of the cytoplasmic RNA was determined using Actinomycin D. Figure 8 shows that the stability of JE mRNA is approximately the same in untransformed and transformed NRK cells. To test whether JE transcription is affected in the transformed cells, nuclear "run-on" analyses were performed, which showed that the JE gene is transcriptionally as active in transformed as in untransformed BMK cells (Figure 9). Thus, the JE gene behaves in Ad5- and Ad12-transformed cells as class I MHC genes in Ad12-transformed cells: expression of JE is suppressed at a post-transcriptional level. Again, the "run-on" data could be explained by transcription of anti-sense RNA or by premature termination.

We are presently studying the expression in the transformed cells of the c-*myc* gene, which reportedly follows kinetics similar to that of JE after stimulation of quiescent cells with PDGF. The results of this experiment may provide information as to whether or not JE and c-*myc* could have functions in the same regulatory pathway.

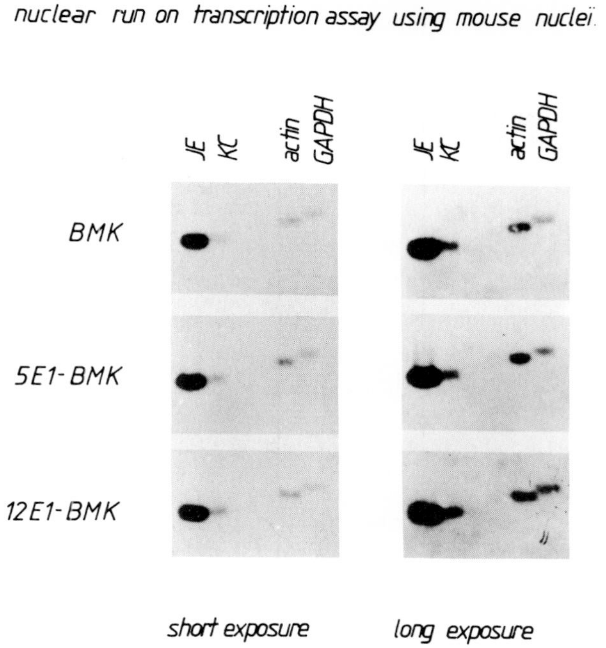

Fig. 9. Run-on transcription assay in isolated nuclei of untransformed, Ad5-transformed and Ad12-transformed BMK cells. The analysis was performed as described in Figure 4. The indicated cDNA inserts of pJE and pKC (Cochran *et al.* 1983), p actin (a mouse γ-actin cDNA clone) and pGAPDH-13, (a rat GAPDH cDNA clone (Fort *et al.* 1985)) were purified from vector sequences, DNA was transferred to Gene Screen Plus filters (NEN) and the filters were used in the hybridizations with labeled RNA.

In our further attempts to explain the suppression of the JE gene in Ad-transformed cells we will have to consider two alternative possibilities: (1) JE is suppressed because the appropriate activating signals that are normally generated by growth factors are lacking.

(2) The activating signals are generated as in normal cells, but induction of JE is prevented, probably as a result of E1A expression.

Experiments similar to those described for JE are carried out in collaboration with Nathans and coworkers, using a set of cDNA clones representing genes that are stimulated as a primary response to serum (Lau *et al.* 1985). Preliminary results have shown that some of the genes respond like the JE gene to adenovirus-transformation, whereas others respond in a different way. Further work with these serum-inducible genes will hopefully provide insight into the mechanism of adenovirus transformation and possibly also on how cell proliferation is regulated in general.

ACKNOWLEDGMENTS

The authors wish to thank Mrs. Rita Veeren for typing the manuscript. This work was supported in part by the Netherlands Organization for the Advancement of Pure Research (ZWO) through the Foundations for Fundamental Medical Research (FUNGO) and Chemical Research in the Netherlands (SON).

REFERENCES

Bernards, R., Houweling, A., Schrier, P. I., Bos, J. L. & Van der Eb, A. J. (1982) Characterization of cells transformed by Ad5/Ad12 hybrid early region 1 plasmids. *Virol. 120,* 422–32.

Bernards, R., Schrier, P. I., Houweling, A., Bos, J. L., Van der Eb, A. J., Zijlstra, M. & Melief, C. J. M. (1983) Tumorigenicity of cells transformed by adenovirus type 12 by evasion of T-cell immunity. *Nature 305,* 776–9.

Bernards, R. & Van der Eb, A. J. (1984) Adenovirus: Transformation and Oncogenicity. *Biochim. Biophys. Acta 783,* 187–204.

Brands, J. H. G. M., Maassen, J. A., Van Hemert, F. J., Amons, R. & Möller, W. (1986) The primary structure of the *a* subunit of human elongation factor 1. *Eur. J. Biochem. 155,* 167–71.

Branton, P. E., Bailey, S. T. & Graham, F. L. (1985) Transformation by human adenoviruses. *Biochim. Biophys. Acta 780,* 67–94.

Cochran, B. H., Reffel, A. C. & Stiles, C. D. (1983) Molecular cloning of gene sequences regulated by platelet derived growth factor. *Cell 33,* 939–47.

Fort, Ph., Marty, L., Piechaczyk, M., el Sabrouty, S., Dani. Ch., Jeanteur, Ph. & Blanchard, J. M. (1985) Various rat adult tissues express only one major mRNA species from the glyceraldehyde-3 phosphate-dehydrogenase multigenic family. *Nucl. Acids Res. 13,* 1431–42.

Friedman, R. L., Manly, S. P., McMahon, M., Kerr, I. M. & Stark, G. R. (1984) Transcriptional and posttranscriptional regulation of interferon-induced gene expression in human cells. *Cell 38,* 745–55.

Jochemsen, A. G., Bernards, R., Van Kranen, H. J., Houweling, A., Bos, J. L. & Van der Eb, A. J. (1986) Different activity of the adenovirus types 5 and 12 region E1A in transformation with the EJ *Ha*-ras oncogene. *J. Virol.* (in press).

Jochemsen, A. G., Bos, J. L. & Van der Eb, A. J. (1984) The first exon of region E1A genes of adenoviruses 5 and 12 encodes a separate functional protein domain. *EMBO J. 3,* 2923–7.

Konecki, D. S., Brennard, J., Fuscoe, J. C. Caskey, C. T. & Chinault, A. G. (1982) Hypoxanthine-guanine phosphoribosyltransferase genes of mouse and chinese hamster: construction and sequence analysis of cDNA recombinants. *Nucl. Acids. Res. 10,* 6763–75.

Kornblihtt, A. R., Umezawa, K., Vibe-Pedersen, K. & Baralle, F. E. (1985) Primary structure of human fibronectin: differential splicing may generate at least 10 polypeptides from a single gene. *EMBO J. 4,* 1755–9.

Lalanne, J-L., Delarbre, C., Gachelin, G. & Kourilsky, P. (1983) A cDNA clone containing the entire coding sequence of a mouse H-2K^d histocompatibility antigen. *Nucl. Acid. Res. 11,* 1567–77.

Lau, L. & Nathans D. (1985) Identification of a set of genes expressed during the G0/G1 transition of cultured mouse cells. *EMBO J. 4,* 3145–51.

McKnight, G. S. & Palmiter, R. D. (1979) Transcriptional regulation of the ovalbumin and conalbumin genes by steroid hormones in chick oviduct. *J. Biol. Chem. 254,* 9050–8.

Quax-Jeuken, Y. E. F. M., Quax, W. J. & Bloemendal, H. (1983) Primary and secondary structure of hamster vimentin predicted from the nucleotide sequence. *Proc. Natl. Acad. Sci. 80,* 3548–52.

Schrier, P. I., Bernards, R., Vaessen, R. T. M. J., Houweling, A. & Van der Eb, A. J. (1983) Expression of class I major histocompatibility antigens switched off by highly oncogenic adenovirus 12 in transformed rat cells. *Nature 305,* 771–5.

Vaessen, R. T. M. J., Houweling, A., Israel, A. Kourilsky, P. & Van der Eb, A. J. (1986) Adenovirus E1A-mediated regulation of class I MHC expression. *EMBO J. 5,* 335–41.

Valenzuela, P., Quiroga, M., Zaldivar, J., Rutter, W. J., Kinscher, M. W. & Cleveland, D. W. (1981) Nucleotide and corresponding amino acid sequences encoded by α and β tubulin. *Nature 289,* 650–5.

Van der Eb, A. J., Bernards, R., Schrier, P. I., Bos, J. L., Vaessen, R. T. M. J., Jochemsen, A. G. & Melief, C. J. M. (1984) Altered expression of cellular genes in adenovirus-transformed cells. In: *Cancer Cells 2/Oncogenes and viral genes,* eds. Van der Woude, G. F., Levine, A. J., Topp, W. C., Watson, J. D., pp. 501–10. Cold Spring Harbor Laboratory, New York.

Van Zoelen, E. J. J., Van Oostwaard, T. M. J., Van der Saag, P. T. & De Laat, S. W. (1985) Phenotypic transformation of normal rat kidney cells in a growth-factor-defined medium: Induction by a neuroblastoma derived transforming growth factor independently of the EGF receptor. *J. Cell. Physiol. 123,* 151–60.

DISCUSSION

WILLUMSEN: When you are suggesting that a mechanism for the difference between the Ad5 and Ad12 infected cells is premature termination of the transcription in one of the lines, are you thinking of a mechanism similar to attenuation described in bacteria, and if yes, how do you interpret your run-on experiments?

VAN DER EB: I think I can interpret the run-on experiments. Say, if the gene is transcribed in the Ad12 transformed cells and prematurely stops for some reason, then there will still be radioactive RNA that can bind to our Southern blots in the run-on assay. So, I do not think that is a problem.

WILLUMSEN: Are you able to estimate the size of the RNA that is being synthetized in the run-on experiments?

VAN DER EB: The RNA is still present in the nucleus, say at 1/10 of the concentration of that present in the nuclei of Ad5 cells. They have the same size as cytoplasmic RNA and there is also precursor RNA, just as in the nuclei of Ad5 transformed cells.

WILLUMSEN: If the run-on RNA is a 50 nucleotide-long piece of RNA, would you be able to see that in your experiment; would you get a sufficient number of counts into that?

VAN DER EB: I think that the transcription rate is high enough, thus I think you might see it. We are now preparing single-stranded probes of the two polarities, both for the 5′ and the 3′ region of the class I gene, so we can get a better idea where the RNA comes from.

CUZIN: Could you comment on the possible role of E1B in cell transformation?

VAN DER EB: That is a difficult question. We have the impression that the major role of E1B in transformation is to cause a high expression of E1A. Region E1B has no detectable transforming activity of its own, as activated *ras,* but it seems to have a role in oncogenecity. This was concluded from experiments with E1B mutants, which still transform normally and in which E1A is expressed at a high level. These transformed cells are not oncogenic in nude mice, despite high expression of E1A. How E1B enhances expression of E1A is not clear yet. It could be a trans effect, by an E1B protein, or an enhancer effect which works only at short distances.

RAPP: Can you reinduce the suppressed early G1 genes, *myc,* or JF in serum responsive cells?

VAN DER EB: We have not yet tested that. These are very recent data. This is one of the things we would like to do. We can reinduce the class I genes, not only by Ad5 E1A, but also by interferon.

LEPPARD: Your Ad12 transformed cells lines are presumably tumorigenic?

VAN DER EB: Yes, others have tested that. They are oncogenic.

LEPPARD: My question was really aimed at asking whether once expression of class 1 genes recovers, is the tumorigenicity of the cell lines lost?

VAN DER EB: Yes, this has been done by Khoury and Jay, and we have done the same in collaboration with Peter Schrier – you can bring back the class I expression by treatment with interferon, and if you do that the oncogenecity drops considerably.

RIGBY: It is very striking that the Ad5 and Ad12 E1A have such different effects on class I expression. Do you have any information from looking at the sequence why such extensively similar proteins have different functions.

VAN DER EB: No, although we have tried. The proteins are different in some areas, but in general they are very homologous. The E1A proteins do not have different effects on JE, for instance, and on expression of a number of the Nathans clone.

YANIV: There is something illogical to me. JE is induced by serum or PDGF, so it is something that relates to growth, and still you suppress it by adenovirus information, where you would expect to synthetize it.

VAN DER EB: Exactly. It was the last thing that we expected. We have not yet looked, however, whether you can induce JE in resting adeno-transformed cells by addition of serum. It looks as if JE, and perhaps also even *myc* are somehow inhibited from being active, either because expression of the gene is inhibited at some stage, or because there is no signal that activates the gene, due to the presence of E1A. So E1A, so to speak, seems to take over the functions of some of the cell cycle genes. This is a situation that may be typical for some DNA viruses.

YANIV: Do you think that it induces the same pathway as gene activation by E1A, or is it a different function of the E1A proteins?

VAN DER EB: I am afraid that I cannot answer that, simply because there are no data.

CORY: There is an old story, which seems backwards to me at least and maybe you can explain it. That is, the Ad12-tranformed cells seemed to be still very dependent on growth factors, but the Ad5-transformed cells, which are not tumorigenic, are independent.

VAN DER EB: Yes, that seems contradictory. I guess that one of the reasons why Ad12 is tumorigenic is the fact that they have no class I antigen or very little of it. That is one explanation. If you bring back the class I antigens by interferon then they suddenly lose their oncogenicity. Furthermore, Ad5-transformed cells are easily killed by cytotoxic T cells and NK cells, so that may explain why they are not oncogenic. In nude mice, however, Ad5-transformed cells are oncogenic. So, oncogenicity of adeno-transformed cells may be determined more by their ability to evade the host immune surveillance than by their independence of growth factors.

LOWY: Isn't Ad5 more oncogenic than Ad12 *in vitro?*

VAN DER EB: You mean transformation *in vitro?* That is right. Ad5 is 10- to 20-fold more efficient in transforming cells *in vitro*. It is a much stronger transforming virus than Ad12. So, if you take all this information together then it becomes rather unclear.

WEINBERG: You said that you delete various portions of the H2 gene and studied their responsiveness of these altered genes, after transfection, to the Ad5 and the Ad12? Could you elaborate on that a little bit?

VAN DER EB: The data that I showed had 4000 bp upstream relative to the H-2 gene. Jochemsen removed most of this 4000 bp, so that only 400 or 100 is remaining upstream. Both deletion mutants were normally active in Ad5-transformed cells, but suppressed in Ad12 cells. You would expect that if regulation is posttranscriptional. Then you would not expect an effect on the promoter, but who knows? There may also be a far upstream sequence that may be activated like Dr. Rigby showed. This is what, at 7 kb upstream?

RIGBY: Between 4 1/2 and 5 kb.

VAN DER EB: So, this would be outside the region we have tested.

RIGBY: Yes, but as you said, we are looking at the short-term response.

VAN DER EB: Yes, that may not be the same.

RIGBY: You are looking at a long-term response which is probably mediated by an entirely different mechanism.

VAN DER EB: If you infect mouse embryo cells with Ad5 and Ad12 virions, this is the Grosveld experiment which we have repeated, then class I expression is activated in both cases. If you then isolate, from the same infected cells, class I is active in the Ad5 cells and not active in the Ad12 cells.

An Adenovirus Oncogene Post-Transcriptionally Modulates mRNA Accumulation

Keith Leppard, Stephen Pilder, Mary Moore, John Logan & Thomas Shenk

Adenoviruses have been shown to express 4 gene products which cooperate in the oncogenic transformation of rodent cells. Two of these products are coded by the E1A transcription unit, and are often referred to by the size of their mRNAs as E1A-13S and E1A-12S polypeptides. The other 2 are encoded by the E1B transcription unit and termed the E1B-55K and E1B-21K products. The E1A products have transcriptional trans-activating (Berk *et al.* 1979, Jones & Shenk 1979, Nevins 1981) and trans-repressing (Borrelli *et al.* 1984, Velcich & Ziff 1985) activities. It is possible that the E1A products effect their role in transformation through transcriptional modification, although there is evidence that the two activities can be unlinked by mutation (Zerler *et al.* 1986). The functions of E1B products are less well established, and we present evidence here for a role of the E1B-55K polypeptide in transport or cytoplasmic stabilization of late viral mRNAs.

1. GENETIC DISSECTION OF THE E1B UNIT

The adenovirus type 5 (Ad5) E1B transcription unit is located immediately 3′ to the E1A unit between 4.5 and 11.5 map units on the viral chromosome (Figure 1). It encodes a variety of mRNAs early after infection. The largest mRNA (22S) can encode both 21K and 55K polypeptides while the smallest (13S) codes only the 21K species (Bos *et al.* 1981). The 55K coding region overlaps the carboxy-terminal 40% of the 21K region in a second reading frame. Anderson *et al.* (1984) have identified an additional E1B-coded polypeptide synthesized from an intermediate-sized mRNA. The 17K moiety is coded in the 55K reading frame

---

Department of Molecular Biology, Princeton University, Princeton, NJ 08544, U.S.A.

VIRAL CARCINOGENESIS, Alfred Benzon Symposium 24,
Editors: N. O. Kjeldgaard, J. Forchhammer, Munksgaard, Copenhagen 1987.

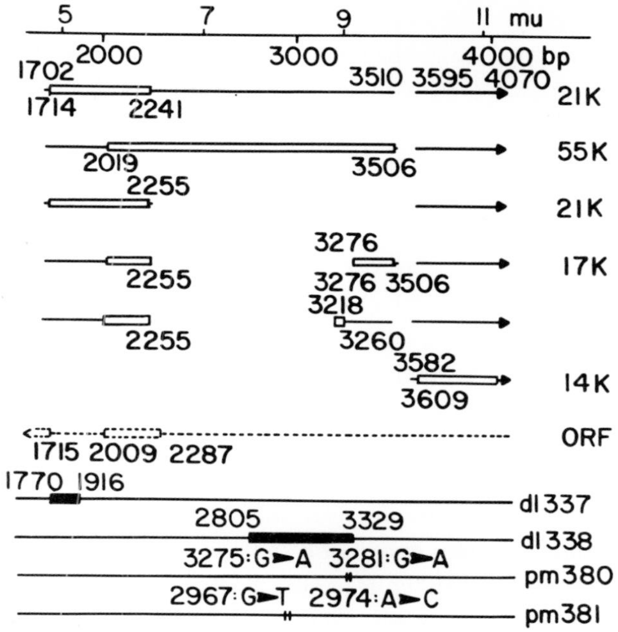

Fig. 1. Diagram of E1B mRNAs, coding regions and mutations. The top of the figure positions the map in terms of map units and nucleotide sequence position relative to the left end of the viral chromosome. mRNA exons are designated by lines, introns by spaces and coding regions by open rectangles. Open reading frames (ORFs) encoded by the opposite DNA strand are designated by open rectangles outlined with a broken line. The segments deleted in *dl*337 and *dl*338 are represented by solid rectangles. Nucleotide sequence numbers identifying the site of substitution mutations or bracketing deletions are indicated.

and comprises the amino- and carboxy-terminal segments of the larger protein. The synthesis of a small (9S), unspliced mRNA is directed by a control element within the E1B region which activates late after infection. This mRNA codes a minor structural polypeptide (IX).

To probe the role of the E1B-55K polypeptide, 2 mutant viruses were constructed. The 1st is a deletion mutant, *dl*338, which lacks 524 base pairs (Figure 1). This alteration prevents synthesis of the related E1B-55K and 17K polypeptides, but not the E1B-21K species. The 2nd mutant, *pm*381, contains a single base-pair substitution at sequence position 2967 which generates a nonsense codon which only affects the E1B-55K polypeptide (Figure 1). These 2 mutants display identical growth properties (data not shown) which suggests that the observed phenotype of *dl*338 is due exclusively to the loss of the 55K moiety. This view gains further support from *pm*380 whose single base-pair change at sequence position 3275 destroys the 3′ splice site required for synthesis of the E1B-17K-coding mRNA (Figure 1) without affecting the amino acid sequence of the 55K polypeptide. This mutant grows indistinguishably from wild-type virus (data not shown). Apparently, the 17K polypeptide is not required for growth of the virus in cultured cells. Consistent with our results, Montell *et al.* (1984) have mutated the 5′ splice site required for synthesis of the mRNA encoding the 17K polypeptide and found the variant to be phenotypically wild-type.

Viruses such as *dl*338, which lack E1B functions essential for viral growth, can be propagated in 293 cells, a human embryonic kidney cell line which contains and expresses the Ad5 E1A and E1B regions (Graham *et al.* 1977). Their growth defect can then be analyzed on a standard transformed cell line such as HeLa. The growth kinetics of *dl*338 and its wild-type parent, *dl*309 were compared in HeLa cells (Figure 2). *dl*338 generates an approximately 100-fold reduced yield as compared to its parent. Transforming capabilities of *dl*338 and its parent were compared (Figure 3) in both primary rat embryo cells and an established rat cell line termed CREF cells (Fisher *et al.* 1982). *dl*338 was markedly defective for transformation of both cell types. We conclude that the E1B-55K polypeptide plays essential roles in both lytic growth and transformation by adenovirus.

2. EARLY GENE EXPRESSION AND DNA REPLICATION ARE NORMAL IN *dl*338-INFECTED HeLa CELLS

To assay early gene expression, RNAs were prepared from HeLa cells at 6 h after infection with either *dl*338 or 309. RNA blot analysis revealed no differences in the variety or levels of early viral mRNAs, and early proteins assayed by immunoprecipitation were also produced in normal quantities in mutant as compared to wild-type virus-infected cells (data not shown). DNA replication was also monitored, and found to be normal in *dl*338-infected HeLa cells (data not shown). Thus, early gene expression and DNA replication proceed normally in the absence of the E1B-55K polypeptide.

3. LATE VIRAL mRNAs ACCUMULATE TO REDUCED LEVELS IN *dl*338-INFECTED HeLa CELLS

Steady-state levels of cytoplasmic mRNAs were monitored by blot analysis at 12, 16 and 24 h after infection. At all times tested, *dl*338-infected cells contained reduced levels of mRNAs coded by the viral late transcription unit. Some late mRNA families were more severely affected than others. For example, the L4

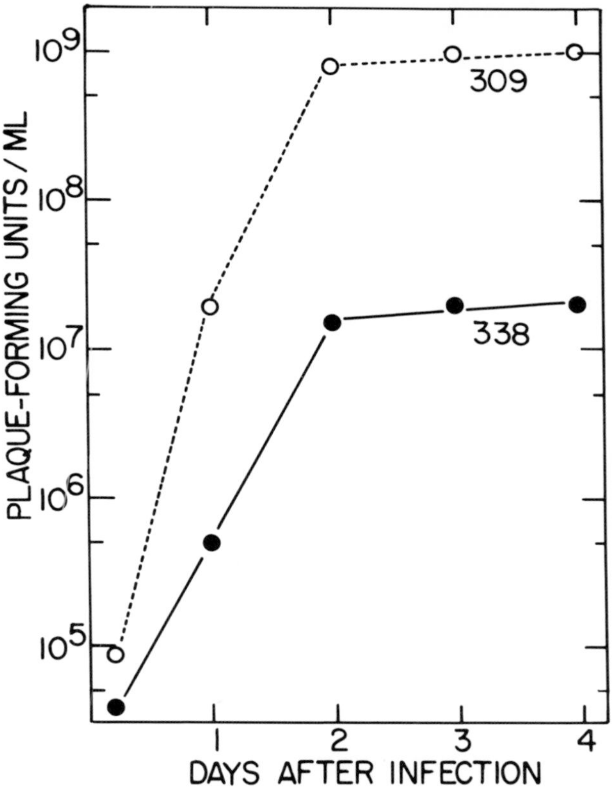

Fig. 2. Growth kinetics of mutant and wild-type viruses in HeLa cells. Cells were infected at a multiplicity of 3 plaque-forming units per cell with either *dl*338 or *dl*309, and the virus yield was assayed at the times indicated by plaque assay on 293 cells. Symbols: ●, *dl*338; ○, *dl*309.

family was reduced only 2- to 3-fold (data not shown) while the L5 family was reduced by a factor of about 15 (Figure 4A).

Transcription rates were measured to determine the basis for abnormal mRNA levels. Nascent RNA synthesized in isolated nuclei was labeled with ^{32}P-UTP, total nuclear RNA was prepared and hybridized to an L5-specific probe DNA (Figure 5A). Transcription rates for the L5 family were identical at 12 and 16 h, but at 20 and 24 h the late RNAs were transcribed at about 4-fold higher levels in wild-type than in mutant virus-infected cells. Reduced transcription rates can at least partially account for the lower mRNA levels in *dl*338-infected cells at 24 h after infection. However, transcription rates are normal at 12 and 16 h when

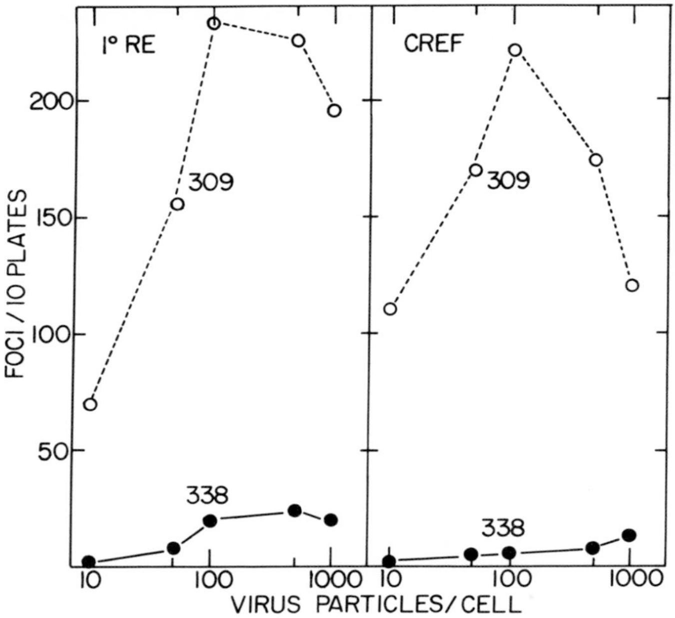

Fig. 3. Transformation of rat cells with mutant or wild-type viruses. Transformations were carried out using equilibrium density gradient-purified virions to infect either primary rat embryo cells or the established cloned rat embryo fibroblast cell line, CREF. Foci were counted 6 wk after infection. Symbols: ●, *dl*338; ○, *dl*309.

steady-state levels are already reduced. Therefore, the earliest pertubation of late mRNA levels in *dl*338-infected cells must be due to a post-transcriptional event.

The rate at which newly synthesized mRNAs accumulated in the cytoplasm of virus-infected cells was monitored to elucidate the nature of the post-transcriptional event. Cultures were labeled continuously with ^{3}H-uridine from 12–14.5 h after infection, and the rate of appearance of labeled L5-specific RNA in the cytoplasm was monitored by hybridization to a specific probe DNA (Figure 5B). Transcription rates were identical for mutant and wild-type viruses at this time (Figure 5A), so the slopes of accumulation plots are a direct measure of relative accumulation rates. L5 mRNAs accumulated in the cytoplasm of *dl*338-infected cells at a 3-fold slower rate than in wild-type virus-infected cells.

The reduced rate of late mRNA accumulation in *dl*338-infected cells could result from a primary defect in nuclear RNA processing. If polyadenylation were compromised, less RNA should be present in the poly A-plus nuclear fraction;

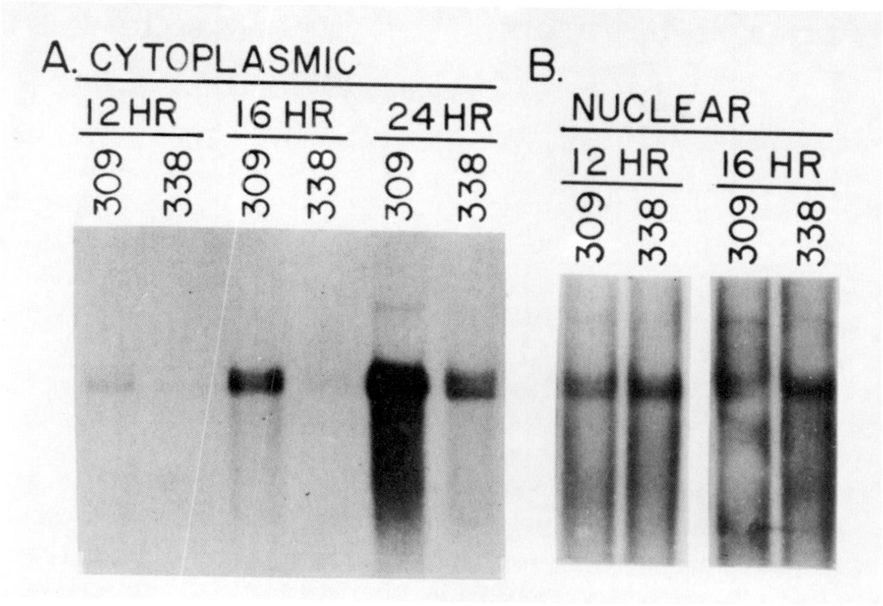

Fig. 4. RNA blot analysis of viral mRNA species present in HeLa cells after infection with mutant or wild-type viruses. Polyadenylated RNA was prepared at various times after infection from either cytoplasm (A) or nucleus (B) and analyzed using an L5-specific probe DNA.

if splicing were inhibited, less mature-sized RNA should be present and large precursors might be evident. Blot analysis of polyadenylated nuclear RNA using an L5-specific probe DNA indicated that similar quantities of identically-sized RNA species were present in mutant as compared to wild-type virus-infected cells (Figure 4B). The presumptive nuclear species cannot be cytoplasmic contaminants since there is little L5-specific cytoplasmic RNA available to contaminate the nuclear fraction of *dl*338-infected cells. Thus, the reduced rate of cytoplasmic accumulation is not due to a processing defect.

The reduction in accumulation rate of L5 mRNAs within *dl*338-infected cells could result from a reduced mRNA half-life. The accumulation plot in Figure 5B does not reach a plateau during the labeling period, suggesting that the mRNAs under study have a half-life in excess of 150 min. However, it is difficult to reach definitive conclusions about viral mRNA stability on the basis of the accumulation data since the rate of viral transcription was increasing during the labeling period. Therefore, a pulse-chase experiment was performed (Figure 5C).

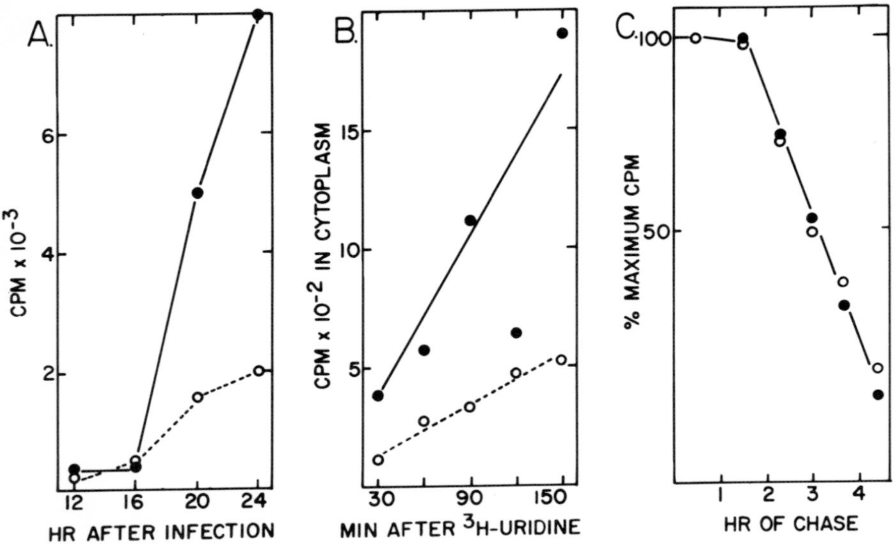

Fig. 5. L5-specific RNA metabolism in mutant or wild-type virus-infected HeLa cells. (A) Transcription rates. Nuclei were prepared from infected cells at the indicated times after infection, incubated for 15 min at 30 °C in the presence of ^{32}P-UTP, and nuclear RNA prepared. (B) Kinetics of mRNA accumulation in the cytoplasm. Cultures were labeled with ^{3}H-uridine beginning at 12 h after infection, portions were harvested every 30 min, and cytoplasmic RNA prepared. (C) mRNA half-life determination. Cells were labeled with ^{3}H-uridine for 30 min at 12 h after infection, chased in the presence of excess uridine and glucosamine, and cytoplasmic, polyadenylated RNA was isolated. All experiments utilized an L5-specific, single-stranded probe DNA. Symbols: ●, *dl*309; ○, *dl*338.

Cells were radiolabeled for 30 min with ^{3}H-uridine at 12 h after infection, and the chase was carried out in medium containing high levels of unlabeled uridine and glucosamine. The stability of L5 mRNA was monitored at 45-min intervals by hybridization of cytoplasmic, polyadenylated RNA to an L5-specific probe DNA. L5 mRNAs displayed the same half-life (about 3.2 h) in *dl*338 and *dl*309-infected cells. Therefore, it is unlikely that the *dl*338 phenotype is due to a primary effect on mRNA half-lives.

Polyadenylation, splicing and cytoplasmic L5 mRNA half-lives are normal in *dl*338-infected cells at 12–16 h after infection. We conclude that abnormal viral mRNA accumulation observed at this time in mutant-infected cells results from a defect in either transport or stabilization of mRNA as it first reaches the cytoplasm. We (Pilder *et al.* 1986) and others (Babiss *et al.* 1985) have recently published additional experiments which support this conclusion.

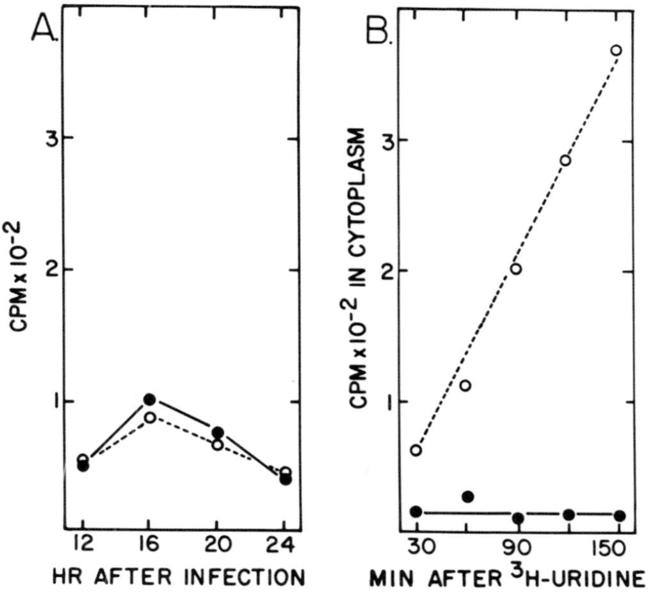

Fig. 6. Actin-specific RNA metabolism in mutant or wild-type virus-infected HeLa cells. (A) Transcription rates. (B) Kinetics of mRNA accumulation in the cytoplasm. Accumulation was measured from 16–18.5 h after infection using an actin-specific probe DNA. Symbols: ●, *dl*309; ○, *dl*338.

4. CELLULAR mRNAs CONTINUE TO ACCUMULATE IN *dl*338-INFECTED HeLa CELLS

Normally, host cell mRNAs continue to be transcribed but fail to accumulate in the cytoplasm of adenovirus-infected cells (Beltz & Flint 1979). As expected, transcription of the cellular actin gene continued in both wild-type and *dl*338-infected cells (Figure 6A). However, actin mRNA continued to accumulate in the cytoplasm of *dl*338-infected cells after its appearance had ceased in wild-type virus-infected cells (Figure 6B). Thus, the E1B-55K polypeptide plays a role in host cell shut-off. This conclusion is consistent with the work of Babiss & Ginsberg (1984) who have demonstrated the failure of E1B mutants to shut off host cell protein synthesis.

5. E1B-55K MUTANTS ARE COLD-SENSITIVE

Both *dl*338 and *pm*381 generate substantially reduced yields at 32°C as compared to 37°C (Table I) with an extended delay in the appearance of infectious progeny at the lower temperature. Their phenotype is not qualitatively altered at the lower temperature. Rather, the magnitude of the defects observed at 37°C is greater at

TABLE I

Replication of E1B-55K mutants is cold-sensitive

Virus	Plaque-Forming Units per ml 32°C	37°C	Ratio 37°/32°
*dl*309	6×10^8	2×10^9	3.3
*dl*338	7×10^6	9×10^7	13
*pm*381	7×10^6	9×10^7	13

HeLa cells were infected at a multiplicity of 3 plaque-forming units/cell and cultures harvested at 90 h after infection. Yields were determined by plaque-assay on 293 cells at 37°C.

32°C (data not shown). The mechanistic basis for the cold-sensitive phenotype is not yet clear. It is interesting to note that a series of cold-sensitive E1B mutants has been isolated previously (Ho *et al.* 1982), and a temperature-dependent phenotype may be a general consequence of E1B-55K mutations.

6. ROLE OF THE E1B-55K POLYPEPTIDE IN ADENOVIRUS-MEDIATED TRANSFORMATION

At present, the mechanistic role of E1B products in adenovirus-mediated transformation is unclear. An appealing view of transformation in which immortalizing and transforming oncogenes cooperate to induce complete transformation (Land *et al.* 1983, Ruley 1983) seems, at first glance, to fit well with the organization of the adenovirus chromosome. E1A has well established immortalizing functions (Houweling *et al.* 1980, Ruley 1983), and mutants with lesions in the E1B region are clearly defective for transformation (e.g. Figure 3). It is possible, however, that E1B products don't impact directly on the cell during transformation, but rather influence the efficiency of E1A expression. This assertion follows from two observations. First, van den Elsen *et al.* (1983) have found that cells containing only the E1A gene express less E1A mRNA than cells which express both E1A and E1B units. Second, Senear & Lewis (1986) have recently reported that a chimeric E1A gene which contains the E1A coding region appended to the mouse metallothionein promoter is able to fully transform rodent cells. The transformants express higher levels of E1A products than cells which contain the gene controlled by its normal promoter region.

Possibly, then, at least one function of E1B in transformation is to enhance expression of the E1A region. Given our present results from the lytic cycle, this effect could be exerted post-transcriptionally by the E1B-55K polypeptide. However, it is not at all clear that the lytic and transforming functions of the

55K moiety are similar. The E1B product exists in a complex with an E4-34K polypeptide during lytic viral growth (Sarnow *et al.* 1984), and this complex mediates efficient accumulation of late mRNAs (Halbert *et al.* 1985). In semi-permissive rodent cells, the 55K moiety is associated with the cellular oncogene p53 (Sarnow *et al.* 1982), and the E4 region is generally not present in transformed cell lines. Nevertheless, it is conceivable that its new association re-directs the post-transcriptional activity of the E1B-55K polypeptide from late viral mRNAs to the E1A species. We are presently testing this notion.

ACKNOWLEDGMENTS

This work was supported by grants from the American Cancer Society (MV-45) and the United States Public Health Service (CA38965). Keith Leppard was the recipient of an EMBO Postdoctoral Fellowship and is presently supported by an SERC Postdoctoral Fellowship, John Logan was supported by a Postdoctoral Fellowship from the Otsuka Pharmaceutical Co. and Thomas Shenk is an American Cancer Society Research Professor.

REFERENCES

Anderson, C. W., Schmitt, R. C., Smart, J. E. & Lewis, J. B. (1984) Early region 1B of adenovirus 2 encodes two coterminal proteins of 495 and 155 amino acid residues. *J. Virology 50*, 387–96.

Babiss, L. E. & Ginsberg, H. S. (1984) Adenovirus type 5 early region 1b gene product is required for efficient shutoff of host protein synthesis. *J. Virology 50*, 202–12.

Babiss, L. E., Ginsberg, H. S. & Darnell, J. E. (1985) Adenovirus E1B proteins are required for accumulation of late viral mRNA and for effects on cellular mRNA translation and transport. *Mol. Cell. Biol. 5*, 2552–8.

Beltz, G. A. & Flint, S. J. (1979) Inhibition of HeLa cell protein synthesis during adenovirus infection: restriction of cellular mRNA sequences to the nucleus. *J. Mol. Biol. 131*, 353–73.

Berk, A. J., Lee, F., Harrison, T., Williams, J. & Sharp, P. A. (1979) Pre-early Ad5 gene product regulates synthesis of early viral mRNAs. *Cell 17*, 935–44.

Borrelli, E., Hen, R. & Chambon, P. (1984) Adenovirus-2 E1A products repress enhancer-induced stimulation of transcription. *Nature 312*, 608–12.

Bos, J. L., Polder, L. J., Bernards, R., Schrier, P. I., van den Elsen, P. J., van der Eb, A. J. & van Ormondt, H. (1981) The 2.2 kb E1b mRNA of human Ad12 and Ad5 codes for two tumor antigens starting at different AUG triplets. *Cell 27*, 121–31.

Fisher, P. B., Babiss, L. E., Weinstein, I. B. & Ginsberg, H. S. (1982) Analysis of type 5 adenovirus transformation with a cloned rat embryo cell line (CREF). *Proc. Natl. Acad. Sci. USA 79*, 3527–31.

Graham, F. L., Smiley, J., Russell, W. C. & Nairu, R. (1977) Characteristics of a human cell line transformed by DNA from human adenovirus type 5. *J. Gen. Virol. 36*, 59–72.

Halbert, D. N., Cutt, J. R. & Shenk, T. (1985) Adenovirus early region 4 encodes functions required for efficient DNA replication, late gene expression and host cell shutoff. *J. Virology 56*, 250–7.

Ho, Y-S., Galos, R. & Williams, J. (1982) Isolation of type 5 adenovirus mutants with a cold-sensitive host range phenotype: genetic evidence of an adenovirus transformation maintenance function. *Virology 122*, 109–24.

Houweling, A., van den Elsen, P. J. & van der Eb, A. J. (1980) Partial transformation of primary rat cells by the leftmost 4.5% fragment of adenovirus 5 DNA. *Virology 105*, 537–50.

Jones, N. & Shenk, T. (1979) An adenovirus type 5 early gene function regulates expression of other early viral genes. *Proc. Natl. Acad. Sci. USA 76*, 3665–9.

Land, H., Parada, L. F. & Weinberg, R. A. (1983) Tumorigenic conversion of primary embryo fibroblasts requires at least two cooperating oncogenes. *Nature 304*, 596–602.

Montell, C., Fisher, E. F., Caruthers, M. H. & Berk, A. J. (1984) Control of adenovirus E1B mRNA synthesis by a shift in the activities of RNA splice sites. *Mol. Cell. Biol. 4*, 966–72.

Nevins, J. R. (1981) Mechanism of activation of early viral transcription by the adenovirus E1A gene product. *Cell 26*, 213–20.

Ruley, H. E. (1983) Adenovirus early region 1A enables viral and cellular transforming genes to transform primary cells in culture. *Nature 304*, 602–6.

Sarnow, P., Hearing, P., Anderson, C. W., Halbert, D. N., Shenk, T. & Levine, A. J. (1984) Adenovirus early region 1B 58,000-dalton tumor antigen is physically associated with an early region 4 25,000-dalton protein in productively infected cells. *J. Virology 49*, 692–700.

Sarnow, P., Ho, Y-S., Williams, J. & Levine, A. J. (1982) Adenovirus E1b-58kd tumor antigen and SV40 large tumor antigen are physically associated with the same 54kd cellular protein in transformed cells *Cell 28*, 387–94.

Senear, A. W. & Lewis, J. B. (1986) Morphological transformation of established rodent cell lines by high-level expression of the adenovirus type 2 E1A gene. *Mol. Cell. Biol. 6*, 1253–60.

van den Elsen, P. J., Houweling, A. & van der Eb, A. J. (1983) Morphological transformation of human adenoviruses is determined to a large extent by gene products of region E1A. *Virology 131*, 242–6.

Velcich, A. & Ziff, E. (1985) Adenovirus E1a proteins repress transcription from the SV40 early promoter. *Cell 40*, 705–16.

Zerler, B., Moran, B., Maruyama, K., Moomaw, J., Grodzicker, T. & Ruley, E. (1986) Adenovirus E1A coding sequences that enable *ras* and *pmt* oncogenes to transform cultured primary cells. *Mol. Cell. Biol. 6*, 887–99.

DISCUSSION

RIGBY: What is known about the subcellular localization of the E1B-55K protein?

LEPPARD: From the work of Sarion (Sarion *et al.* 1982 *Virology 120,* 510–517) it is predominantly nuclear in infected cells.

RIGBY: And is it associated with any discrete structures?

LEPPARD: I could not detect such associations in the pictures I have seen.

VAN DER EB: It is different in transformed cells. Ad5-transformed cells hold the E1B protein in the cytoplasm. It forms a complex with p53 that is present in a small discrete body that you can even see in the electron microscope.

RAPP: Does the 55K actually bind RNA, late messenger RNA?

LEPPART: Not that I know of. We have recently done experiments to examine the composition of hnRNP particles in infected cells using an antibody to one of the component proteins of these particles in uninfected cells, supplied to us by Dr. G. Dreyfuss (Northwestern University). As yet I have not found any differences in the composition of these particles from wild type or mutant infected cells.

VAN DER EB: Do you know whether there is any difference in the activity of 55K in the early and the late phase of infection? This should be in relation to the fact that, early, only 22S RNA is expressed and relatively little 19K, whereas later you have the smaller RNA that only expresses 19K so you get an enormous overexpression of 19K.

LEPPARD: There is no evidence to suggest that those two E1B proteins interact in any way. It is possible, based on the precendent of the E1A+E2 transcription units which each produce a set of proteins with interrelated functions that there may be some interplay at some point in infection. There may be less 55K late in infection because, as you mentioned, splicing of E1B mRNA changes late in infection.

BRUGGE: In the mutant virus infected cells, since the viral RNA cannot be transported, but it is synthesized at the same rate, and there does not seem to be any difference in degradation, don't you expect an accumulation of that RNA in the nucleus?

LEPPARD: Yes, we would expect that. We have to argue that the hnRNA that is not transported is degraded.

BRUGGE: Did you perform a pulse chase experiment in which the cells were labelled with tritiated uridine?

LEPPARD: That experiment examined cytoplasmic and not nuclear RNA.

IV. Gene Activation by Integration

Pim-1 Activation in T-Cell Lymphomas

Anton Berns, H. Theo Cuypers, Gerard Selten & Jos Domen

Moloney murine leukemia virus induces T-cell lymphomas in many mouse strains after a latency of 3 to 6 months. The virus, which does not contain an oncogene, probably exerts its neoplastic potential as an insertional mutagen of cellular proto-oncogenes. Since integration of proviruses occurs relatively randomly, proviral activation of an oncogene will require a large number of integration events. To obtain insight into the clonal composition and sequential selection processes active during progressive oncogenic transformation, we have examined a series of primary and transplanted lymphomas with respect to differentiation-specific and transformation-related markers. One of these markers, the *pim*-1 gene, was characterized in more detail.

Clonal composition of primary and transplanted lymphomas

In a tumor-prone system, each of the steps involved in tumor initiation and progression is expected to occur with a certain probability. Differences in the frequencies with which the various consecutive steps occur to a large extent determines the clonality and the homogeneity of the cell populations present in the outgrown primary tumor. Transplantation of these tumors can serve as a means of revealing the clonal composition and the interdependence of the various cell populations within that particular tumor. The following situations can be encountered:

1. The primary tumor is monoclonal and, upon transplantation, its genotype and phenotype do not change.
2. The primary tumor is oligoclonal (multiple independent transformation events). The different clones do or do not segregate upon transplantation.
3. The primary tumor consists of cells with a common ancestor, but several

Section of Molecular Genetics of the Netherlands Cancer Institute and the Department of Biochemistry of the University of Amsterdam, Plesmanlaan 121, 1066 CX Amsterdam, The Netherlands.

VIRAL CARCINOGENESIS, Alfred Benzon Symposium 24,
Editors: N. O. Kjeldgaard, J. Forchhammer, Munksgaard, Copenhagen 1987.

subclones grow out independently, some of which have a strong selective advantage upon transplantation.

To reveal the clonal composition of the lymphomas, we used the following markers: i) The proviral integration pattern; ii) The somatic rearrangements of the beta T-cell receptor and immunoglobulin heavy chain; and iii) The structural alteration of the c-*myc* and *pim*-1 genes. Each of the categories described above was found in Moloney MuLV-induced T-cell lymphomas. Oligoclonal tumors were predominantly found in mice with a relatively short latency of disease. In a number of tumors some of the subsequent events during tumor outgrowth could be traced back (Cuypers *et al.* 1986b). An instructive example is shown in Figure 1. Primary lymphomatous spleen cells of mouse #9 transplanted into syngeneic hosts gave rise to the development of a monoclonal tumor in the mesenteric lymph nodes and spleens of recipient mice 1–4. The monoclonality of these lymphomas was evident from the molecular markers used (equal hybridization intensities of all proviral integrations, and of the rearranged c-*myc* and *pim*-1 genes as compared to the germ-line fragments; rearrangement of the beta T-cell receptor genes at both alleles (absence of germ-line allele), and a single allele immunoglobulin heavy chain D-J rearrangement). The rearrangements observed near c-*myc* and *pim*-1 were caused by proviral integrations. This resulted in the enhanced expression of both genes and suggests that the subsequent activation of these genes each contributed to the proliferative capacity of the tumor. Furthermore, this transplantation experiment shows that an additional selection took place during progression of this lymphoma. Upon transplantation of this primary tumor, a subclone, marked by an additional proviral integration site, grew out in independent transplantation experiments (marked by a triangle in Figure 1). This subclone, which was below detection level in the primary tumor, appeared to be the only constituent of the outgrown transplant supporting the notion that this clone was probably composed of a more malignant cell type. Since MuLV in this system acts as a highly efficient insertional mutagen, it is possible that the additional proviruses observed after transplantation are physically linked to genes directly involved in tumor progression. We are currently testing this hypothesis by characterizing the subset of proviral integration sites emerging after transplantation of the primary tumor.

THE *pim*-1 GENE

Frequency of proviral integrations near pim-*1*

The *pim*-1 locus was identified as a preferred integration site in MuLV-induced T-cell lymphomas (Cuypers *et al.* 1984). Proviral integrations near *pim*-1 are

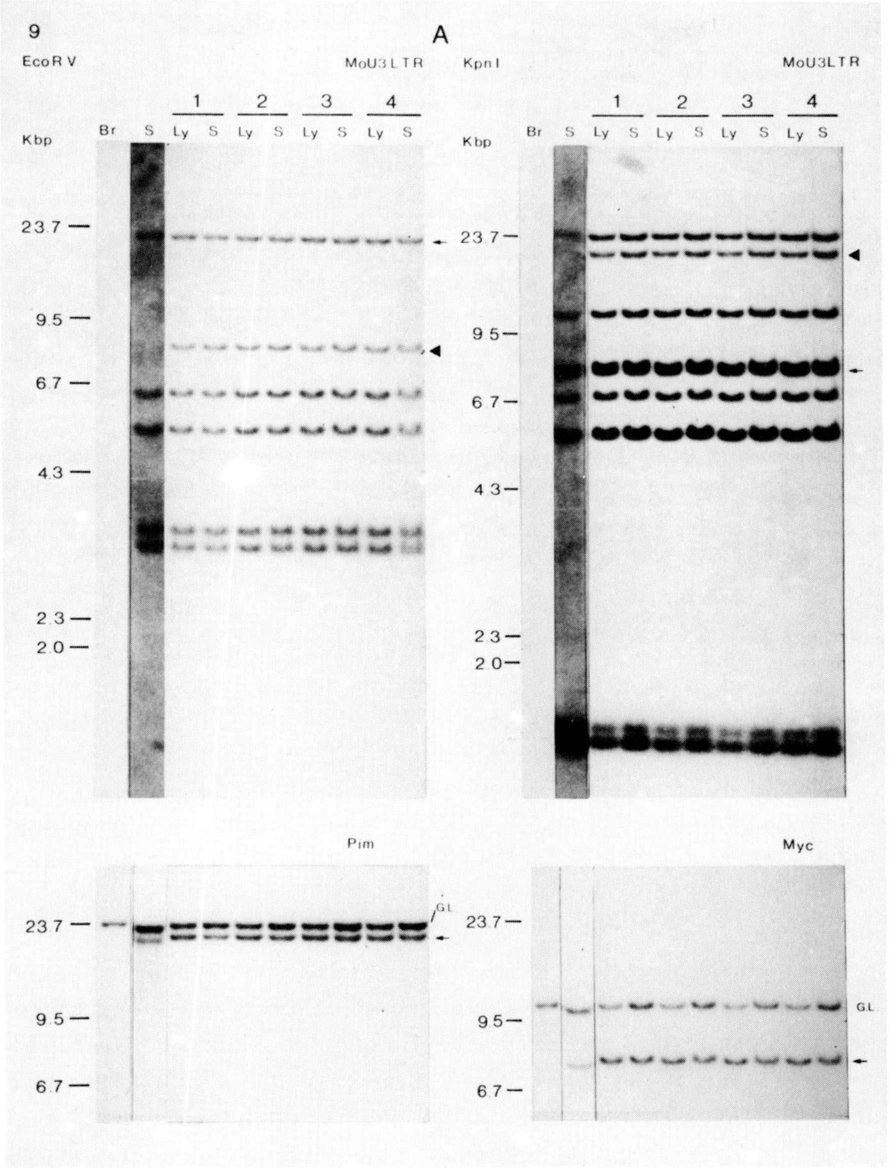

Fig. 1. Analysis of lymphoma DNA from primary and transplanted tumors of BALB/c mouse #9. Southern blot analysis of DNA from brain (Br), spleen, (S) and lymph nodes (Ly). Second lane of each panel: Primary tumor (S). Numbers 1–4 refer to recipient mice #1–4. Probes and digests are indicated in the figure.

TABLE I

*Frequency (MuLV) proviral integrations near c-*myc *and* pim-*1*

virus strain	mouse strain	induction mode	latency (months)	number tested	%int pim	%int myc	%int pim+myc	%int total
MoMuLV	C57Bl							
MCF247	BALB/c		<6	63	46%	37%	19%	63%
	AKR	NBI						
MCF1233	BALB/c	NBI						
	C57Bl	NBI	7–12	45	11%	13%	0%	24%
AKV	AKR	GT						
MoMuLV	Mov-1	GT						
MCF	C57Bl	NBI						
	BALB/c	NBI	>12	16	0%	0%	0%	0%
C57/MLV	C57Bl	MT						

Abbreviations: NBI, newborn infected; GT, germline transmitted; MT, milk transmitted. The tumors positive for *pim*-1 and *myc* comprise oligoclonal tumors as well as tumors with *pim*-1 and *myc* activated within the same cell.

found with a frequency similar to that of integrations near c-*myc* (Selten *et al.* 1984). The frequency of *pim*-1 (and c-*myc*) activation appears to be correlated with the latency of the disease (Table I). The shorter the latency, the higher the frequency of integrations near c-*myc* and *pim*-1. The observed frequency of approximately 50% for *pim*-1 activations in early T-cell lymphomas might be an overestimation as a substantial number of these tumors are oligoclonal in origin (Cupyers *et al.* 1986b), and often only one of the constituent clones harbors an activated *pim*-1.

Chromosomal location of pim-*1*

The chromosomal location of the *pim*-1 gene was determined both in mouse and human. In the mouse the *pim*-1 gene was localized on chromosome 17 by Southern blot analyses of DNAs obtained from mouse-hamster somatic cell hybrids (Hilkens *et al.* 1986). Using restriction fragment length polymorphisms between various recombinant inbred mouse strains, C. Blatt (personal communication) localized the *pim*-1 gene close to the mouse MHC. Recently, Ark and Bennet (personal communication), using their t-haplotype recombinants, mapped *pim*-1 betweem the alpha crystallin and tw12 marker. Fine mapping placed *pim*-1 0.59 cM proximal to tw12 at approximately 5 cM from H-2.

Using human-mouse and human-hamster somatic cell hybrids, the *pim*-1 gene

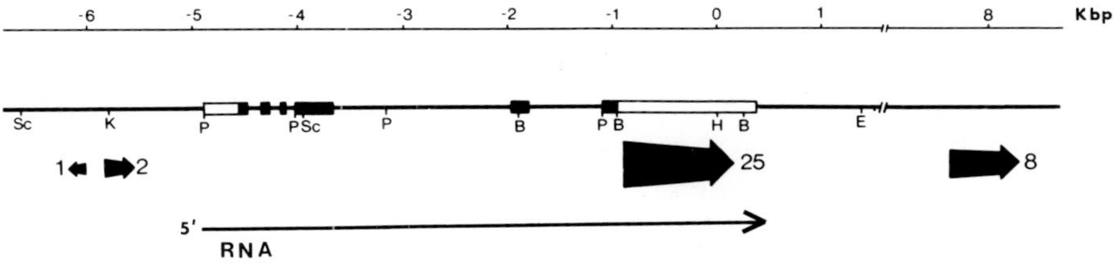

Fig. 2. Proviral integration sites within the *pim*-1 domain. Exons are indicated by blocks, coding sequences are filled in. Large arrows denote regions in which proviral integrations were detected. Figures beneath arrows give the number of lymphomas with an integrated provirus in that region.

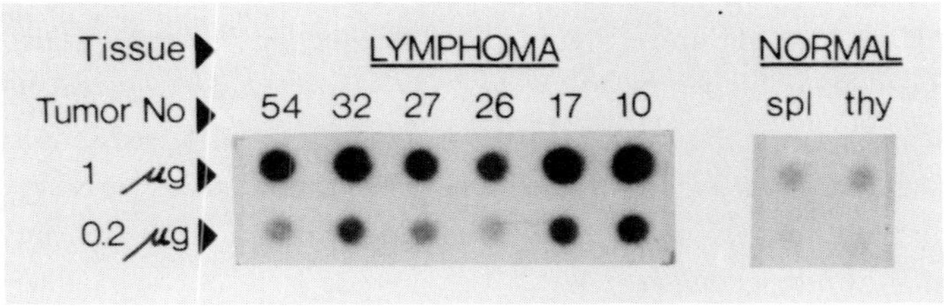

Fig. 3. Expression level of *pim*-1 in tumor and non-tumor tissues.

was assigned to the pter-q12 region of human chromosome 6 (Cuypers *et al.* 1986a). Croce and colleagues recently assigned the human *pim*-1 gene to the p12 region of chromosome 6 using *in situ* hybridization. They observed chromosomal translocations to the region in some acute leukemias with the concomitant enhanced levels of *pim*-1 mRNA (Nagarajan *et al.* 1986). Experiments are in progress to compare the mouse and human *pim*-1 genes, and to map the chromosomal breakpoint relative to the *pim*-1 gene.

Mode of activation

Pim-1 activation seems to occur almost exclusively by enhancement (Selten *et al.* 1985). As depicted in Figure 2, the majority of proviruses are found to be integrated in the 3' region of the *pim*-1 transcription unit. All the proviruses in this region have the same transcriptional orientation as *pim*-1. Furthermore, when 5 randomly chosen tumors, with integrations in the *pim*-1 transcription unit, were analyzed in detail with respect to the position of integration, a high preference for distinct nucleotide positions was found: 3 of 5 proviruses were integrated within 3 nucleotides (2 at the same position), whereas the other 2 were inserted 5 nucleotides apart and 50 nucleotides from the first 3. Integration within the *pim*-1 transcription unit results in a truncated *pim*-1 mRNA, which is terminated at the polyA addition signal of the 5' LTR of the provirus (Selten *et al.* 1985). Proviruses at the 5' end of the *pim*-1 transcription unit were found in both orientations. The proviruses in the "promoter insertion" orientation were internally deleted, while the 5' and 3' LTRs were intact. The provirus (a single example) with the opposite transcriptional orientation seemed structurally intact. We have been unable to show the covalent linkage between U5-LTR sequences and *pim*-1 sequences in tumors with integrations 5' of *pim*-1.

The expression levels of *pim*-1 were highest when integration occurred within the *pim*-1 transcription unit (Figure 3). Whether this is caused by a higher transcription rate or by a higher stability of the mRNA has not yet been determined. The high *pim*-1 expression level seen in tumors with proviruses in the 3' region of *pim*-1 (Selten *et al.* 1986) might have provided the driving force for their selective outgrowth, and therefore be responsible for the high incidence of proviral integrations in this region.

Expression pattern of pim-*1*

The expression of *pim*-1 was monitored in pre- and postnatal tissues. *Pim*-1 was expressed as a 2.8 kb mRNA in hematopoietic tissues throughout development. The highest levels were found during fetal life in liver, and later in spleen and thymus (Figure 4). After birth the level of expression decreased. In the adult a moderate level of the 2.8 kb *pim*-1 mRNA was found in bone marrow, thymus and spleen. High amounts of a smaller sized *pim*-1 mRNA were detected in adult testis and in placenta early in gestation. *Pim*-1 expression is also seen in a number of cell lines, both of lymphoid and myeloid origin. In normal T-cells, *pim*-1 could be induced by concanavalin-A, suggesting that the *pim*-1 gene product is involved in T-cell proliferation (Mally *et al.* 1985).

Structure of the pim-*1 gene*

The *pim*-1 gene contains 6 coding exons (see Figure 2), as determined by sequencing genomic and cDNA clones (Selten *et al.* 1986). The total length of these 6 exons is approximately 2.6 kbp, close to the mRNA size of 2.8 kb, which includes the polyA tail. There are 2 polyA addition signals, approximately 50 nucleotides apart, of which only the second signal appears to be used, as was determined by sequencing a series of cDNA clones. A similar situation has been observed for *int*-1 (van Ooyen & Nusse 1984).

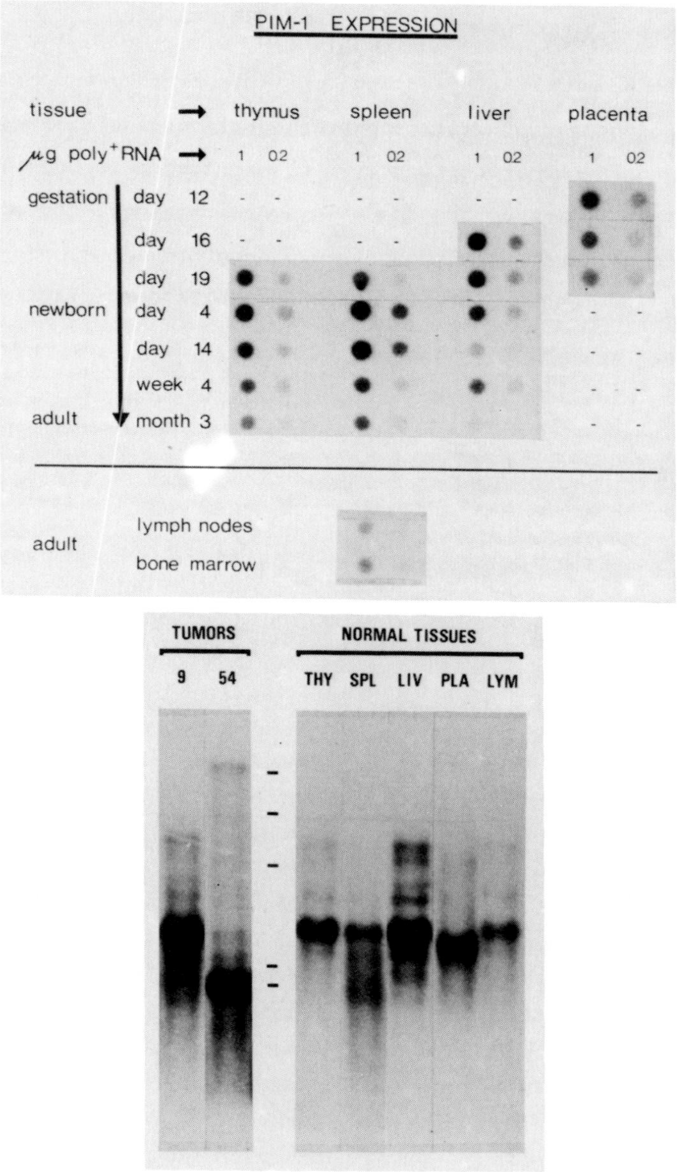

Fig. 4. Expression of *pim*-1 during development. Level and size of *pim*-1 mRNAs. A. The dot-blot only shows the positive hybridizing tissues. No expression was detected in Brain, lung, heart, muscle, salivary gland, uterus, testis (up to 3 weeks). B. Northern blot analysis shows the sizes of *pim*-1 transcripts in normal (embryonic) tissues and in tumors. In tumor 54 proviral integration occurred within the *pim*-1 transcription unit.

PUTATIVE PIM-1 PROMOTER REGION

```
ggcggggcccgcaacgccttatgcaaatagggcgtcttctggttggctgaagcgcgcccg  60
agggcggggccggggagaggggtgtagccgcgagggggcggagcggagggggag-ggccc 120
tggtcccgccgcctccccgccccctctccgccctccgc-gccgagccagcccggctcccc 180
acctcg-c-tcccggagaggccccgccccgtccccgcccgccgccgccctccccagcggc 240
gcgtccgcccccttactcctggcggcggggcgagccgggcgtctgcagcagcggccgtgg 300
cGGCGGAGGAGGCTGGAGGGGAGTCGGCAGTGCCGGCGGCGGGATCGGCAGTAGCAGCAG 360
 -------->
 Exon 1
```

Fig. 5. Sequence of the *pim*-1 promoter region. The start of exon 1 corresponds with the 5' position of 2 independently obtained cDNA clones (corresponds with 5'exon of 419 bp). S-1 analysis (not shown) indicated various initiation sites. SP-1 binding sites and lymphoid-specific octamer motif are underlined.

S-1 analysis showed a heterogeneous pattern for the 1st exon with protected fragments of 450, 430, and 360 nucleotides. Although the presence of an additional 5' exon cannot be excluded, transcription can most likely be initiated at different positions with this region. No TATA box sequences are found upstream from the putative transcriptional initiation sites. However, this region displays resemblance with other genes lacking TATA box sequences (Figure 5). The GC-rich motif CCGCCC, which has been shown to bind specifically to the transcription factor Sp1 (Dynan & Tjian 1985), is repeated eight times in this region, which also contains the lymphoid-specific octamer sequence ATGCAAAT (Singh *et al.* 1986).

The predicted pim-*1 protein*

The nucleotide sequence of the *pim*-1 gene revealed a single open reading frame, encoding 313 amino acids. A single AUG initiator codon in exon 1 was preceded by stopcodons in all 3 reading frames. Long untranslated regions are present both at the 3' (1300 nucleotides) and the 5' ends (approx. 300 nucleotides). *In vitro* translation of pSP6-produced *pim*-1 mRNA confirmed the size of the *pim*-1 protein. A polypeptide of approximately 33 kd was produced close to the calculated mol. wt. of 35.5 kd. The proposed primary structure of the *pim*-1 protein does not reveal a hydrophobic amino terminus or any other signal sequence which might direct the *pim*-1 gene product to a membrane within the cell. However, the primary structure of *pim*-1 shows remarkable amino acid

```
PREDICTED PIM-1 AMINO ACID SEQUENCE

MLLSKINSLAHLRARPCNDLHATKLAPGKEKEPLESQYQVGPLLGSGGFGSVYSGIRVAD  60
                                           LG G FG V

NLPVAIKHVEKDRISDWGELPNGTRVPMEVVLLKKVSSDFSGVIRLLDWFERPDSFVLIL 120
   VAIK                                    I LLD

ERPEPVQDLFDFITERGALQEDLARGFFWQVLEAVRHCHNCGVLHRDIKDENILIDLSRG 180
                                            HRD    NIL

EIKLIDFGSGALLKDTVYTDFDGTRVYSPPEWIRYHRYHGRSAAVWSLGILLYDMVCGDI 240
  K  DFG   L                *PE   Y        *VWS GILL *   G

PFEHDEEIIKGQVFFRQTVSSECQHLIKWCLSLRPSDRPSFEEIRNHPWMQGDLLPQAAS 300
PF                       L   C     P DRP F

EIHLHSLSPGSSK*                                               313
```

Fig. 6. Amino acid sequence of *pim*-1 and regions of homology with protein kinases. Conserved regions are denoted under the *pim*-1 sequence.

sequence homology with protein kinases (Hunter & Cooper 1985, Van Beveren & Verma 1985) (Figure 6). The highest overall homology was observed with the gamma-subunit of phosphorylase kinase (Reiman *et al.* 1984). In the NH_2-terminal region of the catalytic domain the homology was high with *mos,* whereas at the carboxyl terminus close resemblance with *abl* was observed. All the domains conserved in protein kinases are found in *pim*-1 (Figure 6). A single consensus sequence for N-linked glycosylation was found (Pesciotta *et al.* 1981). At present we cannot predict whether *pim*-1 is a serine/threonine or a tyrosine kinase. The tyrosine present at position 416 in v-*src*, which can serve as an autophosphorylation site in all tyrosine kinases, is also conserved in *pim*-1. Probably, *pim*-1 belongs to a class of protein kinases which fulfills a role in the intracellular transduction of signals involved in the proliferative stimulation of lymphoid and myeloid cells. Since proviral integrations leave the protein-encoding domain of *pim*-1 intact, the increased expression of *pim*-1 is apparently sufficient to confer a selective (growth) advantage to the cell. Experiments are in progress to identify the enzymatic activity and the transforming activity of the *pim*-1 gene.

REFERENCES

Cuypers, H. Th., Selten, A., Berns, A. & Geurts van Kessel, A. H. M. (1986a) Assignment of the human homologue of *pim*-1, a mouse gene implicated in leukemogenesis, to the pter-q12 region of chromosome 6. *Hum. Genet. 72,* 262–5.

Cuypers, H. T., Selten, G., Quint, W., Zijlstra, M., Robanus-Maandag, E., Boelens, W., van Wezenbeek,

P., Melief, C. & Berns, A. (1984) Murine leukemia virus-induced T-cell lymphomagenesis: Integration of proviruses in a distinct chromosomal region. *Cell 37,* 141–50.

Cuypers, H. Th., Selten, G., Zijlstra, M., de Goede, R., Melief, C. J. & Berns, A. J. (1986b) Tumor progression in MuLV induced T-cell lymphomas: Monitoring of clonal selections with viral and cellular probes. *J. Virol.* (in press).

Dynan, W. S. & Tjian, R. (1985) Control of eukaryotic mRNA synthesis by sequence-specific DNA-binding proteins. *Nature 316,* 774–8.

Hilkens, J., H. Th. Cuypers, Selten, G., Kroezen, V., Hilgers, J. & Berns, A. (1986) Genetic mapping of the *pim*-1 putative oncogene to mouse chromosome 17. *Somat. Cell. Mol. Genet. 12,* 81–8.

Hunter, T. & Cooper, J. A. (1985) Protein-tyrosine kinases. *Ann. Rev. Biochem. 54,* 897–930.

Mally, M., Vogth, M., Swift, S. E. & Haase, M. (1985) Oncogene expression in murine splenic T-cells and in murine T-cell neoplasms. *Virology 144,* 115–26.

Nagarajan, L., Louie, E., Tsujimoto, Y., ar-Rushdi, A., Huebner, K. & Croce, C. M. (1986) Localization of the human *pim* oncogene (PIM) to a region of chromosome 6 involved in translocations in acute leukemias. *Proc. Natl. Acad. Sci. USA 83,* 2556–60.

Pesciotta, D. M., Dickson, L. A., Showalter, A. M., Eikenberry, E. F. De Crombrugghe, B., Fietzek, P. P. & Olsen, B. R. (1981) Primary structure of the carbohydrate-containing regions of the carboxyl propepetides of type I procollagen. *FEBS Letts. 125,* 170–4.

Reimann, E. M., Titani, K., Ericsson, L. H., Wade, R. D., Fischer, E. H. & Walsh, K. A. (1984) Homology of the gamma subunit of Phosphorylase b kinase with c-AMP dependent protein kinase. *Biochem. 23,* 4185–92.

Selten, G., Cuypers, H. T. & Berns, A. (1985) Proviral activation of the putative oncogene *pim*-1 in MuLV-induced T-cell lymphomas. *EMBO J. 4,* 1793–8.

Selten, G., Cuypers, H. T., Boelens, W., Robanus-Maandag, E., Verbeek, J., Domen, J., Van Beveren, C. & Berns, A. (1986) The primary structure of the putative oncogene *pim*-1 shows extensive homology with Protein Kinases. *Cell 46*, 603–11.

Selten, G., Cuypers, H. T., Zijlstra, M., Melief, C. & Berns, A. (1984) Involvement of c-*myc* in MuLV-induced T-cell lymphomas in mice: frequency and mechanisms of activation. *EMBO J. 3,* 3215–22.

Singh, H., Sen, R., Baltimore, D. & Sharp, P. A. (1986) A nuclear factor that binds to a conserved sequence motif in transcriptional control elements of immunoglobulin genes. *Nature 319,* 154–8.

Van Beveren, C. & Verma, I. M. (1985) Homology among oncogenes. *Curr. Top. Microbiol. Immunol. 123,* 73–98.

Van Ooyen, A. & Nusse, R. (1984) Structure and nucleotide sequence of the putative mammary oncogene *int*-1, a mouse gene implicated in mammary tumorigenesis. *Cell 39,* 233–40.

Voronova, A. F. & Sefton, B. M. (1986) Expression of a new tyrosine protein kinase is stimulated by retrovirus promoter insertion. *Nature 319,* 682–5.

DISCUSSION

YANIV: What kind of cells did you transfect with the *pim* gene?

BERNS: Rat embryo fibroblast, and NIH 3T3, the current transformation assay. We tried with the different constructs, actually, both with LTR upstream in the region of the *pim*-1 promoter and downstream, and none of them would give any transformation. The next experiments we did was to cotransfect NIH 3T3 with *ras* and *pim* and screened for transformation, because of the *ras*, and looked for the *pim* gene expression. It would not give any reasonably sized messenger or anything like that. We have to go back now using a cDNA clone and see if we get a different result, or go back with a retrovirus in lymphoid cell and see if we get effects there.

YANIV: Did you try to introduce it into T cells?

BERNS: No, we did not. We are just at the moment trying to do that, using retroviruses.

BRUGGE: According to your model for the tumorigenicity of murine leukemia viruses, the cells that would be derived from the first clonal expansion, i.e., the first oncogene activation, might be expected to be resistant to infection by the resident virus, because they would be expressing gp70. So, the second virus that you invoke to be the second transforming event might be a different virus, say, a recombinant virus and able to infect those cells. Have you looked at the relatedness of the two resident proviruses where you find integration upstream from both *myc* and *pim?*

BERNS: First of all, the fact that you find a number of proviruses in one cell type makes it clear that you can get more than one infection round. Secondly, it is amazing, but many of these integrated viruses are defective in their glycoprotein. We have cloned three or four and checked their biological activity. They appeared biologically inactive, and we could assign that to a mutation within the envelop region. If that happens frequently that would circumvent the problem. Thirdly, it is not only the MCF which is activating these oncogenes. For example, 50% of the proviruses near the *myc* oncogene are of ecotropic origin, whereas at the *pim* locus only 20% are ecotropic.

WEINBERG: There actually is a mechanism whereby you can get multiple proviruses of the same type into cells in spite of the existence of the interference barrier. First you put one good provirus into a cell. That provirus will start making virus particles which can leave the cell and reinfect that cell immediately prior to the establishment of interference. Eventually, you cannot get any more viruses in, but there is a window of time where virus can still enter. That could explain your observations. What I wanted to ask or suggest was that the precedents with most of the kinases, but not all of them, have been that structural alterations in the protein are especially effective in creating oncogenes, and I am wondering whether it would not be to your advantage to see whether certain kinds of truncations might enhance the oncogenicity of the protein, this protein having only relatively weak transforming effects in the absence of such structural change.

BERNS: The protein is not altered in the tumor, because the sequence is exactly the same as the genomic germ line sequence.

WEINBERG: But you might have more potent effects if you were to change its structure. There is great precedent for that with all the other kinases: by changing them structurally, often with C-terminal deletions, they acquire biological activity.

BERNS: Yes, we do not have such a domain like for *scr,* where a tyrosine might be regulatory for the rest of the gene. This is only a 35 K protein and there is actually little more left than the protein kinase domain. But it certainly would be an approach. On the other hand, we still did not test the proper system to see the activity of the *pim* gene. I think we should first go ahead and try to get it expressed properly in transgenic mice and in lymphoid cells and then see what the effect is, because at least *in vivo* it has to be a selective advantage, otherwise we would not find it.

LOWY: Have you considered the possibility that down at the 3′ end of your gene there might be a suppressor analogous to the one which occurs in *fos?*

BERNS: Yes, that might be, but we do not know yet.

LOWY: It is just that the integration sites are striking in their location.

BERNS: Right, and it is clearly associated with a higher expression level. But that could also be simply a stabilization. As far as I remember, Irwin Wagner has *fos* transgenic mice, from which the ones where transcription is terminated in the LTR have significant levels of *fos* mRNA in different organs.

LOWY: Would you speculate on why you do not tend to see common integration sites going on in long latency tumors?

BERNS: Maybe other people should speculate on that. I think some of these very late tumors might not even carry proviral integrations. The main mechanism there may be a proliferative stimulus by the virus, which gives a higher level of proliferation of several cell types which finally results in transformation. And that would be quite different from what we see here. So, that might have nothing to do with an integration event, but be the result of a long-term proliferative stimulation of the cells.

WONG-STAAL: Along that line, could it also be a hit and run. Are the *pim* and *myc* genes also activated anyway in those tumors?

BERNS: *Pim*-1 is always activated if we do see a proviral integration. There is only one exception where we see an activated *pim* without proviral integration. For *myc* the same holds true except that we see many activated *myc*'s without proviral integration. I showed you in one slide that 37% of the lymphomas had a proviral integration near *myc;* in that same group of tumors, 63% of the tumors had a significantly higher level of *myc* expression. So, for *myc* there are more ways to be highly expressed. But with *pim* there is a very good correlation between proviral integration and enhanced expression.

SHERR: To my knowledge, most if not all the tyrosine kinases which have been described to have transforming activity have undergone some sort of mutational change which has affected the activity of the enzyme. But I have the impression listening to you that the cDNA sequence you have determined is identical to the genomic sequence?

BERNS: That is correct.

SHERR: How many cDNAs have actually been sequenced? Have you sequenced more than one?

BERNS: We have sequenced two cDNAs and we have sequenced one genomic clone with the provirus, and one germ line copy.

SHERR: And they are always identical?

BERNS: They are identical. On the other hand, *pim*-1 could well be a serine kinase. So, it might be better to refer to it as an analogy with *mos,* for example, than with the tyrosine kinases.

RAPP: I wonder what you mean by activation in tumor. You have cells which are presumably progenitors of the tumor cells, expressing the RNAs. Did you not have expression in normal lymphoid tissue?

BERNS: Sure, at low level.

RAPP: So what is the pattern. You would say the cells which are growing in tumors have a much higher level, (and) how much?

BERNS: That is slightly complicated, because not all these tumors are monoclonal. But if you have the clean cases you might think about 5, 10, 20, in that range. The problem is that we do not know the real tissue to compare to. We have examples of lymphomas in this series, where you do not have a proviral integration, and absolutely have no *pim*-1 expressions. So, we are simply using the dimes and the spears as a reference, but why should that be a reference for the tumor cells? We do not have a pure bottle with progenitor cells.

RAPP: Is the *pim* gene actually a growth factor-regulated gene?

BERNS: If you add T-cell mitogenes you will enhance the *pim*-1 levels. With ConA you will get high levels of *pim*.

RAPP: And your tumor cells, do they have different growth properties?

BERNS: We do not know.

Activation of the Cellular Oncogenes *int*-1 and *int*-2 by Proviral Insertion of the Mouse Mammary Tumor Virus

Roel Nusse[1], Albert van Ooyen[1,2] & Arnoud Sonnenberg[3]

Mammary tumors can be induced in experimental animals by a variety of methods: hormones, chemical carcinogens, irradiation, retroviruses, and by the introduction of oncogenes into the germ line (Stewart *et al.* 1984). Some of these systems offer the opportunity to examine genetic alterations in mammary tumors at the molecular level. Mutations in the *ras* oncogenes have been implicated in some forms of chemical carcinogenesis (Sukumar *et al.* 1983), but it is as yet unclear how hormones, which are thought to play an important role in human breast cancer, act at the genomic level. Finding more and novel oncogenes could be of great value in the elucidation of the mechanism of hormonal carcinogenesis.

The large majority of the cellular oncogenes that we know of have been discovered with the aid of retroviruses and therefore the Mouse Mammary Tumor Virus (MMTV) could be the system of choice to define novel mammary oncogenes. MMTV is a replicating retrovirus of B-type morphology. The virus is found in the milk of high mammary tumor incidence mouse strains such as C3H. It is, as an endogenous provirus, present in the germ line of every inbred strain of mice, but most endogenous proviruses, except for the Mtv-2 locus in the GR strain, are expressed at a low level and are not thought to play a role in tumorigenesis.

MMTV does not transform cells in culture, and it induces tumors only after a long period of latency, ranging from 4 to 12 months. The tumors are clonal with respect to the integrated proviral copies, a finding that suggested that the tumors were outgrowths of cells with proviruses integrated at sites that predispose

---

Divisions of Molecular Biology[1] and Tumor Biology[3], The Netherlands Cancer Institute (Antoni van Leewenhoekhuis), Plesmanlaan 121, 1066 CX Amsterdam, The Netherlands. [2]Present Address: Gist Brocades, Postbus 1, Delft, The Netherlands.

VIRAL CARCINOGENESIS, Alfred Benzon Symposium 24,
Editors: N. O. Kjeldgaard, J. Forchhammer, Munksgaard, Copenhagen 1987.

to tumorigenesis. Experimental support for this model came from the startling discovery that another slowly oncogenic retrovirus, Avian Leukosis Virus (ALV), integrates near the cellular *myc* gene, and activates *myc* transcription (Hayward *et al.* 1981). The proposal that retroviruses lacking viral oncogenes induce tumors by activating host cell oncogenes suggests a general stategy to uncover novel oncogenes (Varmus 1984). Host cell DNA adjacent to an integrated provirus, preferentially from a tumor with a single aquired provirus is cloned as recombinant DNA in *E. coli.* The host genomic sequences are then examined for more independent proviral integrations. This would indicate that cells with a provirus in that particular domain have a selective growth advantage over cells with proviruses at other regions. The model also predicts that a gene within the common integration domain is transcriptionally active as a consequence of proviral activation. Ultimately, a biological assay in which the gene of interest is introduced into other cells may tell us whether a gene with oncogenic potential has been identified.

In this paper we shall review our efforts to isolate oncogenes activated by MMTV in murine mammary cancers. We show that a gene, called *int*-1 (Nusse & Varmus 1982), has been found that conforms to many of the predictions listed above, and we report on the structure of the gene. We also present some data on activation of the *int*-2 gene, isolated by Peters *et al.* (1983) in a set of cell lines from a mammary tumor, and discuss the implications of our findings.

FINDING A COMMON INTEGRATION SITE FOR MMTV PROVIRUSES IN MAMMARY TUMORS

Mammary tumors induced by MMTV contain variable numbers of integrated proviruses, and only rarely a single insertion. The right half of such a single provirus was, with its adjacent cell DNA, cloned into a bacteriophage vector. From the cellular sequences, we selected other DNA fragments and used them as probes to isolate a number of overlapping phage clones from a library of normal cell DNA. Thus we obtained over 30 Kb of cloned DNA of the locus, which we called *int*-1 (Nusse & Varmus 1982). Single copy DNA probes were prepared by subcloning appropriate fragments, and were used to screen mammary tumors for disruption of this locus as a consequence of proviral insertion. It appeared that many tumors contained novel restriction fragments hybridizing to *int*-1 probes. The fragments also annealed to probes from the MMTV provirus, indicating that they arose due to proviral insertion.

To assign *int*-1 to a specific mouse chromosome, we used a probe from the

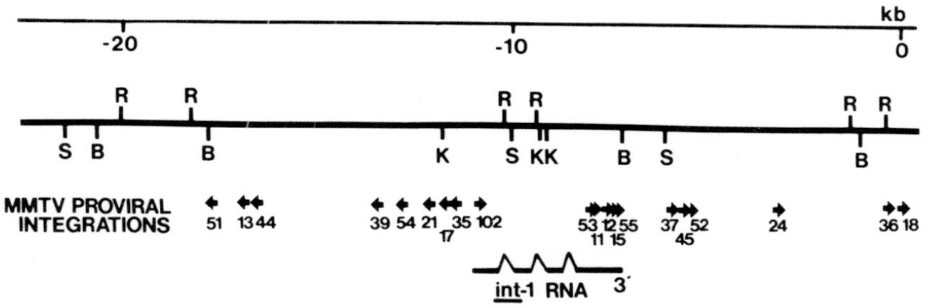

Fig. 1. Map of the *int*-1 region and integrated proviruses in different tumors. The restriction map is compiled from previous studies with minor modifications. Transcription is from left to right. Sites of proviral integrations and direction of transcription of the proviruses are indicated. R: EcoRI; B: BglII; S: SacI; K: KpnI.

region to detect restriction enzyme fragments containing *int*-1 sequences in Chinese hamster-mouse somatic cell hybrids segregating mouse chromosomes. Segregation analysis of murine *int*-1 restriction fragments with isoenzyme markers revealed 96% concordant segregation with *int*-1 with aryl sulphatase 1, an enzyme marker on chromosome 15 (Nusse *et al.* 1984).

In Figure 1 we present the *int*-1 restriction map, including the position and orientation of the integrated MMTV proviruses. The insertions are found over a distance of about 20 Kb, surrounding the portion of the locus that is transcribed into polyadenylated RNA in mammary tumors. Proviruses integrated downstream from the transcriptional unit, which is from left to right on this map, are all in the same orientation as the *int*-1 gene; proviruses found upstream from the gene are in the opposite orientation (Nusse *et al.* 1984). One exception has been found: tumor 102 contains an MMTV provirus upstream from *int*-1 in the same transcriptional orientation (Figure 1).

TRANSCRIPTION OF *int*-1 AND THE MECHANISM OF ACTIVATION

Most tumors with proviral integrations surrounding the *int*-1 gene contain a polyadenylated transcript of 2.6 Kb. This RNA has not been found in most normal tissues of the mouse, including normal mammary glands. The mechanism of transcriptional activation of *int*-1 seems to be enhancement rather than promoter insertion, because the MMTV insertions at *int*-1 are, in the large majority of tumors, always pointing away from the gene. This configuration may also be relevant for the mechanism of enhancement: If the enhancer is located in the U3 region of the viral LTR, the orientation away from the *int*-1 gene would avoid

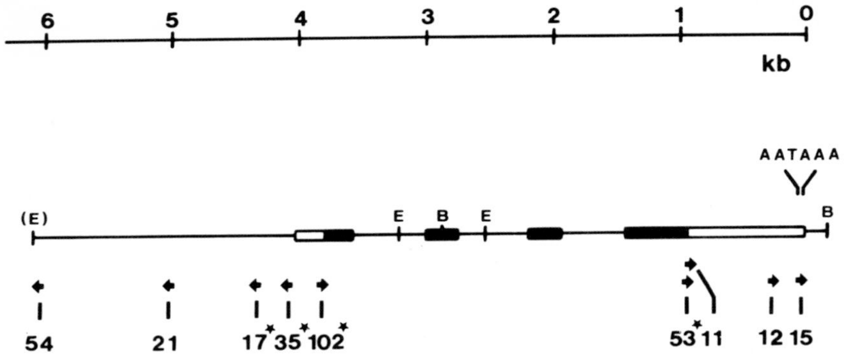

Fig. 2. Structure of the *int*-1 gene. The intron-exon structure as determined by nuclease S1 mapping and DNA sequencing is shown (Wan Ooyen & Nusse 1984). Exons are indicated by blocks; coding sequences are black. The putative polyadenylation signal (AATAAA) at the 3' end of the gene and a putative RNA polymerase initiation signal (TATAA) are indicated. Sites and orientations of proviral integrations in tumors having an integration in the depicted area are represented by arrows (cf. Figure 1). Tumor numbers are below arrows. B: BamHI; R: EcoRI. The EcoRI site in brackets is not present in the chromosomal DNA but derived from DNA cloning.

the interposition of the MMTV promoter between the viral enhancer and the *int*-1 promoter. Since enhancers are thought to act only on a proximal promoter, a proviral insertion pointing towards *int*-1 cannot active the gene. The insertion in tumor 102, as it replaces the *int*-1 promoter, is compatible with this notion.

THE STRUCTURE OF THE GENE

The transcriptionally active region is located in the middle of the proviral insertion clusters, giving an indication of the position of the gene. The exact structure of *int*-1 was determined by DNA sequencing, nuclease S1 mapping and by isolating cDNA clones (Figure 2) (Van Ooyen & Nusse 1984, Fung *et al.* 1985, Rijsewijk *et al.* 1986).

The position of the 5' end of *int*-1 is not yet clear. Two overlapping exons have been found, sharing a splice donor site at the 3' end but differing at their 5' end. The smaller variant is bounded by a splice acceptor consensus sequence, but is also preceded by a TATA box 30 nucleotides upstream from an adenine that might serve as a cap site of *int*-1 RNA.

THE *int*-1 PROTEIN

In order to deduce the amino acid sequence of the protein encoded by *int*-1 we aligned the nucleotide sequence of the 4 exons, searched for translational start

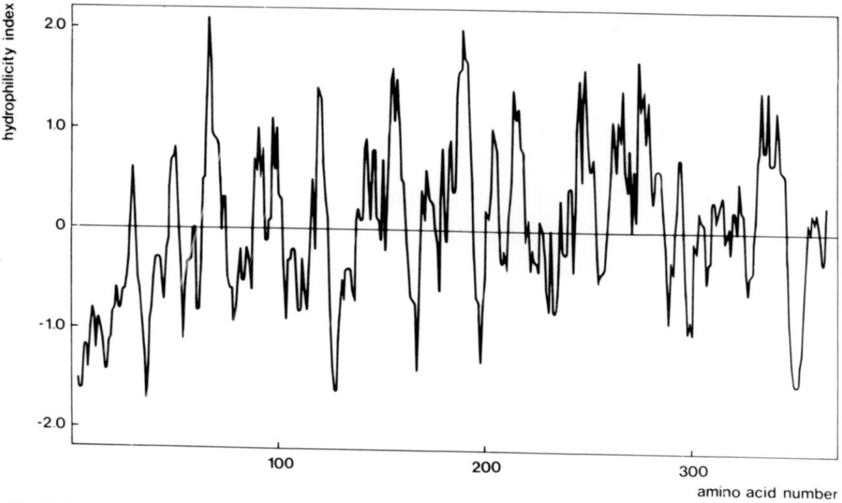

Fig. 3. Hydropathicity profile of *int*-1 protein.

and stop signals and translated the nucleotide sequence in the 3 possible open reading frames. The first AUG found in the derived mRNA sequence is followed immediately by a stop codon. The second AUG, is followed by a large open reading frame traversing the 4 aligned exons. This frame encodes a protein of 370 amino acids, stops at a UGA stop codon within the last exon and is followed by a relatively large untranslated trailer sequence (Figure 2). By *in vitro* transcription and translation of the available *int*-1 cDNA clones, a protein of 37000 daltons has been synthesized, which can be immunoprecipitated by antisera raised against peptides corresponding to the *int*-1 amino acid sequence (Rijsewijk *et al.* 1986, Van de Vijver, personal communication). These sera are currently being used to search for *int*-1 protein expression in cells and tissues.

Comparison of the amino acid sequence of the deduced protein with over 2000 sequences present in the database of the University of California at San Diego did not reveal any significant similarities (R. F. Doolittle, personal communication), so that little can be predicted about the function of the *int*-1 protein. The recently published sequence of the *int*-2 gene also has no homology to *int*-1 (Moore *et al.* 1986). A salient feature of the *int*-1 protein sequence is the strong hydrophobic leader (Figure 3), and a high cysteine content of the carboxy terminal half. These two properties are shared with growth factors and their receptors.

PROVIRAL INTEGRATIONS NEAR THE *int*-1 TRANSCRIPTIONAL UNIT

Near the 5' end of the transcriptional unit proviruses are most proximal to the gene in tumors 17, 35 and 102. Restriction enzyme fragments containing part of the MMTV provirus and neighboring *int*-1 gene sequences were cloned from these tumors and the precise insertion sites were determined. The integrated proviruses of tumors 17 and 35 were located in front of the 1 exon (Figure 2). The proviral integration in tumor 102, which is in the same transcriptional orientation as the *int*-1 gene, was found within the 1 exon, but before the AUG start codon of the *int*-1 protein. Many mammary tumors have a provirus integrated close to the 3' end of the *int*-1 transcripts, all in the same orientation as the gene. In addition to the tumors shown in Figure 2, we have found five others with an insertion within the last exon. Restriction site mapping showed that, of all these tumors, no. 53 has its provirus closest to the central portion of the *int*-1 gene. DNA sequencing of the cloned host-proviral DNA junction showed that integration of MMTV in this tumor was only 5 nucleotides downstream from the TGA stop codon of the *int*-1 protein.

MOLECULAR CLONING OF HUMAN *int*-1

Probes from the transcriptional domain of *int*-1 detect, at reduced hybridization stringency, homologous sequences in various other organisms including man and Drosophila (Nusse *et al.* 1984). We have employed this high degree of conservation to isolate a molecular clone of the human *int*-1 homolog and to determine its nucleotide sequence (Van 't Veer *et al.* 1984, Van Ooyen *et al.* 1985). The conservation between mouse and man is, as expected, maximal in the protein encoding domain of the gene. Only four amino acid changes were found, all in the hydrophobic leader of the *int*-1 protein. None of these substitutions, however, influence the hydrophobic nature of the leader domain. Surprisingly, the conservation of *int*-1 is not limited to the coding domain. Parts of the introns are more than 75% conserved, and so is the region upstream from the gene (Figure 4). Evolutionary conservation on non-coding DNA may indicate some function in regulation. For *int*-1, this prediction cannot be tested unless the normal function of the gene in normal cells has been elucidated. By means of somatic cell hybrids, the human *int*-1 homolog has been mapped on chromosome 12 (Van 't Veer *et al.* 1984).

THE MECHANISM OF TRANSCRIPTIONAL ACTIVATION OF THE *int* GENES

The transcriptional orientation of MMTV proviruses with respect to the *int* genes in tumors has been taken as evidence that an enhancer on the MMTV genome

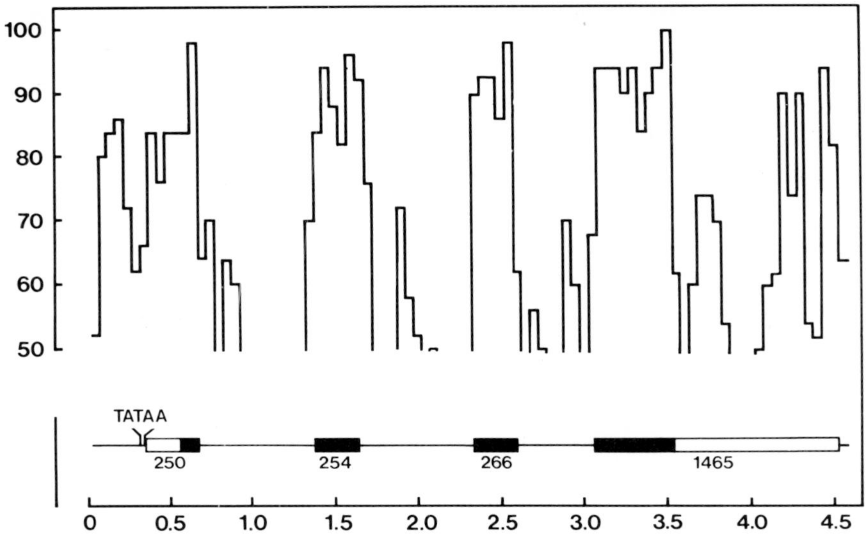

Fig. 4. Overall homology between the human and the mouse *int*-1. The human and the mouse genes were compared in blocks of 50 nucleotides with the aid of a computer, without introduction of gaps in either of the sequences. Regions with > 50% homology are shown. Exons are indicated by blocks, coding region in black.

is responsible for transcriptional activation of the *int* promoter. In other systems, it has been shown that the MMTV LTR contains a transcriptional enhancer whose the activity is dependent on steroid hormones (Chandler *et al.* 1983, Hynes *et al.* 1983, Majors & Varmus 1983). This raised the possibility that transcription of the *int* genes is also under the control of steroid hormones. We have investigated this in a set of cell lines established from a mammary tumor containing an MMTV insertion upstream and in the opposite orientation of the *int*-2 gene (Peters *et al.* 1983, Dickson *et al.* 1984). These cell lines, called RAC lines (Sonnenberg *et al.* submitted), show morphological differences upon *in vitro* culture, despite being of clonal origin. They can also serve as an *in vitro* model for progression of mammary tumors, as the original polygonal from the RAC cells induces moderately differentiated adenocarcinomas in syngeneic animals, whereas the elongated variants produce highly malignant carcinosarcomas. There is even a third cell type, of large cuboidal morphology, which is not tumorigenic.

Polygonal cells have retained many characteristics of epithelial cells: they express keratins amd milkfat globule membrane antigens, contain surface microvilli and desmosomes and form domes when grown in the presence of glucocorticoid hormones.

EXPRESSION OF *int*-2 DURING PROGRESSION *IN VITRO*

The presence of *int*-2 transcripts in our cell lines was checked by isolating polyadenylated RNA from cells cultured under normal conditions. Four RNA species hybridizing to the *int*-2 probe were detected, of 3.2, 3.0, 1.8 and 1.4 kb, but, surprisingly, only in the polygonal cell line. Neither the cuboidal cells nor the elongated cells showed any detectable amounts of RNA hybridizing to the *int*-2 probe. The loss of expression was not due to deletion of the provirus or the gene, as we had not detected alterations in the restriction pattern.

We assessed whether the synthetic glucocorticoid hormone dexamethasone would elevate *int*-2 and MMTV transcription in the polygonal and in the elongated cells. Hormone was added to cells which had grown for 48 h in medium depleted of steroids, and total RNA was isolated after 3 h of steroid treatment, followed by RNA blotting. The levels of the *int*-2 transcripts were not influenced by the hormone; they remained at the same level in polygonal cells and were still undetectable in the elongated cells. MMTV transcription, however, could be stimulated by dexamethasone in all lines, most strikingly in elongated cells where levels slightly lower than in untreated polygonal cells were reached.

These results suggest that the expression of *int*-2 and MMTV in the absence of steroid hormone is restricted to mammary tumor cells which retain differentiation characteristics of epithelial cells. In other cells, MMTV can be induced by hormone but this does not suffice to activate the adjacent *int*-2 gene. The high constitutive expression of MMTV and *int*-2 in polygonal cells may be caused by a viral enhancer which responds to a factor specific for epithelial cells of certain tissues, such as the mammary gland. This factor and the enhancer domain on the DNA could be different from the dexamethasone receptor and its binding site on the viral LTR. Alternatively, these components could be identical but not dependent on exogenously added hormone. This explantation, however, cannot account for the lack of *int*-2 expression in steroid-treated elongated cells. The presence of a cell type- and tissue-specific enhancer on the MMTV LTR can also account for the relatively high expression in mammary tissue of MMTV-LTR fusion genes in transgenic mice (Stewart *et al.* 1984, Ross & Solter 1985).

THE ROLE OF *int*-2 IN TUMORIGENESIS

The restricted activity of the MMTV enhancer has profound consequences on transcription of *int*-2; it is turned off when polygonal cells convert to cuboidal cells. Hence, expression of *int*-2 is correlated with tumorigenicity, as polygonal cells produce adenocarcinomas while the cuboidal cells are poorly oncogenic. Nevertheless, when the cuboidal cells switch to the highly malignant elongated cells, *int*-2 is not re-expressed, and is therefore not involved in progression. Peters *et al.* (1984) have also shown that this gene is involved in early stages of hormone-dependent mammary tumorigenesis.

CONCLUDING REMARKS

The results presented in this paper show that the transposon tagging strategy has been fruitful in finding genes activated by proviral insertion of MMTV. The 2 *int* genes, while unrelated to each other, share some intriguing properties: they are highly conserved in evolution, are not expressed in most normal mouse tissues, and are unique among the oncogene family. The latter poses some problems in designing experiments to assay for the biological property of these genes; most known oncogenes transform fibroblasts, whereas the *int*-1 gene does not (Rijsewijk *et al.* 1986). Possibly, the *int* genes are specific for mammary epithelial cells in their oncogenic action, but transformation parameters for epithelial cells are poorly defined. Moreover, we have shown that expression of *int*-2 is not required, and perhaps is even inhibitory, for progression of mammary tumor cells. Another interesting observations regarding the function of these genes had been made by Peters *et al.* (1986) who showed that, quite often, both *int*-1 and *int*-2 are provirally activated in mammary tumors of the BR6 strain.

A further elucidation of the mechanism of action of the *int* genes in tumorigenesis will need the availability of highly specific antibodies to their protein products, an *in vitro* model system such as the RAC cells described here, and a better understanding of the role of these genes during normal development. Perhaps the high degree of evolutionary conservation of the *int*-1 gene will then be helpful; by isolating homologous genes from the 'pet' organisms of developmental biologists – *C. elegans* and *Drosophila* – we may get some clues as to the normal function of genes whose oncogenic action is so uniquely mammalian.

REFERENCES

Chandler, V. L., Mater, B. A. & Yamamoto, K. R. (1983) DNA sequences bound specifically by glucocorticoid receptor in vitro render a heterologous promoter hormone responsive in vivo. *Cell 33,* 489–99.

Dickson, C., Smith, R., Brookes, S. & Peters, G. (1984) Tumorigenesis by mouse mammary tumor virus: Proviral activation of a cellular gene in the common integration region *int*-2. *Cell 37,* 529–36.

Fung, Y. K. T., Shackleford, G. M., Brown, A. M. C., Sanders, G. S. & Varmus, H. E. (1985) Nucleotide sequence and expression in vitro of cDNA derived from mRNA of *int*-1, a provirally activated mouse mammary oncogene. *Mol. Cell Biol. 5,* 3337–44.

Hayward, W. S., Neel, B. G. & Astrin, S. M. (1981) Activation of a cellular oncogene by promoter insertion in ALV-induced lymphoid leukosis. *Nature 290,* 475–80.

Hynes, N., Van Ooyen, A. J. J., Kennedy, N., Herrlich, P., Ponta, H. & Groner, B. (1983) Subfragments of the large terminal repeat cause glucocorticoid responsive expression of mouse mammary tumor virus and of an adjacent gene. *Proc. Natl. Acad. Sci. USA 80,* 3637–41.

Majors, J. & Varmus, H. E. (1983) A small region of the mouse mammary tumor virus long terminal repeat confers glucocorticoid hormone regulation on a linked heterologous gene. *Proc. Natl. Acad. Sci. USA 80,* 5866–70.

Moore, R., Casey, G., Brookes, S., Dixon, M., Peters, G. & Dickson, C. (1986) Sequence, topography and protein coding potential of mouse *int*-2: a putative oncogene activated by mouse mammary tumor virus. *EMBO J. 5,* 919–24.

Nusse, R., Van Ooyen, A., Cox, D., Fung, Y. K. T. & Varmus, H. E. (1984) Mode of proviral activation of a putative mammary oncogene (*int*-1) on mouse chromosome 15. *Nature 307,* 131–6.

Nusse, R. & Varmus, H. E. (1982) Many tumors induced by the mouse mammary tumor virus contain a provirus integrated in the same region of the host genome. *Cell 31,* 99–109.

Peters, G., Brookes, S., Smith, R. & Dickson, C. (1983) Tumorigenesis by mouse mammary tumor virus, evidence for a common region for provirus integration in mammary tumors. *Cell 33,* 369–77.

Peters, G., Lee, A. E. & Dickson, C. (1984) Activation of cellular gene by mouse mammary tumour virus may occur early in mammary tumour development. *Nature 309,* 273–5.

Peters, G., Lee, A. L. & Dickson, C. (1986) Concerted activation of two potential proto-oncogenes in carcinomas induced by mouse mammary tumor virus. *Nature 320,* 628–631.

Rijsewijk, F. A. M., van Lohuizen, M., van Ooyen, A. & Nusse, R. (1986) Construction of a retroviral cDNA version of the *int*-1 mammary oncogene and int expression *in vitro. Nuc. Acids Res. 14,* 693–702.

Ross, S. R. & Solter, D. (1985) Glucocorticoid regulation of mouse mammary tumor virus sequences in transgenic mice. *Proc. Natl. Acad. Sci. USA 82,* 5880–4.

Stewart, T. A., Pattengale, P. K. & Leder, P. (1984) Spontaneous mammary adenocarcinomas in transgeneic mice that carry and express MTV/*myc* fusion genes. *Cell 38,* 627–37.

Sukumar, S., Notario, V., Martin-Zanca, D. & Barbacid, M. (1983) Induction of mammary carcinomas in rats by nitroso-methylurea involves malignant activation of H-ras-1 locus by single point mutations. *Nature 306,* 658–61.

Van Ooyen, A., Kwee, V. & Nusse, R. (1985) The nucleotide sequence of the human *int*-1 mammary oncogene; evolutionary conservation of coding and non-coding sequences. *EMBO J. 11,* 2905–9.

Van Ooyen, A. & Nusse, R. (1984) Structure and nucleotide sequence of the putative mammary oncogene *int*-1; proviral insertions leave the protein-encoding domain intact. *Cell 39*, 233–40.

Van 't Veer, L. J., Geurts van Kessel, A., van Heerikhuizen, H., van Ooyen, A. & Nusse, R. (1984) Molecular cloning and chromosomal assignment of the human homolog of *int*-1 a mouse gene implicated in mammary tumorigenesis. *Mol. Cell Biol. 4,* 2532.

Varmus, H. E. (1984) The Molecular Genetics of Cellular Oncogenes. *Ann. Rev. Genet. 18,* 553–612.

DISCUSSION

WILLUMSEN: As far as I remember, the hormone-induced tumors of the GR mouse are not very transplantable. In the experiments I know of, each tumor goes into four mice, and that is the amount you have to use in order to get successful transplantation. Have you done any experiments of this type? How many of these cells does it take to keep the tumor growing?

NUSSE: We have not really done that. But that is true, you need quite high numbers of cells to keep the cells growing even in the presence of hormones.

WILLUMSEN: Do these cells express the endogenous mammary tumor virus?

NUSSE: Yes, they do.

RIGBY: Can you say anything more about the expression of *int*-1 and *int*-2 during embryogenesis, do you know anything about timing or cell type specificity?

NUSSE: Yes, but it is not my own data mostly. We have seen *int*-1 expression in testis, but as Don Blair just pointed out almost any oncogene is expressed in the testis. And in embryos *int*-1 is expressed between day 9 and day 12 of development and more recently it is found to be expressed in developing brain. *int*-2 is expressed in embryos around day 7 1/2, and also in differentiating F9 cells that go into the endodermal lineage. We are going to do *in situ* hybridization experiments to see where exactly it is expressed in the testis and also in the embryos.

VENNSTRÖM: You said that the *int*-1 protein is potentially glycosylated and it has a signal peptide-like sequence. Can you make the protein in normal cells and get it secreted?

NUSSE: The amounts of protein that are found in the tumor cells and also in the cell lines are probably quite low, and we have not seen the protein yet in a normal cell. I guess we have to overexpress the protein in order to detect it and to see whether it is processed and glycosylated. I would like to do it in these cell lines that are derived from mammary tumors, because that is really where *int* exerts its action.

YANIV: Is there any activation of *int*-1 or *int*-2 in human breast tumors?

NUSSE: We have looked quite extensively at human breast tumors and we have not seen expression of *int*-1. Either it is not involved at all in human breast cancer, or one can almost postulate, in parallel to the mouse model, that it is activated early in tumorigenesis and not necessarily expressed any more in the later phases.

FORCHHAMMER: Except for the parallel that the hormone dependency is also present in many of the human mammary carcinomas. So there is no *a priori* reason to believe that you are in such a late stage in the human situation, because one out of three cases is hormone-dependent.

NUSSE: Yes, but on the other hand, human mammary tumor cells are not much like mouse mammary tumors. The mouse tumors are derived from alveolar cells and arise during pregnancy, whereas most of the human tumors are derived from the ductal cells.

CORY: You made retroviruses containing *int*-1. Can you remind me of what happened when you did infections with that virus?

NUSSE: We have infected 3T3 cells which do not become transformed. We have infected some mouse mammary gland cells, more in particular the cuboidal cells that I showed you, also with the idea that maybe *int*-1 and *int*-2 can complement each other. We have not really seen good effects except for a morphological change. So the cuboidal cells, if they acquire *int*-1 by infection of a virus, become more elongated, but I doubt whether that is the real mechanism of action in view of the fact that during normal progression into those elongated cells the *int*-2 gene expression is lost. What we would like to do is take the primary mammary gland cells and then infect them with the virus, but those experiments are difficult.

LEPPARD: If the *int*-1 protein is a secreted protein and you have not been able to detect it by other methods, does the culture medium from *int*-1-transformed cells have any specific growth-conferring properties for untransformed mammary cells?

NUSSE: Those are experiments that we are also trying to do. The cells that I showed you, the polygonal cells, can be grown in serum-free medium, so we are trying to collect medium from cells to see whether we can stimulate the growth of other cells. But then, of course, the controls are essential in that experiment and antibodies are very important to have. We are also making the protein in all kinds of expression vector systems like yeast and more recently Baculovirus to see whether we can isolate it and then do those types of assays.

V. Gene Activation by Translocation

Chromosome Aberration and Oncogene Activation in Two Histologically Related Human and Rat B-Cell Tumors

Janos Sümegi, Warren Pear, Stanley Nelson, Gunilla Wahlström, Sigurdur Ingvarsson, Cecilia Melani, Anna Szeles, Francis Wiener & George Klein

Activation of a cellular oncogene by chromosomal translocation which brings an oncogene under the influence of a highly active chromosome region appears to play a pivotal role in the genesis of certain hematopoetic and lymphoid tumors (Klein 1983). This is most consistently seen in mouse plasmacytoma (MPC) where the MPC-associated specific chromosomal translocations bring the *c-myc* oncogene under the influence of the immunglobulin heavy (IgH) or, more rarely, the kappa light chain gene (Klein 1983). Closely homologous translocation activates the *c-myc* oncogene in human Burkitt lymphoma (BL) and less frequently in other human B-cell lymphomas (Leder *et al.* 1983, Croce & Klein 1985). The *c-myc* is transcriptionally activated by the integrated proviral DNA LTR region in avian bursal lymphoma (Hayward *et al.* 1981). The critical effect of the translocation and proviral DNA integration is the constitutive expression of the *c-myc* oncogene. The purpose of our work over the past 2 years has been to investigate how stringent and general is the connection between DNA rearrangement, *c-myc* activation and the genesis of B-cell derived tumors. Our specific question was; is the perturbation of *c-myc* expression a consistent feature in the genesis of B-cell tumors of the same histological type. The nearest counterpart of the MPC is the human multiple myeloma (MM), plasma cell leukemia (PCL) and the rat spontaneous immunocytoma (RIC).

Department of Tumor Biology, Karolinska Institute, S-104 01 Stockholm, Sweden. Supported by PHS grant No 2RO1 CA 14054-12 awarded by National Cancer Institute, The Swedish Cancer Society and the Balthzar von Platen Foundation.

VIRAL CARCINOGENESIS, Alfred Benzon Symposium 24,
Editors: N. O. Kjeldgaard, J. Forchhammer, Munksgaard, Copenhagen 1987.

CYTOGENETICS

In human MM and PCL the pattern of chromosomal aberration is quite variable. The loss or the translocation of chromosome 8 appears to be a frequent chromosomal change (Itani *et al.* 1970, Ferti *et al.* 1984). In F. Mittelman's catalogue on chromosomal translocation in human tumors, the translocation between the chromosomes 11q21 and 14q32 occurs in approximately 30–40% of MMs and PCLs (Mittelman 1983). This translocation is also seen in human chronic lymphatic leukemia (CLL) and both small and large cell lymphoma. The DNA sequence involved in the translocation has been cloned and partially characterized by C. Croce's group at the Wistar Institute (Tsujimoto *et al.* 1985).

The other tumor we analyzed was the spontaneous immunocytoma which occurs at an unusually high frequency in the LOU/Ws1 inbred rat strain (Brazin *et al.* 1972). The tumors originate from the ileocecal lymph nodes. The incidence of RIC is 30% for the male and 16% for the female. The tumors appear in the affected animals in 9th month after birth. About 60% of the tumors secrete monoclonal IgM, IgG, IgD, IgA, IgE and Bence-Jones proteins. 10% of the tumors secrete only Bence-Jones protein (Bazin 1985). It is noteworthy that the majority of tumors secrete IgG or IgE. The non-random chromosomal aberration observed in the RIC is an elongated chromosome 6. The presence of the elongated chromosome 6 is consistent with a translocation between chromosomes 6 and 7 (Wiener *et al.* 1982). The distal segment of the q-arm of chromosome 7 breaks in band 3.2–3.3 and the telomeric part is translocated onto the deleted telomeric region of chromosome 6.

THE *C-MYC* GENE IS AMPLIFIED AND HIGHLY EXPRESSED IN SOME CASES OF PLASMA CELL LEUKEMIA

DNAs from bone marrow and/or peripheral lymphocytes of 26 patients with MM or PCL were cleaved with BamHI restriction enzyme, subjected to electrophoresis on 0.7% agarose gel and transferred to nitrocellulose filters. The nitrocellulose filters were hybridized with 32P-labeled human *c-myc* cDNA, Ryc 7.4 (Erikson *et al.* 1983). Cleavage with BamHI, known to cut the DNA outside of the *c-myc* coding region, resulted in only 1 *myc*-specific band, upon autoradiography, corresponding to the expected 27 kb germ-line size. In 2 DNA samples (PCL-B8 and PCL-B16) obtained from leukemic cells of plasma cell leukemia patients, a strongly labeled band was apparent (Sumegi *et al.* 1985). This suggested an amplification of the *c-myc* oncogene in 2 of the 5 plasma cell leukemia-derived DNA samples. DNA amplification is usually accompanied by elevated expression

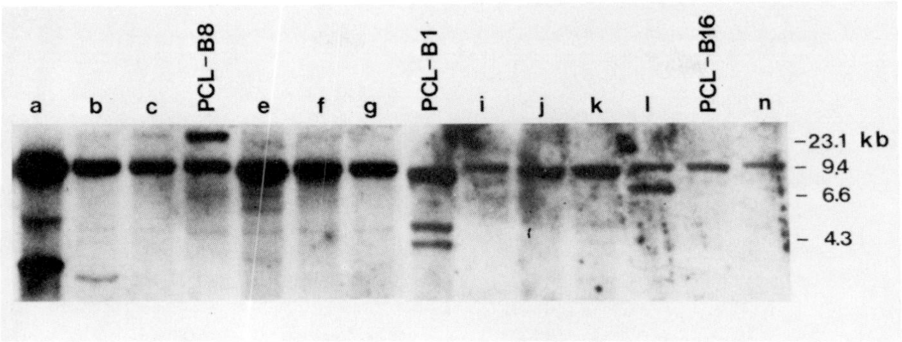

Fig. 1. Southern blot analysis of MM and PCL DNAs. DNAs were cleaved with Bcl-1 restriction enzyme and run on 0.8% agarose gel. The Southern blot filter was hybridized with 32P-labeled Bcl-1 DNA probe B (Tsujomoto 1984).

of genes within the amplified DNA region (Schimke 1984). The level of *c-myc* mRNA, determined by Northern-blot hybridization, was about 20- to 30-fold higher in PCL-B8 and 5-fold higher in PCL-B16 cells compared to that in peripheral blood lymphocytes and mouse fibroblasts (Sumegi *et al.* 1985). Slightly elevated levels of *c-myc* mRNA were seen in some of the MMs and PCls without obvious *c-myc* gene amplification.

THE t(11:14)(q13:q32) TRANSLOCATION INVOLVES THE *BCL-1* REGION IN HUMAN MULTIPLE MYELOMA AND PLASMA CELL LEUKEMIA

The t(11:14) translocation has been observed in diffuse small and large cell lymphoma and chronic lymphocytic leukemia (Yunis 1982), as well as in multiple myeloma (Van der Berghe *et al.* 1984). The chromosome 11 specific DNA sequence which contains the chromosomal breakpoint has been cloned from DNA of 2 CLL patients (Tsujimoto *et al.* 1984). The chromosome 11 breakpoints in the 2 DNAs have been shown to occur only 7 nucleotides away from each other (Tsujimoto *et al.* 1985). The chromosome 11 specific DNA region which carries the chromosomal breakpoints has been designated *bcl-1* (Tsujimoto *et al.* 1984). We screened 26 human MM and PCL DNAs for the rearrangement of *bc1-1* locus. The DNAs were cleaved with Bcl-1 or HindIII restriction enzymes; 5 DNAs showed an additional rearranged fragment beside the germ-line *bcl-1* fragment (Figure 1). Two rearranged *bcl-1* fragments have been cloned in lambda vectors and analyzed. Southern hybridization revealed that the chromosome 11 specific DNA sequence has been juxtaposed into the J region of the immunglobulin heavy chain gene (data not shown).

TABLE I

Chromosomal translocations, immunoglobulin production and the status of the c-myc *gene in 13 RICs*

Tumor	Translocation type	Ig secreted	MYC status
IR9	t(6:7)	BJ	NR
IR27	t(6:7)	gamma/kappa	r>23kb
IR33	t(6:7)	gamma/kappa	r>23kb
IR49	t(6:7)	BJ/kappa	r>23 kb
IR50	ND	epsilon/kappa	r>23 kb
IR72	t(6:7)	epsilon/kappa	r>23 kb
IR74	t(6:7)	none	r>23 kb
IR75	t(6:7)	epsilon/kappa	r=12 kb
IR88	t(6:7)	epsilon/kappa	r>23 kb
IR89	t(6:7)	epsilon	r>23 kb
IR209	t(6:7)	gamma/kappa	r=12 kb
IR222	t(6:7)	epsilon	r=18 kb
IR223	t(6:7)	BJ	r=20 kb
IR241	ND	gamma/kappa	r>23 kb
IR304	ND	gamma/kappa	NR

CHROMOSOMAL TRANSLOCATION IN RIC INVOLVES THE CHROMOSOMES BEARING THE *C-MYC* AND IMMUNOGLOBULIN HEAVY CHAIN GENES

We have karyotyped 20 spontaneous rat immunocytomas and found that all contained the 6:7 chromosomal translocation. The different immunocytomas showed an identical translocation pattern, which closely resembles the translocation found in the majority of MPCs (Wiener *et al.* 1982). The banding homologies between the chromosomes involved in these translocations suggested that the identity of translocation patterns cannot be accidental. We hypothesized that the translocation in the RIC would involve the same loci (*c-myc* and IgH chain gene) as in the t(12:15) MPC translocation. By utilizing rat/mouse somatic cell hybrids we have assigned the *c-myc* oncogene to chromosome 7 (Sümegi *et al.* 1983) and the immunglobulin heavy chain gene to chromosome 6 (Pear *et al.* 1986). Cytogenetic studies of 20 RICs have not identified any tumor which carries translocation between chromosome 7 and the kappa light chain gene bearing chromosome 4 (Perlmann *et al.* 1985) or the lambda light chain gene carrying chromosome 11 (G. Wahlström, unpublished). Another remarkable chromosomal aberration observed in some of the RICs is the duplication of the rearranged chromosome 6/7 (A. Szeles, unpublished).

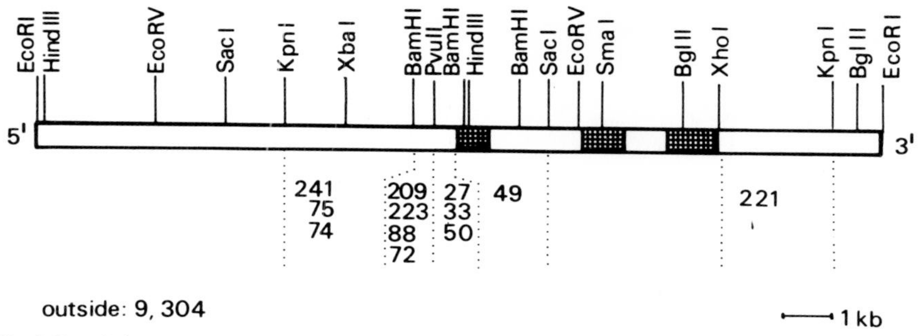

Fig. 2. Restriction enzyme mapping of the *c-myc* locus in different RICs. Rat DNAs from different RICs were cleaved with restriction enzymes, subjected to electrophoresis on 0.8% agarose gels, transferred to nitrocellulose filters and hybridized to 32P-labeled DNA fragments derived from the 1st exon of the normal rat *c-myc* gene (Pear *et al.* 1986).

THE RAT *C-MYC* GENE

The rat *c-myc* gene with its 3 exons and 5′- and 3′-flanking sequences resides on a 17 kb long EcoRI fragment (Sümegi *et al.* 1983). In 13 of the 15 RICs, an additional *myc*-specific fragment is detectable by Southern analysis. As shown in Table I, the rearranged *c-myc*-specified EcoRI fragment is greater than 23 kb in 10 of the 13 RICs. The rearrangement of the oncogene is accompanied by elevated levels of *c-myc* mRNA in the tumor cells (W. Pear, in preparation). The *c-myc* mRNA is shorter than the normal 2.4 kb transcript in only one tumor, IR49, suggesting that the chromosomal breakpoints fall outside of the 3 exons in the majority of the RICs.

THE CHROMOSOMAL BREAKPOINTS CLUSTER AT THE 5′ END OF THE *C-MYC* GENE

Using suitable DNA fragments as hybridization probes, Southern-blot hybridization has revealed that in 12 of the 13 RICs the recombination interrupted the 5′ end of the *c-myc* transcriptional unit (Figure 2). In one tumor, IR221, the recombination occurred at the 3′-end of the gene but left the 3 coding exons intact (figure 2). As shown in Figure 2, similar to Burkitt lymphoma and MPC, no translocation breakpoint is found in the protein coding the 2nd and 3rd exon sequences. Two RICs, IR9 and IR304, do not show rearrangement within the 17 kb long EcoRI fragment. Restriction enzyme analysis demonstrated that in these 2 tumors the *c-myc* gene locus is not rearranged within at least 23 kb of the 1st exon. As shown in Figure 2, the *c-myc*-associated breakpoints are clustered within a 0.9 kb long BamHI fragment, which contains the 2 promoters of the *c-myc* gene.

TABLE II

Correlation between immunoglobulin production and the location of chromosome 6 breakpoint in 8 different RICs

tumor	IG secreted	IgH target
IR50	epsilon/K	IgG epsilon
IR72	epsilon/K	IgH epsilon
IR74	none	IgH epsilon
IR 88	epsilon/K	IgH epsilon
IR222	epsilon	IgH epsilon
IR223	BJ	IgH epsilon
IR75	epsilon/K	IgH Sμ
IR241	gamma/K	IgH Sμ

In MPC and BL several sites in the IgH locus can provide breakpoints for translocation. In RICs, 6 out of 8 tumors investigated show rearrangement with the epsilon gene (Table II). Four of these 6 tumors produce IgE, suggesting that the recombination mechanism for normal IgH switch renders the chromosome temporarily susceptible to illegitimate recombination with other DNA sequences. Three RICs show unusual recombination between the *c-myc* gene and chromosome 6-derived sequences:

IR75 tumor carries a 6:7 translocation. As shown in Figure 3, the rearranged *c-myc* gene has been juxtaposed to the IgH region by multiple events. Genomic Southern blotting and sequence analysis of the cloned *c-myc* gene revealed that the chromosome 6 sequences lie 850 bp 5′ of the rearranged *c-myc*. The sequences originate from the switch μ region (Figure 3). The switch μ sequences, however, do not continue into the Cμ gene. Instead of Cμ sequences, a region upstream of the switch gamma-1 locus (5′ SWIGI) is joined to the switch μ sequences. The 5′ SWIGI sequences are in the same transcriptional orientation as the *c-myc* gene. At least two events are necessary to explain the juxtaposition of the switch μ and 5′ SWIGI regions in opposite orientations; deletion in the IgH cluster downstreams of the switch μ, combined with inversion, or a second translocation combined with inversion.

In the IR209 tumor, a 12 kb long EcoRI fragment accomodates the rearranged translocated *c-myc* gene (figure 4). Analysis of the cloned *c-myc* gene showed that the chromosomal breakpoint is 850 bp ustream of the 1st promoter of the gene. In order to identify the juxtaposed DNA sequences, the DNA fragment upstream of the chromosomal breakpoint in IR209 has been hybridized to cloned DNA sequences originating from the IgH chain gene. No hybridization was

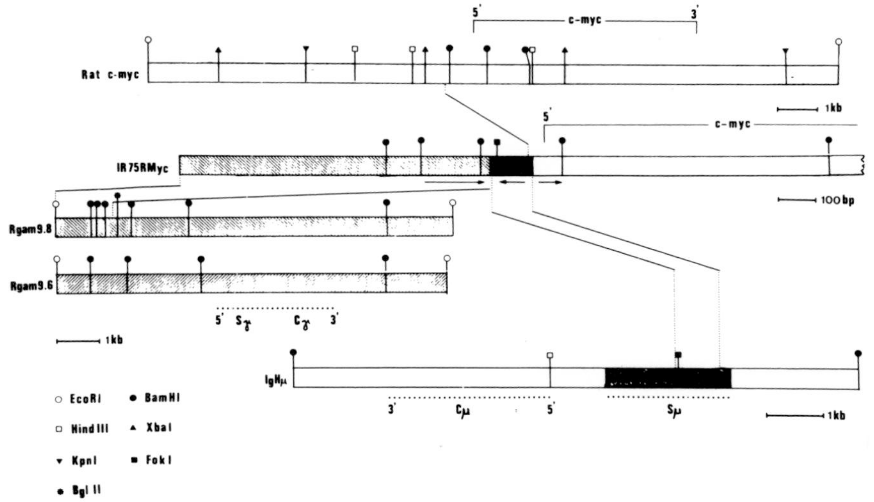

Fig. 3. Restriction enzyme analysis of the rearranged *c-myc* gene from IR 75 tumor. Structure of the normal and re-arranged *c-myc* and immunglobulin genes has been deduced by analyzing recombinant clones obtained from lambda libraries (Pear *et al.* 1986).

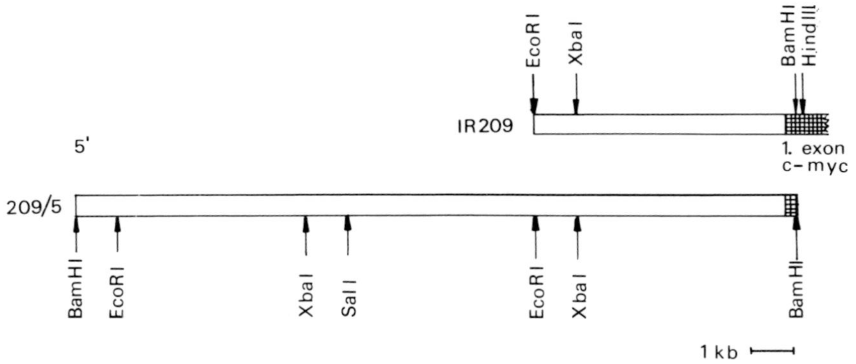

Fig. 4. Restriction enzyme analysis of the rearranged *c-myc* gene in IR 209 tumor. The rearranged *c-myc* gene has been isolated from a lambda library.

found between the IgH chain gene sequences and the DNA fragment upstream of the breakpoint (data not shown). When hybridized to EcoRI, BamHI or HindIII cleaved rat genomic DNA, it identified a class of highly repetitive DNA (Figure 5). On the basis of the unique pattern of the location of BamHI, EcoRI and HidIII restriction enzyme sites within the repeats, a specific intense banding

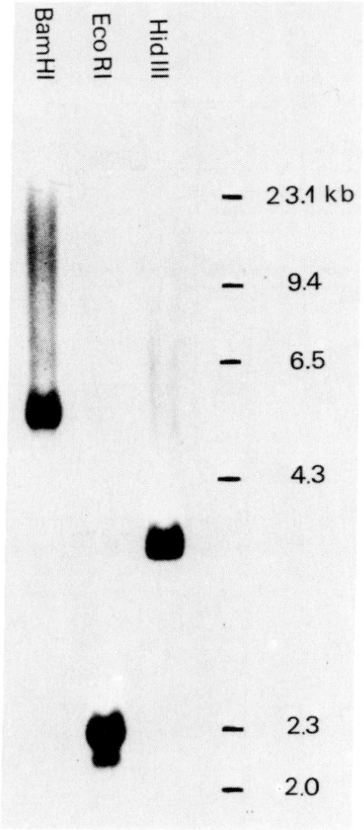

Fig. 5. Southern analysis of the LINE sequences in rat genomic DNA. Rat liver DNA was cleaved with restriction enzymes BamHI, EcoRI and HindIII, separated on 0.8% agarose gel electrophoresis and transferred to nitrocellulose filter and hybridized with a 32P-labeled 1 kb long BamHI/EcoRI fragment isolated from 209/5 lambda clone (see Figure 4).

pattern was apparent (Figure 5). This specific pattern became visible in autoradiography after a very short (5–10 min) exposure. Exposing the X-ray film longer to the blot resulted in a complete blackness. On the basis of the EcoRI, BamHI and HindIII pattern, we could identify the cloned DNA sequence as a member of the LINE family of interspersed long repeated DNA (Witney *et al.* 1984).

The 3rd tumor, IR221, does not show rearrangement at the 5′-end of the *c-myc* gene. Genomic Southern analysis showed that the chromosomal breakpoint is distal from the 3rd coding exon of the *c-myc* gene (data no shown).

DISCUSSION

The regular occurrence of the *myc/IgH* juxtaposition in MPCs and RICs can only be interpreted as meaning that the translocation represents a crucial event in the genesis of these tumors. The close homology between MPC and RIC is striking. There are only minor differences in detail. The comparative aspects of the immunglobulin gene target in the translocation is interesting. In MPCs, switch alpha is the most frequently involved translocation acceptor site, whereas in RICs it is switch epsilon. Most mouse plasmacytomas produce IgA, while the majority of RICs make IgE. As the translocation effects the non-productively rearranged IgH chain gene in both MPC and RIC, and explanation of this phenomenon is not evident. In a recent review, Cory (1986) has classified the *c-myc/IgH* translocation breakpoints. They can occur in three regions of the *c-myc*-carrying chromosome; i.) within the normal *myc* transcriptional unit, ii.) within a 2 or 3 kb interval 5′ to the gene, ii.) at a large, unknown distance 5′ to the gene. In all cases of MPC so far studied the 2 coding exons are not affected by the break. In murine tumors the breakpoints fall within a 1.5 kb region spanning exon 1 and the first half of intron 1. In RICs, there is a very strong preference for breaks within a 1.0 kb region 5′ of the 1st exon. Among 13 tumor translocation breakpoints characterized, only one breaks the 1st exon. In MPCs there is a preference for the *c-myc/IgH* juxtaposition (typical translocation) over the rearrangement between the oncogene and 1 of the 2 light chain genes (variant translocation). In RICs, we could not find a single "variant" translocation among 20 tumors karyotyped.

We have found no *myc* rearrangement in 20 human multiple myeloma and plasma cell leukemia cases. The *c-myc* was amplified and highly expressed, however, in 2 plasma cell leukemia cases. This is of considerable interest, in view of the well demonstrated occurrence of *myc* amplification in small cell lung cancer (Little *et al.* 1984), colorectal carcinoma cell lines (Schwab *et al.* 1984) and breast cancer lines (Kozbor *et al.* 1984). The *c-myc* amplification in small cell lung carcinoma has been related to the progression of the disease (Little *et al.* 1984). Plasma cell leukemia is rare among malignancies of B-cell origin. They can occur terminally in MM, but may also appear directly, without previously existing myeloma. By analogy with small cell lung carcinoma, it is conceivable that the rare *myc* amplification (2 out of 5) occurs in some of the tumors during progression. The most common chromosomal translocation in MMs is the t(11q13:14q32). The translocation involves the region denoted *bcl-1*. What is the role of the *bcl-1* rearrangement in tumorigenesis? Does the immunoglobulin gene region activate an oncogene (*bcl-1*?) on chromosome 11? The *bcl-1* is not homologous to any known retroviral oncogenes, nor is it well conserved among species (J. Sümegi, unpublished). We and others have failed to detect complementary transcripts in normal and tumor cells. It is tempting to speculate that the *bcl-1* has a function similar to that of *pvt-1* on murine chr. 15 (Cory 1986). Both *pvt-1* and *bcl-1* alterations may act in *cis* via alterations of long-range chromatin folding.

REFERENCES

Bazin, H. (1985) In: *Mechanism of B-cell Neoplasia,* pp. 208–16. Roche, Basle.

Bazin, H., Deckers, C., Beckers, A. & Heremans, J. F. (1972) Immunglobulin secreting rat tumors. *Int. J. Cancer. 10,* 568–80.

Cory, S. (1986) Activation of cellular oncogenes in hemopoetic cells by chromosome translocation. *Adv. Cancer Res. 47,* 189–234.

Croce, M. C. & Klein, G. (1985) Chromosome translocations and human cancer. *Scientific American 252,* 54–60.

Erikson, J., Ar-Rushdi, A., Drwinga, H. L., Nowell, P. C. & Croce, C. M. (1983) transcriptional activation of the translocated c-myc oncogene in Burkitt lymphoma. *Proc. Natl. Acad. Sci. USA 80,* 820–4.

Ferti, A., Pananai, A., Arapabis, G. & Raptis, S. (1984) Cytogenetic study in multiple myeloma. *Cancer Genet. Cytogenet. 12,* 247–53.

Hayward, W. S., Neel, B. G. & Astrin, S. M. (1983) Activation of a cellular onc gene by promoter insertion in ALV-induced lymphoid leukosis. *Nature 209,* 475–9.

Itani, S., Hoshino, T., Kawasaki, S. & Nakayama, S. (1970) Chromosome abnormality and its significance in human multiple myeloma. *Acta Haemat. 33,* 54–66.

Leder, P., Battey, J., Lenoir, G., Moulding, C., Murphy, W., Potter, H., Stewart, T. & Taub, R. (1983) Translocation among antibody genes in human cancer. *Science 222,* 765–71.

Little, C. D., Nau, M. M., Carney, D. N. & Minna, J. D. (1983) Amplification and expression of the c-myc oncogene in human lung cancer cell lines. *Nature 306,* 194–6.

Klein, G. (1983) Specific chromosomal translocations and the genesis of B-cell derived tumors in moce and man. *Cell 32,* 311–5.

Kozbor, D. & Croce, C. M. (1984) Oncogene in one of five human breast carcinoma cell lines. *Cancer Res. 44,* 438–41.

Mittelman, F. (1983) catalogue of chromosome aberrations. *Cancer Cytogenet. Cell Genet. 36,* 1–511.

Pear, W., Ingvarsson, I., Steffen, S., Munke, M., Francke, U., Bazin, H., Klein, G. & Sumegi, J. (1986) Multiple chromosomal rearrangements in a spontaneously arising t(6:7) rat immunocytoma juxtapose c-myc and immunglobulin heavy chain sequences. *Proc. Natl. Acad. Sci. USA* (in press).

Perlmann, C., Sumegi, J., Szpirer, C., Levan, G. & Klein, G. (1985) The rat immunglobulin light chain locus is on chromosome 4. *Immunogenet. 22,* 97–100.

Schimke, R. T. (1984) Gene amplification in cultured animal cells. *Cell 37,* 705–13.

Schwab, M., Alitalo, K., Varmus, H. E. & Bishop, J. M. (1984) Amplification of cellular oncogenes in tumor cells. In: *Cancer cells. Oncogenes and viral genes,* eds. Vande Woude, F. G., Levine, J. A., Topp, C. W. & Watson, J. M., pp. 216–20. Cold Spring Harbor Laboratory.

Sümegi, J., Hedberg, T., Björkholm, M., Godal, T., Mellstedt, H. & Klein, G. (1985) Amplification of the c-myc oncogene in human plasma-cell leukemia. *Int. J. Cancer 36,* 367–72.

Sümegi, J., Spira, J., Bazin, H., Szpirer, J., Levan, G. & Klein, G. (1983) Rat c-myc oncogene is located on chromosome 7 and rearranges in immunocytomas with t(6:7) chromosomal translocation. *Nature 306*, 497–8.

Tsujimoto, Y., Jaffe, E., Cossman, J., Gorham, F., Nowell, P. C. & Croce, C. M. (1985) Clustering of breakpoints on chromosome 11 in human B-cell neoplasms with t(11:14) chromosome translocation. *Nature 315,* 340–3.

Tsujimoto, Y., Yunis, J., Onorato-Showe, L., Erikson, J., Nowell, P. C. & Croce, C. M. (1984) Molecular cloning of the chromosomal breakpoint of B cell lymphomas and leukemias with the t(11:14) chromosome translocation. *Science 224*, 1403–1406.

Van den Berghe, H. & Louwagie, A. (1979) Philadelphia chromosome in human multiple myeloma. *J. Natl. Cancer Inst. 63*, 11–6.

Wiener, F., Babonits, M., Spira, J., Klein, G. & Bazin, H. (1982) Non-random chromosomal changes involving chromosome 6 and 7 in spontaneous rat immunocytoma. *Int. J. Cancer 29,* 431–2.

Witney, F. R. & Furano, A. V. (1984) Long interspersed repeat sequences in mammalian genomes. *J. Biol. Chem. 259*, 10481–92.

Yunis, J. J., Oken, M. M., Kaplan, N. E., Enstrud, K. M., Howe, R. R. & Theologides, A. (1982) Distinctive chromosomal abnormalities in histologic subtypes of non-Hodgkin lymphoma. *N. Engl. J. Med. 307*, 1231–6.

Constitutive c-*myc* Expression and Lymphoid Neoplasia

Suzanne Cory, W. Y. Langdon, A. W. Harris, M. W. Graham, W. S. Alexander & J. M. Adams

It seems likely that the proto-oncogene c-*myc* plays a central role in the complex chain of events emanating from the interaction of a growth factor with its receptor on the cell surface. In general, the level of c-*myc* expression correlates with the rate of cell division (Stewart *et al.* 1984) (Pfeifer-Ohlsson *et al.* 1984). Moreover, there is a rapid increase in the amount of c-*myc* RNA in several cell types in response to mitogens and growth factors (Kelly *et al.* 1983, Makino *et al.* 1984, Reed *et al.* 1985). While differentiation appears to be associated with down-regulation of *myc* (Reitsma *et al.* 1983, Gonda & Metcalf 1984, Campisi *et al.* 1984, McCormack *et al.* 1984, Lachman & Skoultchi 1984, Filmus & Buick 1985), it has been unclear whether this is cause or effect. The phosphoprotein encoded by c-*myc* is located in the nucleus and, although *in vitro* studies suggest that it binds to DNA (Donner *et al.* 1982), its function remains unknown.

Dysregulated expression of the c-*myc* gene has been strongly implicated in lymphoid neoplasia. In retrovirus-induced avian B lymphomas (Hayward *et al.* 1981), and a proportion of T lymphomas (Corcoran *et al.* 1984), proviral insertion near c-*myc* brings its expression under the influence of the promoter and/or enhancer within the retroviral LTR. In contrast, in most Burkitt lymphomas and murine plasmacytomas, deregulation is achieved by translocation to the immunoglobulin heavy chain locus (for reviews, see Leder *et al.* 1983, Klein & Klein 1985, Adams & Cory 1985, Cory 1986). Although translocation often excises the 5′ untranslated segment, the *myc* coding region remains unperturbed. Most strikingly, *myc* transcription in the tumors derives only from the translocated allele, the normal allele being silent (Bernard *et al.* 1983, Nishikhira *et al.*

The Walter and Eliza Hall Institute of Medical Research, Post Office, Royal Melbourne Hospital, Victoria 3050, Australia.

VIRAL CARCINOGENESIS, Alfred Benzon Symposium 24,
Editors: N. O. Kjeldgaard, J. Forchhammer, Munksgaard, Copenhagen 1987.

1983). The major consequence of the translocation is therefore not a structural change in the *myc* polypeptide but constitutive expression of the normal gene. It is reasonable to postulate that Ig regulatory elements are involved in imposing constitutive expression. The transcriptional enhancer (Eμ) associated with the constant region (C_H) locus is an attractive candidate. Indeed, in one remarkable plasmacytoma, the *myc* gene has been activated by a novel mechanism in which the Eμ element has "jumped" from chromosome 12 (Corcoran *et al.* 1985).

TRANSGENIC MICE PROVE THE TUMORIGENIC POTENTIAL OF c-*myc*

While the structural alterations in lymphomas provided strong circumstantial evidence that dysregulated c-*myc* expression is tumorigenic, transgenic mice produced by injecting various c-*myc* constructs into the pronuclei of fertilized oocytes have provided direct proof. Mice bearing c-*myc* linked to the regulatory region within the LTR of the murine mammary tumor virus were found to have increased susceptibility to mammary carcinomas (Stewart *et al.* 1984b). In collaboration with R. Brinster, C. Pinkert and R. Palmiter, we tested a series of c-*myc* constructs in transgenic mice. Linkage to the SV40 promoter/enhancer provoked a lymphosarcoma, a renal carcinoma and a fibrosarcoma, but the tumor incidence was relatively low (Adams *et al.* 1985). Subjugation of c-*myc* to immunoglobulin enhancer elements was strikingly effective (Adams *et al.* 1985): 13 or 14 of 15 primary transgenic animals bearing c-*myc* linked to the heavy chain enhancer (Eμ) developed lymphomas, as did 6 of 17 with c-*myc* linked to the kappa enhancer (E_κ). In contrast, a normal c-*myc* gene was innocuous as a transgene, as was a *myc* gene which lacked exon 1 and the putative 5′ regulatory region, but which retained the cryptic intron 1 promoters from which *myc* transcripts derive in most plasmacytomas.

It seems clear from these results that the Ig loci play a vital role in the B-cell tumorigenesis promoted by c-*myc* chromosome translocations. The efficacy of the Eμ-*myc* construct leaves little doubt that Eμ is involved in those translocations that directly link Eμ to c-*myc*. However, in at least 3/26 Burkitt lymphomas, Eμ is more than 18 kb 5′ to c-*myc*. Moreover, in almost all plasmacytomas and some 45% of Burkitt lymphomas, Eμ is *not* linked to c-*myc* but lies on the reciprocal chromosome product. Perhaps in these tumors *myc* is governed by an additional unknown IgH enhancer element. On the other hand, Eμ may still be the principal actor if its crucial role is played before chromosome translocation takes place. If one postulates that deregulation of c-*myc* is effected simply by conjunction with an active C_H locus, Eμ may be needed for the initial establishment of this state but not for its subsequent maintenance. In support of this notion, IgH genes remain active *in vivo* even when deletions remove Eμ (Wabl & Burrows 1984, Klein *et al.* 1984).

The transgenic mouse studies clearly establish that E$_\kappa$-controlled *myc* expression is tumorigenic. While some 15% of translocations in Burkitt lymphomas and murine plasmacytomas involve a light chain locus, in most of these tumors recombination occurs far 3′ to c-*myc*. The major recombination site in the variant plasmacytomas, *pvt*-1, maps more than 94 kb 3′ to c-*myc* (Cory *et al.* 1985) and is weakly homologous (Graham & Adams 1986) to the region that recombined near C$_\kappa$ in a Burkitt lymphoma (Taub *et al.* 1984). All the variant plasmacytoma breakpoints occur 5′ to E$_\kappa$ (Cory *et al.* 1985) and the structure of the 15q$^+$ chromosome can be represented as centromere/myc/pvt-1/E$_\kappa$C$_\kappa$ (Banerjee *et al.* 1985). If the variant tumors also result from deregulation of c-*myc*, rather than activation of another oncogene (for a discussion, see Cory 1986), the influence of E$_\kappa$ and/or the activated C$_\kappa$ locus must be conveyed to c-*myc* over distances greater than 100 kb. Since *pvt*-1 is co-amplified with c-*myc* in the Abelson fibroblast line ANN-1, and most amplicons are of the order of 100–1000 kb, the separation may be several hundred kb (Graham & Adams 1986).

Eμ-*myc* INDUCES AGGRESSIVE B-CELL NEOPLASIA

The fatal predisposition of Eμ-*myc* transgenic mice to develop tumors has enabled us to analyze the course of the disease in detail (Adams *et al.* 1985, Harris *et al.* in preparation). The typical diagnosis is a disseminated lymphoma, usually accompanied by leukemia, and involving most of the lymph nodes and often the thymus. In some mice the lymph nodes are not grossly affected and the animals succumb to a thymoma or to bowel obstruction caused by proliferation of tumor cells in the intestinal wall. Some 50 different tumors from 20 mice have been analyzed for cell surface markers and DNA rearrangement. All have proved to be B-lymphoid in origin, even those derived from the thymus. Some 60% were initially pre-B cells but some have evolved in culture to sIg$^+$ B cells. Thus Eμ-*myc*-induced tumorigenesis does not totally prevent further differentiation.

As anticipated, the tumor cells exhibited relatively abundant transcription of the Eμ-*myc* transgene. Moreover, as in tumors bearing a *myc*-IgH translocation, *no* normal c-*myc* transcripts could be detected (Adams *et al.* 1985). These observations support the model (Leder *et al.* 1983, Rabbitts *et al.* 1984) that normal c-*myc* regulation involves a negative feedback loop. The rearranged allele is apparently refractory to repression and continues to produce c-*myc* protein which, either directly or indirectly, silences the normal allele. We believe that the inability to turn off *myc* transcription is crucial to the transformation process (see below).

The nearly invariant tumors in Eμ-*myc* mice might suggest that deregulation of c-*myc* is the sole factor required. However, several aspects of the disease suggest that additional event(s) are required for the emergence of a fully malignant clone. First, the disease in individual mice usually involves only one clonal cohort (Adams *et al.* 1985, Cory *et al.*, in preparation), even though all B-lineage cells express Eμ-*myc*. Secondly, the onset of tumors is highly variable and can occur anywhere between 3 weeks and more than 6 months of age. Thirdly, and most compelling, while the tumor cells are aggressively malignant when injected into syngeneic recipients (Harris *et al.*, in preparation), the lymphoid cells from young animals lacking overt signs of disease usually fail to induce tumors, even when injected in large numbers (Langdon *et al.* 1986). Thus, even though the Eμ enhancer is turned on early in B-cell ontogeny, Eμ-*myc* mice exhibit a true pre-neoplastic phase.

Eμ-*myc* MICE INITIALLY UNDERGO A BENIGN EXPANSION OF EARLY B-LINEAGE CELLS

The pre-neoplastic phase in young Eμ-*myc* mice is characterized by a profound disturbance of B-cell differentiation, which provides important clues about c-*myc* function (Langdon *et al.* 1986). Phenotype analysis revealed a remarkable expansion of pre-B cells that is evident as early as 18 days of gestation. In young adults, the expansion dominates the bone marrow and also involves the spleen. Significantly, there are no detectable pre-B cells in the lymph nodes or thymus, even though these organs are ultimately the major sites for tumor growth. Overall, prelymphomatous mice exhibit a 4- to 5-fold increase in pre-B cells and a concomitant ~30% reduction of sIg^+ B cells. Analysis of bone marrow DNA for J_H rearrangement indicated that the expansion is polyclonal and probably includes a considerable number of pro-B cells which have not yet commenced J_H rearrangement.

The B-lineage cells in Eμ-*myc* mice all appear to be in an "activated" state. The small resting B cell and its immediate precursor, the small $B220^+ThB^+$ cell, are undetectable, and all $B220^+$, ThB^+ and sIg^+ cells are large. Moreover, at least half the pre-B and B cells are actively cycling. We conclude that constitutive *myc* expression maintains B-lineage cells in cycle and may indeed preclude a G_0 state. Surprisingly, many of the Eμ-*myc* pre-B cells express the Ia antigen. In normal mice, this antigen is expressed on B cells, and its level increases after stimulation by mitogens or by antigen plus growth factors (Dasch & Jones 1986). The significance of the premature Ia expression is not clear but may mean that constitutive *myc* expression partly short-circuits the normal pathway responsive to mitogenic signals at the cell surface.

The role of *myc* in mitogenesis has been addressed by several groups (e.g. Kelly *et al.* 1983, Campisi *et al.* 1984). Our observations on B-cell differentiation in pre-lymphomatous mice strongly indicate that *myc* also plays an important role in differentiation. We propose (Langdon *et al.* 1986) that the level of *myc* expression is an important factor in setting the probability of self-renewal versus maturation during differentiation, with increased *myc* expression favoring self-renewal.

In summary, our studies with transgenic mice have amply vindicated the hypothesis that constitutive *myc* expression highly predisposes to lymphoid malignancy and have underlined the important role of the Ig loci in the deregulation process. A major consequence of constitutive *myc* expression is that self-renewal is favored over maturation and this results in a substantial polyclonal expansion of early cells. Emergence of a fully malignant clone apparently requires additional genetic change(s). The increased proliferative potential of Eμ-*myc*-driven cells undoubtedly increases the likelihood of such a genetic accident, as does the expansion of a vulnerable pool of pre-B cells which actively undergo DNA rearrangement and may therefore be particularly susceptible.

ACKNOWLEDGMENTS

We gratefully acknowledge the collaboration of Drs. R. Brinster, C. Pinkert and R. Palmiter in the production of the *myc* transgenic mice; informative discussions with Drs. D. Metcalf, B. Pike and T. Mandel; and dedicated technical assistance from M. Crawford, L. Gibson, J. Mitchell and S. Kyvetos. This work was supported in part by National Institutes of Health Grant CA-12421, the American Heart Association, the Drakensberg Trust, and the National Health and Medical Research Council (Canberra).

REFERENCES

Adams, J. M. & Cory, S. (1985) *Myc* oncogene activation in B and T lymphoid tumours. In: *Oncogenes: Their Role in Normal and Malignant Growth*, eds. Bodmer, W. F., Weiss, R. & Wyke, J., pp. 59–72. The Royal Society, London.

Adams, J. M., Harris, A. W., Pinkert, C. A., Corcoran, L. M., Alexander, W. S., Cory, S., Palmiter, R. D. & Brinster, R. L. (1985) The c-*myc* oncogene driven by immunoglobulin enhancers induces lymphoid malignancy in transgenic mice. *Nature (London) 318*, 533–8.

Banerjee, M., Wiener, F., Spira, J., Babonits, M., Nilsson, M.-G., Sumegi, J. & Klein, G. (1985) Mapping of the c-*myc*, *pvt*-1 and immunoglobulin kappa genes in relation to the mouse plasmacytoma-associated variant (6;15) translocation breakpoint. *EMBO J. 4*, 3183–8.

Bernard, O., Cory, S., Gerondakis, S., Webb, E. & Adams, J. M. (1983) Sequence of the murine and human cellular *myc* oncogenes and two modes of transcription resulting from chromosome translocation in B lymphoid tumours. *EMBO J. 2*, 2375–83.

Campisi, J., Gray, H. E., Pardee, A. B., Dean, M. & Sonenshein, G. E. (1984) Cell-cycle control of c-*myc* but not c-*ras* expression is lost following chemical transformation. *Cell 36*, 241–7.

Corcoran, L. M., Adams, J. M., Dunn, A. R. & Cory, S. (1984) Murine T lymphomas in which the cellular *myc* oncogene has been activated by retroviral insertion. *Cell 37*, 113–22.

Corcoran, L. M., Cory, S. & Adams, J. M. (1985) Transposition of the immunoglobulin heavy chain enhancer to the *myc* oncogene in a murine plasmacytoma. *Cell 40*, 71–9.

Cory, S. (1986) Activation of cellular oncogenes in hemopoietic cells by chromosome translocation. In: *Advances in Cancer Research*, eds. Klein, G. & Weinhouse, S., Vol. 47, pp. 189–234. Academic Press, New York.

Cory, S., Graham, M., Webb, E., Corcoran, L. & Adams, J. M. (1985) Variant (6;15) translocations in murine plasmacytomas involve a chromosome 15 locus at least 72 kb from the c-*myc* oncogene. *EMBO J. 4*, 675–81.

Dasch, J. R. & Jones, P. P. (1986) Independent regulation of IgM, IgD and Ia antigen expression in cultured immature B lymphocytes. *J. Exp. Med. 163*, 938–51.

Donner, P., Greiser-Wilke, I. & Moelling, K. (1982) Nuclear localization and DNA binding of the transforming gene product of avian myelocytomatosis virus. *Nature (London) 296*, 262–6.

Filmus, J. & Buick, R. N. (1985) Relationship of c-*myc* expression to differentiation and expression of HL-60 cells. *Cancer Res. 45*, 822–5.

Gonda, T. J. & Metcalf, D. (1984) Expression of myb, myc and fos protooncogenes during the differentiation of a murine myeloid leukaemia. *Nature (London) 310*, 249–51.

Graham, M. & Adams, J. M. (1986) Chromosome 8 breakpoint far 3′ of the c-*myc* oncogene in a Burkitt lymphoma 2;8 variant translocation is equivalent to the murine *pvt*-1 locus. *EMBO J. 5* (in press).

Hayward, W., Neel, B. G. & Astrin, S. (1981) Activation of a cellular *onc* gene by promoter insertion in ALV-induced lymphoid leukosis. *Nature (London) 290*, 475–80.

Kelly, K., Cochran, B. H., Stiles, C. D. & Leder, P. (1983) Cell-specific regulation of the c-*myc* gene by lymphocyte mitogens and platelet-derived growth factor. *Cell 35*, 603–10.

Klein, G. & Klein, E. (1985) Evolution of tumours and the impact of molecular oncology. *Nature (London) 315*, 190–5.

Klein, S., Sablitzky, F. & Radbruch, A. (1984) Deletion of the IgH enhancer does not reduce immunoglobulin heavy chain production of a hybridoma IgD class switch variant. *EMBO J. 3*, 2473–6.

Lachman, H. M. & Skoultchi, A. J. (1984) Expression of c-*myc* changes during differentiation of mouse erythroleukaemia cells. *Nature (London) 310*, 592–4.

Land, H., Parada, L. & Weinberg, R. (1983) Tumorigenic conversion of primary embryo fibroblasts requires at least two co-operating oncogenes. *Nature (London) 304*, 596–602.

Langdon, W. Y., Harris, A. W., Cory, S. & Adams, J. M. (1986) Perturbed B cell development induced by the c-*myc* oncogene in Eμ-*myc* transgenic mice. *Cell 47* (in press).

Leder, P., Battey, J., Lenoir, G., Moulding, C., Murphy, W., Potter, H., Stewart, T. & Taub, R. (1983) Translocations among antibody genes in human cancer. *Science 222*, 765–71.

Makino, B., Hayashi, K. & Sugimura, T. (1984) c-*myc* transcript is induced in rat liver at a very early stage of regeneration or by cycloheximide treatment. *Nature (London) 310*, 697–8.

McCormack, J. E., Pepe, V. H., Kent, R. B., Dean, M., Marshak-Rothstein, A. & Sonenshein, G. (1984)

Specific regulation of c-*myc* oncogene expression in a murine B-cell lymphoma. *Proc. Natl. Acad. Sci. USA 81*, 5546–50.

Nishikura, K., ar-Rushdi, A., Erikson, J., Watt, R., Rovera, G. & Croce, C. M. (1983) Differential expression of the normal and of the translocated human c-*myc* oncogenes in B cells. *Proc. Natl. Acad. Sci. USA 80*, 4822–6.

Pfeifer-Ohlsson, S., Goustin, A. S., Rydnert, J., Wahlström, T., Bjersing, L., Stehelin, D. & Ohlsson, R. (1984) Spatial and temporal pattern of cellular *myc* oncogene expression in developing human placenta: implications for embryonic cell proliferation. *Cell 38*, 585–96.

Rabbitts, T. H., Forster, A., Hamlyn, P. & Baer, R. (1984) Effect of somatic mutation within translocated c-*myc* genes in Burkitt's lymphoma. *Nature (London) 309*, 592–7.

Reed, J. C., Nowell, P. C. & Hoover, R. G. (1985) Regulation of c-*myc* mRNA levels in normal human lymphocytes by modulators of cell proliferation. *Proc. Natl. Acad. Sci. USA 82*, 4221–4.

Reitsma, P. H., Rothbert, P. G., Astrin, S. M., Trial, J., Bar-Shavit, Z., Hall, A., Teitelbaum, S. L. & Kahn, A. J. (1983) Regulation of *myc* gene expression in HL-60 leukaemia cells by a vitamin D metabolite. *Nature (London) 306*, 492–3.

Stewart, T. A., Bellvé, A. R. & Leder, P. (1984a) Transcription and promoter usage of the c-*myc* gene in normal somatic and spermatogenic cells. *Science 226*, 707–10.

Stewart, T. A., Pattengale, P. K. & Leder, P. (1984b) Spontaneous mammary adrenocarcinomas in transgenic mice that carry and express MTV/myc fusion genes. *Cell 38*, 627–37.

Taub, R., Kelly, K., Battey, J., Latt, S., Lenoir, G. M., Tantravahi, U., Tu, Z. & Leder, P. (1984) A novel alteration in the structure of an activated c-*myc* gene in a variant t(2;8) Burkitt lymphoma. *Cell 37*, 511–20.

Wabl, M. R. & Burrows, P. D. (1984) Expression of immunoglobulin heavy chain at a high level in the absence of a proposed immunoglobulin enhancer element in *cis*. *Proc. Natl. Acad. Sci. USA 81*, 2452–5.

DISCUSSION

HANAHAN: Could you elaborate a little on the possibility that there is a locus in SJL which affects thymic abnormalities or something like this, which could be part of the early onset you described?

CORY: It is certainly true that SJL mice are tumor-prone, but only very late in their life. It seems conceivable that the gene(s) controlling this might also influence the early onset of Eμ-*myc* tumors.

WONG-STAAL: Are there any karyotypic abnormalities associated with the Eμ-tumors that are not found in the preneoplastic cells?

CORY: We are collaborating with G. Webb in Melbourne on this question and so far do not have enough data on enough different tumors to be draw any conclusions.

YANIV: Is there any difference in the amount of *myc* expressed in the preneoplastic and the neoplastic cells?

CORY: No, it is essentially the same.

CUZIN: Did you try to establish cell lines from this prelymphoma stage?

CORY: We have tried, and so far we have failed. I think that simply means that they represent early cells that do not grow readily under standard tissue culture conditions. We plan next to try bone marrow stromal feeder layers.

VI. Transformation by BPV

Multiple Skin Pathologies in Transgenic Mice Harboring the Bovine Papilloma Virus Genome

Douglas Hanahan, Susan Alpert & Mark Lacey

The ability to transfer new genetic information in the mouse germ line is providing a novel approach to studying the control and consequences of gene expression. In this situation mice inherit a new 'transgene' from a parent and, therefore, harbor it in every cell of their body, manifesting any phenotype it endows (Palmiter & Brinster 1986). The stable introduction of new genes in mice is proving to be particularly applicable to the study of carcinogenesis (Hanahan 1986). A variety of oncogenes have been established in lines of transgenic mice; in some cases the mice develop specific tumors as a consequence of the expression of the oncogene. The transfer of viral genomes into the mouse germ line can provide information on the specificity of the viral regulatory elements, on the effectiveness of viral oncogenes on different cell types, and on the requirements for additional events in the genesis of any tumors or other abnormalities produced by the viral genomes. For example, when the early (transforming) region of simian virus 40 was established in transgenic mice, tumors of the choroid plexus, a layer of cells which lines the ventricles of the brain, were heritably produced (Brinster *et al.* 1984).

Bovine papilloma virus type 1 produces fibroepithelial tumors upon infection of cutaneous tissue in cattle. These skin tumors are about equally composed of dermal fibroblasts and epidermal keratinocytes. BPV is also capable of morphologically transforming certain cultured mouse cells. This *in vitro* genetic assay has been used to study the functional organization of the BPV DNA, in conjunction with the molecularly cloned BPV genome (an ~8 kb circular double stranded DNA molecule). Elements in the BPV genome involved in viral replication,

---

Cold Spring Harbor Laboratory, P.O. Box 100, Cold Spring Harbor, N.Y. 11724, U.S.A.

VIRAL CARCINOGENESIS, Alfred Benzon Symposium 24,
Editors: N. O. Kjeldgaard, J. Forchhammer, Munksgaard, Copenhagen 1987.

transcriptional control, and transformation have been identified. (The biology, molecular biology, and genetics of papillomaviruses are reviewed in Lancaster & Olsen 1982, Pfister 1984, Howley & Schlegel 1986.) These types of genetic analysis can now be expanded into the analysis of the control and consequences of BPV gene expression in a situation where the BPV genome is vertically transmitted as a heritable genetic element in a transgenic mouse.

A LINE OF TRANSGENIC MICE HARBORING THE BPV GENOME

The complete BPV genome has been established in a line of transgenic mice (Lacey *et al.* 1986). This was accomplished by microinjecting a plasmid clone of the BPV-I genome into fertilized, one-cell mouse embryos which were implanted into the oviducts of pseudopregnant female mice and allowed to develop. One mouse, derived from these microinjections, carried the plasmid as an integrated head-to-tail concatenate, as is generally observed in this method of gene transfer (Palmiter & Brinster 1986). This transgenic mouse transmitted the BPV DNA to its progeny in a Mendelian fashion, and mice homozygous for this insertion were derived by intercrossing transgenic siblings. Both heterozygous and homozygous mice which inherited the BPV genome were phenotypically normal during their early lives, suggesting that throughout this period the viral genome was either inactive or ineffective.

The plasmid clone of the bovine papilloma virus genome which was established in this line of mice contained a complete, redundant copy of the normally circular genome (Lacey *et al.* 1986). This was accomplished with a plasmid (pBPV1.69) which contained a partial tandem duplication of the complete genome. BPV can be subdivided into two domains: the 69% transforming region, sufficient for cell transformation *in vitro*, and the remaining 31% "late region". In this plasmid, the 69% region was duplicated to create a 169% copy of the viral genome, thus mimicking a circular topology in the linear molecule, which was created during integration into a mouse chromosome (see Figure 1 of Lacey *et al.* 1986).

HERITABLE DEVELOPMENT OF SKIN TUMORS

As the transgenic mice harboring the BPV genome aged, skin tumors began to arise with high frequency. Although protuberant tumors were the first recognizable abnormality, areas of abnormal skin became apparent as well. The development of these abnormalities was unique to the line of transgenic mice harboring the BPV genome, and have never been observed in normal mice in this institution, nor in other transgenic mice. Thus a correspondence existed between the presence

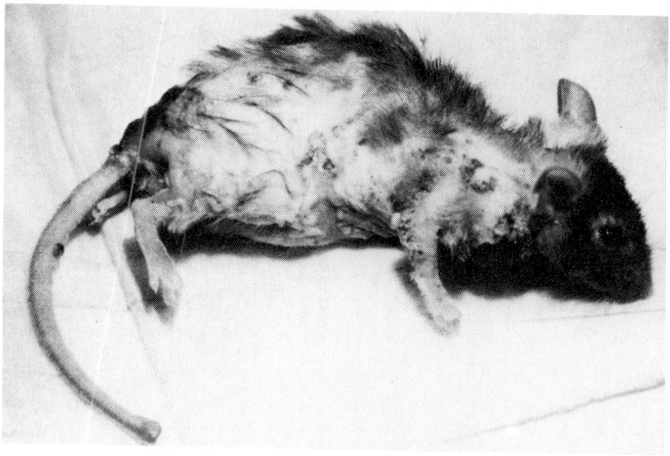

Fig. 1. A BPV transgenic mouse showing multiple skin pathologies. This mouse, (#806), was about 12 months old when this photograph was taken. Most of the body skin is grossly abnormal, showing marked hardening and loss of hair. The abnormal skin contains scattered, small protuberant tumors and blue nevi, and the mouse carries a large tumor on its neck, and two tumors on the tail.

of the BPV genome and the development of abnormalities of the skin, which is the tissue affected by BPV in cattle. Skin tumors eventually appear on most of the BPV1.69 transgenic mice, and individual mice can show multiple pathologies, including large areas of abnormal skin and multiple, separate protuberant tumors. A severly affected mouse is shown in Figure 1.

The analysis of a considerable number of affected mice has revealed several distinguishable types of abnormality, as is shown in Figure 2. One frequent type is described as abnormal skin, which shows radical hair loss, black, wart-like regions which are called blue nevi, and dramatic thickening and hardening of the skin layer. A second type, also frequent, consists of slow-growing protuberant tumors. These protuberant tumors most frequently arise around the head and neck, and, in heterozygous mice, on the tips of their tails, which are clipped at 3 weeks of age for DNA analysis. A third class of pathology is fast-growing tumors that can appear in a region of otherwise normal tissue. These aggressive tumors are quite rare, and only 2–3 have been observed in the $\sim$100 mice which have been studied. There have been no reproducible examples of tumors or other abnormalities occurring in internal tissues of these transgenic mice.

There are two notable aspects to the development of these abnormal skin pathologies. The first is the long latency period. In spite of the fact that every cell in the body of these transgenic mice harbors the BPV genome, the skin abnormalities do not begin to appear until 8–9 months of age. The second point of interest is that tumors first appear in sites that are prone to irritation and wounding – in particular, around the head, face, and neck.

HISTOPATHOLOGY OF AFFECTED SKIN

Normal skin is composed of two distinct tissue layers – the dermis and the epidermis. The epidermis consists of a layer of basal stem cells which underly layers of progressively differentiated keratinocytes. The basal epithelium contacts the dermis, which is composed of a layer of fibroblasts that supports the growth and regulation of the epidermis. Embedded in the dermis are various glands, hair follicles, and clusters of melanocytes. A standard thin section of normal skin is shown in Figure 3A, for comparison with the various types of affected skin in the BPV transgenic mice.

Abnormal skin is characterized by a loss of hair, general thickening of the skin layer, and the presence of small wart-like protrusions. The histochemical analyses show that this tissue has undergone a dramatic hyperplasia of the dermal fibroblasts, resulting in a significant thickening of the dermis (Figure 3B). The epidermal layer is characteristically atrophic, generally being only 1–2 cells thick. The dermal proliferation is apparently accompanied by the disorganization of the sweat glands, and the loss of hair follicles. In addition, there are often focal proliferations of melanocytes, which produce the small, black wart-like protrusions that are found scattered throughout regions of abnormal skin.

The protuberant tumors show hyperplasia of both dermal and epidermal tissue, as is shown in Figure 3C. In these cases there is again dramatic proliferation of the dermal fibroblasts, which are observed in dense sworls of cells. In contrast to abnormal skin, proliferation of epidermal tissue is observed. The hyperplasia of the epidermis is clearly abnormal with respect to both the number and alignment of the keratinocytes. It is possible that the types and sequential progression of the differentiating keratinocytes are abnormal as well. As in the abnormal skin, there are few, if any, normal glands or hair follicles. However, in contrast to abnormal skin, focal proliferations of melanocytes are not observed. Thus these neoplasms can be classified as fibropapillomas composed of dermal fibroblasts and squamous epithelial cells.

The third class of abnormality is one of rapidly growing protuberant tumors. As shown in Figure 3D, the cells of this tumor type are anaplastic. The cells are multinucleated, show frequent mitotic spindles, and do not have a characteristic

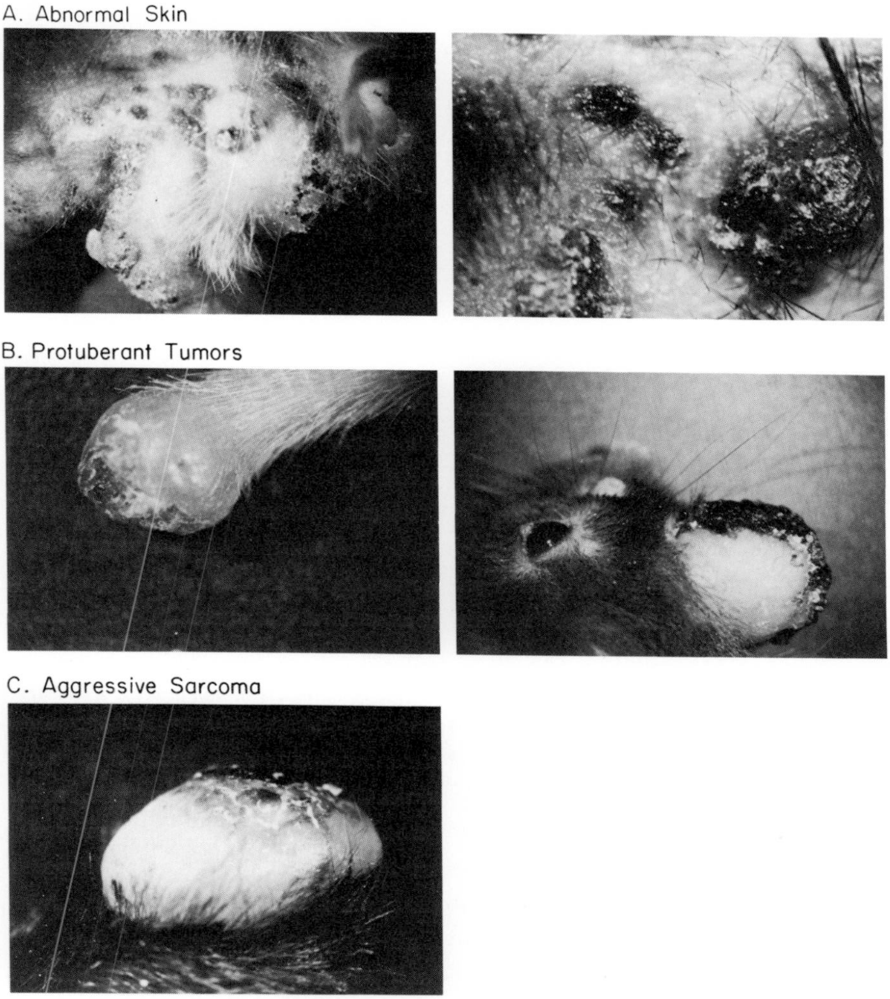

Fig. 2. Examples of the three characteristic tumor pathologies observed. A. Abnormal skin, with crusted sections, and small wart-like protrusions (called blue nevi). B. (Left) A tumor growing out from the tip of a tail, which had been clipped at 3 wk of age. (Right) A protuberant tumor which encapsulates the right eye. C. A sarcoma or fibrosarcoma that burst out of an area of otherwise normal skin.

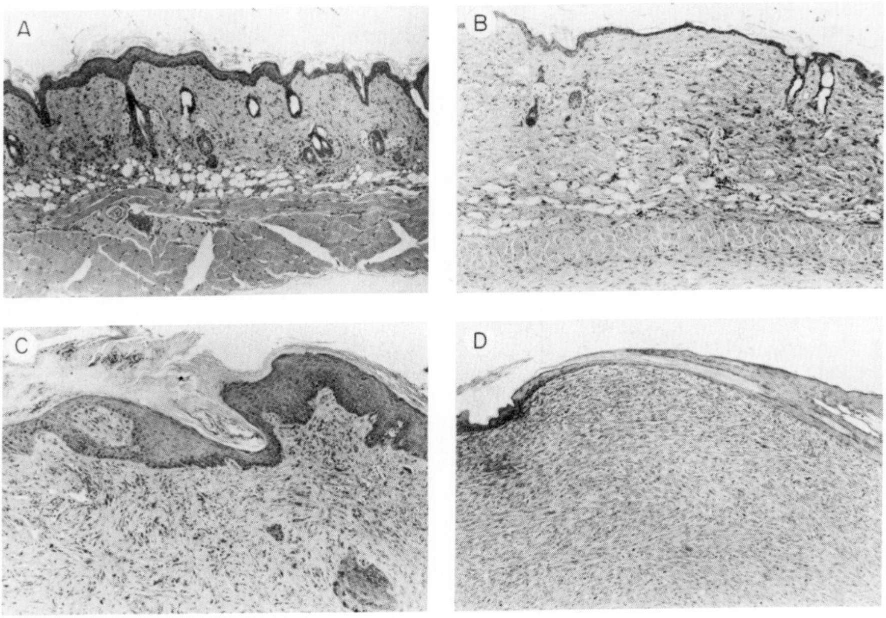

Fig. 3. Histopathology of the tumor types. Thin sections of formalin-fixed tissue were stained with hematoxylin and eosin (×40). A. Normal skin, showing the epidermis and dermis overlying a layer of fat. Hair follicles and glands are embedded in the dermis. B. Abnormal skin, with a very thin epidermal layer, enlarged dermal layer, and loss of glands and hair follicles. C. Protuberant tumor from the tip of a tail, where both dermis and epidermis are enlarged and disorganized. D. An anaplastic tumor, with dense multinucleated cells not readily identified as dermal or epidermal cell type.

dermal (or epidermal) appearance. These tumors are rare, and can be classified not as fibropapillomas but as aggressive sarcomas or fibrosarcomas.

DISCUSSION

The vertical transmission of the bovine papilloma virus genome through the mouse germ line produces a phenotype analogous to that observed in the natural horizontal infection of cattle. The development of various types of abnormalities in skin tissue suggests that there is a tissue specificity inherent in the viral genome, one that is separable from its ability to infect a particular cell type. The distinct tissue specificity demonstrates that transgenic mice can provide a model system for studying certain aspects of BPV-induced cell transformtion and carcinogenesis in the context of a whole animal.

An important quality of the oncogenesis which emerges from the analysis of

TABLE I
Characteristics of affected skin tissue

Tumor classification:	Abnormal Skin	Fibroepithelial Tumors	Sarcomas
Occurrence:	frequent	frequent	rare
Cell types:	dermal fibroblasts + focal clusters of melanocytes	dermal fibroblasts + epidermal keratinocytes	anaplastic cells (probably dermal in origin)
Notable features:	atrophic epidermis; loss of hair; disorganization of glands; thickening and hardening of skin; wart-like protrusions (blue nevi)	loss and disorganization of hair follicles and glands; lack of melanocytes and of blue nevi	aggressive; locally invasive; multinucleated

a considerable number of transgenic mice in this family is that the process is slow. Affected skin tissue is first apparent at 8–9 months of age, before which the transgenic mice are apparently normal. The facts of slow oncogenesis, and of the reproducible development of distinct types of tumors suggest that multiple events are required to produce these neoplastic conditions, and that distinct sets of events may be responsible for producing each type of tumor. The characteristics of the three classes of abnormalities are summarized in Table 1.

In all three classes of skin lesion induced by BPV we observe rearrangements in the BPV genome, in particular, the appearance of extrachromosomal copies of the unit length BPV genome (Lacey *et al.* 1986). It is unclear whether this is causal, or merely consequential, to the development of the transformed cells. DNA amplification occurs in all three classes of tumor and cannot, therefore, be used to distinguish the three pathologies. In addition, preliminary results indicate that all three types of tumor tissue express BPV RNA (M. Sippola-Thiele, D. Hanahan and P. Howley, unpublished observations), whereas unaffected tissues do not. Thus, although transcriptional activation of the BPV genome is clearly necessary for oncogenesis, it is unlikely to be responsible for the observed difference in the three types of affected skin tissue.

Initial analyses of the three tumor pathologies suggest that their differences most likely lie in the pattern of secondary changes which have occurred during oncogenesis. Support for this theory comes from ongoing experiments to establish each of the affected cell types in culture. Dermal fibroblasts established from abnormal skin and from protuberant tumors show cell-heritable differences. Cultured fibroblasts derived from abnormal skin are contact-inhibited, whereas fibroblasts originally populating fibroepithelial tumors spontaneously form foci in culture (M. Sippola-Thiele, D. Hanahan and P. Howley, unpublished observations). These preliminary results suggest that secondary mutations are necessary for tumorigenesis, and that these different events (or sets of events) may dictate the distinct tumor types. In light of this possibility, it notable that an apparent progression in transformed phenotype has been observed in cultured rodent cells harboring the BPV genome (Law *et al.* 1983, Cuzin *et al.* 1985). This is consistent with the occurrence of cell-heritable changes which improve cell proliferation. An additional characteristic of BPV-1 transformation *in vitro* relevant to this multistep model is the inability of BPV-1 to efficiently immortalize primary rodent cells to growth in culture (Cuzin *et al.* 1985). One can therefore postulate that the development of an "immortal" proliferative state in the skin cells of a transgenic mouse requires establishment events to complement the direct actions of the BPV oncogenes. In summary, this transgenic mouse model of skin carcinogenesis provides a system to study the regulation and activities of the BPV oncogenes, as well as to investigate the nature of the other changes which occur *in vivo* in order to convert a normal cell into a cancer cell.

ACKNOWLEDGMENTS

This research was funded by a grant from Monsanto Company, as well as by funds from the National Foundation for Cancer Research and the Leukemia Society of America. We wish to thank Jean Roberts and Dave Green for artwork and photography.

REFERENCES

Brinster, R. L., Chen, H. Y., Messing, A., van Dyke, T., Levine, A. J. & Palmiter, R. D. (1984) Transgenic mice harboring SV40 T-antigen genes develop characteristic brain tumors. *Cell 37*, 367–79.

Cuzin, F., Meneguzzi, G., Binetray, B., Cerni, C., Connan, G., Grisoni, M. & de Lapeyrieve, O. (1985) Stepwise tumoral progression in rodent fibroblasts transformed with bovine papilloma virus type 1 DNA. In: *Papillomaviruses: Molecular and Clinical Aspects*, pp. 473-86. Alan R. Liss, New York.

Hanahan, D. (1986) Oncogenesis in transgenic mice. In: *Oncogenes and Growth Control*, eds. Kahn, P. & Graf, T. Springer-Verlag, Heidelberg (in press).

Howley, P. M. & Schlegel, R. (1986) Papillomavirus transformation. In: *Papilloma viruses*, eds. Howley, P. M. & Salzman, N. Plenum Press, New York (in press).

Lacey, M., Alpert, S. & Hanahan, D. (1986) Bovine papillomavirus elicits skin tumors in transgenic mice. *Nature 322*, 609–12.

Lancaster, W. D. & Olson, C. (1982) Animal papillomaviruses. *Microbiol. Rev. 46*, 191–207.

Law, M. F., Byrne, J. C. & Howley, P. M. (1983) A stable bovine papilloma virus hybrid plasmid that expresses a dominant selective trait. *Mol. Cell. Biol. 3*, 2110–5.

Palmiter, R. D. & Brinster, R. L. (1986) Germline transformation of mice. *Ann. Rev. Genet. 20* (in press).

Pfister, H. (1984) Biology and biochemistry of papillomaviruses. *Rev. Physiol. Biochem. Pharmacol. 99*, 111–81.

DISCUSSION

YANIV: Did you analyze the three forms that you get in the tumor cells? What part of the genome did they delete?

HANAHAN: The analysis we have done is perhaps not sufficient to truly answer that. It appears that the primary form we have is a precise excision of the 100% BPV plasmid. We see different mobility changes on those blots because of the mass of DNA. We have not cloned the extra chromosomal BPV forms and subjected them to detailed analysis.

YANIV: Can you assume that the excision of the circular viral genome will switch on tumor formation?

HANAHAN: I do not know. You can argue that it is a cause or a consequence. I think that the precedent of the human papilloma viruses might argue that it is a consequence, such that the proliferating tissue is what is permissive for BPV replication and therefore what happens is simply a consequence of this proliferative condition. We are currently addressing this question, as we have derived transgenic mice that are carrying replication-defective transformation complement BPV genomes.

ZUR HAUSEN: You showed that the kidney cells *in vivo* do not transcribe the BPV RNA. Did you have a chance to look at the kidney cells kept in tissue culture? Was there any transcription or not?

HANAHAN: We have not done this experiment, to see whether one can get spontaneous activation in culture. I think it is important as well in terms of the possibility that activation is a rate-limiting event in tumor progression and could result in the formation of internal tumors as well. I might add though that we have not observed with any frequency reproducible occurrence of internal tumors. These mice get very old and abnormalities occur in them all, so we occasionally get tumors, but we have not been able to ascribe those to BPV.

WEINBERG: There is a small, but perhaps significant difference between what you are doing and what Balmain seems to be doing. As I understand it, in Balmain's work it is clear that the initiating event in these chemically-induced tumors would appear to be the acquisition of a *ras* oncogene. In fact, one can mimic initiation by putting a *ras* oncogene into keratinocytes. Those cells would seem to behave much differently from the ones that you are working with. One might almost argue that the genes that you are putting into cells represent, not initiating events in Balmain's system, but rather a set of genetic functions that are required late in Balmain's protocol of tumorigenesis. So, it may well be that the chronological order of events affecting the alteration of the tumor cells is exactly the opposite of what Balmain's (group) is seeing.

HANAHAN: I think that is quite possible. One of the opportunities here may be to try to correlate our results to what is known about classical skin carcinogenesis and of progression. We are trying classical initiators and promoters of skin cancer to see whether they will accelerate the rate of tumor formation. It may be that oncogenes and chemicals can play different roles in different places. But I would suspect that maybe the BPV oncogenes are involved at a later stage as well as initiation.

BALTIMORE: Have you tried if any of the compounds which the literature suggests are able to increase recombination or gene amplification. Or have you tried UV-radiation?

HANAHAN: We tried one experiment with UV-radiation which did not do anything. However, we have not yet done a sufficient study, which is certainly warranted. I would like to try gamma or X-radiation as well.

LOWY: I would like to comment about the biology of the skin carcinogenesis model versus what is known about BPV biology in rodents. The skin carcinogenesis model is for epidermal tumors. The animals initially develop papillomas, which then go on to develop into squamous cell carcinomas. It is well-recognized if you manipulate rodents with bovine papilloma virus that the animals develop dermal (rather than epidermal) tumors. In those instances with the transgenic mice the dermal component is much more apparent than the epidermal component. It is not clear whether the epidermal component actually represents neoplasia or whether it represents some kind of reactive hyperplasia due to the dermal proliferation that is underneath. In trying to relate the experience of skin carcinogenesis to BPV, I think it is important to keep in mind that one is probably primarily an epidermal tumor and the other a dermal tumor.

HANAHAN: Yes, I think that is a very good point.

LOWY: I think a particular utility of your system, if you could indeed show epidermal proliferation, would be if exposure of the mice to ultraviolet light resulted in epidermal tumors, not dermal tumors.

HANAHAN: I might add that the precedent of this does motivate an attempt to do the same thing directly to the epidermis. We have derived mice now that have several human papilloma viruses which produce epithelial tumors in humans (not fibroepithelial). I think there we can ask the same sort of questions and perhaps begin to compare and contrast the viral oncogene model with the classical skin model. Obviously, in this case, addressing the epidermal component is really important.

ZUR HAUSEN: You mentioned that wounding had some effect on the development of the tumors. Just to mention it, in the two-stage mouse carcinogenesis model wounding has a promoting effect.

HANAHAN: Yes, and in fact there is even some evidence of this in internal tumors. People tie sutures around the liver and things like that which seems to provoke a carcinogenesis. So, there certainly is precedent for that in the classical literature.

PEDERSEN: Have you looked for viral particles in any of the tumors or the tissues in the animals?

HANAHAN: Yes, first of all we have seen no transmission of this phenotype to normal mice following extended cohabitation. We have analyzed a couple of tumors in a virus transformation assay, where we mince the tumor and spread it on C 127 cells, and finally Peter Howley has looked very extensively with antiserum that is specific with capsid proteins and was unable to detect them. So, we believe that the virus particles are not being produced, nor is infectious virus.

WONG-STAAL: Have you looked at the expression of other oncogenes, particularly *sis,* in these tumors?

HANAHAN: No, that is something that we will try to do.

Multiple Bovine Papillomavirus Genes Influence Transformation of Mouse Cells

Douglas R. Lowy, William C. Vass, Elliot J. Androphy & John T. Schiller

Papillomaviruses (PV) induce benign tumors (warts) on cutaneous and squamous mucosal surfaces of many vertebrates, including humans. A subset of the benign lesions induced by some non-human PV have been shown to undergo malignant conversion, and interest in PV has has been stimulated further by the close association in humans between certain epithelial malignancies and infection with particular human (H) PV types (Gissmann 1984, Pfister 1984, Jablonska & Orth 1985, zur Hausen *et al.*, this volume). *In vitro* studies of HPV have been limited because these viruses have still not been propagated in cultured cells, although there are preliminary reports of *in vitro* cell transformation by HPVs (Watts *et al.* 1984, Yasumoto *et al.* 1986).

Bovine papillomavirus type 1 (BPV) is the prototype of a group of PV that induce fibropapillomas in their natural host, in contrast to the purely epithelial growths induced by HPVs (Lancaster & Olson 1984). As is true of all PV lesions, full BPV replication is limited to the differentiating layers of the infected squamous epithelium. The additional capacity of BPV to induce non-productive transformation of the dermal fibroblasts located under these epithelial cells distinguishes BPV biologically from HPV. This expanded host range of BPV extends to cultured cells, where BPV, in contrast to HPV, readily induces non-productive tumorigenic transformation of certain cell lines, such as mouse NIH 3T3 and C127 (Dvoretzky *et al.* 1980). Consequently, BPV is the PV whose genetics has been studied in greatest detail, and certain specific functions have begun to be assigned to defined regions within the BPV genome. Since DNA sequence analysis suggests that all PV possess a common genetic organization, it seems likely that, despite the greater host range of BPV, an understanding of BPV genetics will be relevant to other PV including HPVs.

Laboratory of Cellular Oncology, Building 37, Room 1B-26, National Cancer Institute, Bethesda, Maryland 20892, USA.

VIRAL CARCINOGENESIS, Alfred Benzon Symposium 24,
Editors: N. O. Kjeldgaard, J. Forchhammer, Munksgaard, Copenhagen 1987.

BPV TRANSFORMATION

This report summarizes genetic aspects of BPV transformation *in vitro*. Infection of NIH 3T3 and C127 mouse cells by BPV virions or the full-length 8 kb BPV DNA genome induces focally transformed tumorigenic cells in which the BPV genome is maintained as a multicopy episome (Law *et al.* 1981). Cellular transformation by BPV is a complex phenomenon that depends upon multiple viral factors, only some of which have been defined. The results obtained to date indicate that BPV contains 2 genes which can independently transform mouse cells. Other viral genes encode *trans* acting functions that positively or negatively influence BPV transformation. In addition to these viral determinants, it is likely that as yet undefined viral and cell factors will also be found to influence BPV transformation.

The 8 kb BPV DNA genome contains at least 10 open reading frames (ORF) greater than 100 nucleotides in length (Figure 1); the same strand contains all of these ORFs, and the only detectable viral RNAs are transcribed from this strand (in a left to right orientation in Figure 1; Danos *et al.* 1983, Stenlund *et al.* 1985). The L1 and L2 ORFs, which encode the major structural viral proteins, are not expressed as RNA in the dermal portion of BPV-induced fibropapillomas or in mouse cells transformed by BPV (Engel *et al.* 1983). These results correlate well with the earlier finding that the viral sequences required for full BPV transformation and episomal maintenance of the viral DNA were localized to a 5.4 kb viral DNA fragment (called 69T because it contains 69% of the viral genome) that lacked L1 and L2 but contained the other 8 ORFs (E1–8) (Lowy *et al.* 1980).

TRANSFORMATION BY E6

To investigate the potential transforming activity of specific ORFs within 69T, we have chosen to study their biological potential by insertion and deletion analyses of relatively small subgenomic BPV DNA fragments that have been activated with a retroviral long terminal repeat (Schiller *et al.* 1984), rather than study the effect of mutants within the context of the full-length BPV genome. The viral functions that control replication and expression of the viral genome are complex and poorly understood. Since they might indirectly affect BPV transformation, it seemed prudent initially to examine the transforming activity of the individual ORFs independently of BPV regulatory elements.

A clone (pXH800) that contained the entire E6 and E7 ORFs, but not E1–E5, was found to transform C127 cells (Figure 1). In contrast to the full length BPV

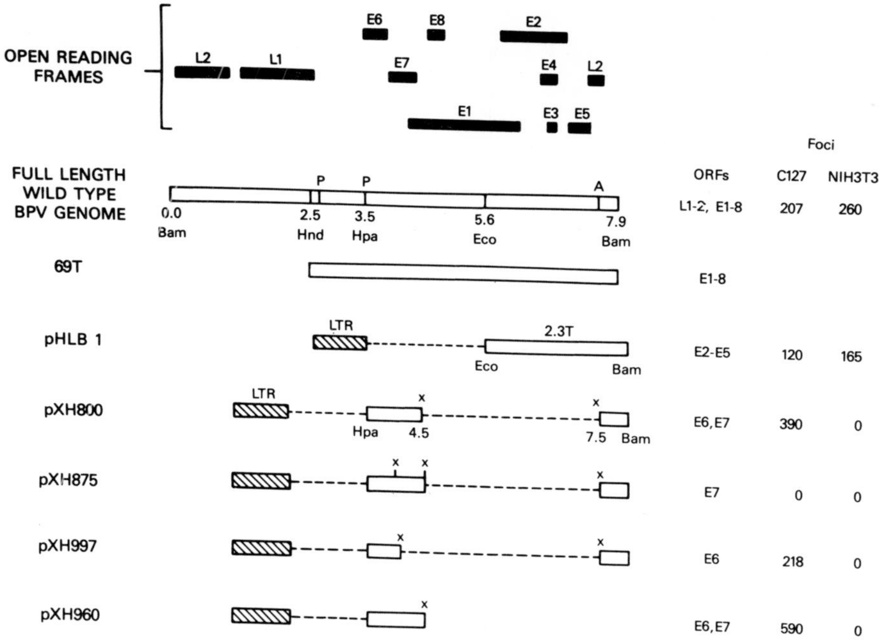

Fig. 1. BPV genome and transformation by E6. The genome of BPV-1 linearized at the unique BamHI site in which it was cloned into the pBR322 derivative pML2d is shown (Sarver *et al.* 1982). The numbers below the BPV genome indicate the distance in kb from the BamHI site at 0.0. The locations of putative promoters and polyadenylation signal are designated "P" and "A" respectively. Open bars represent BPV sequences included in the clones, and dashed lines indicate deleted sequences. Sites of XhoI linker insertions are denoted by "X". Bam = BamHI; Hnd = HindIII; Hpa = HpaI; and Eco = EcoRI recognition sites. (From Schiller *et al.* 1984).

genome, this construction did not induce focal transformation of NIH 3T3 cells – although the murine LTR functions efficiently in this latter cell line – suggesting that another BPV gene mediates BPV transformation of NIH 3T3 cells. A series of XhoI linker insertion and deletion mutants were then employed to determine if the transforming activity resulted from E6, E7, or from sequences in both ORFs, since an E6–7 fusion gene product has been predicted from analysis of BPV mRNA in C127 cells transformed by the full length genome (Yang *et al.* 1985). These mutants localized the transforming function to E6, since lesions in E7 had little effect on transformation but mutations in E6 abolished this activity (Figure 1). The E6 ORF of the HPVs detected in human tumor cell lines is selectively retained and expressed, suggesting an important role for E6 in the establishment or maintenance of these tumors (Schwarz *et al.* 1985).

THE E6 GENE PRODUCT

Having demonstrated the transforming capacity of E6, we then sought to identify the protein product of this gene in transformed cells (Androphy *et al.* 1985). No non-structural PV-encoded protein had been identified previously, presumably because, even in morphologically transformed cells, BPV mRNA levels (and their translation products) are quite low (Engel *et al.* 1983). We reasoned that it should be easier to detect the E6 product in cells transformed by the LTR activated E6 gene, since they expressed E6 RNA levels that were 5–10 times higher than cells transformed by full-length BPV DNA.

Large quantities of an E6 polypeptide that could be used to immunize rabbits were prepared by expressing the E6 ORF as a fusion protein in bacteria. The rabbit sera were first shown to react specifically with the E6 portion of the fusion protein and then tested for their ability to immunoprecipitate proteins from C127 cells transformed by the LTR-activated E6 ORF (clones pXH800 and pXH997 in Figure 1).

When extracts from these transformed cells were tested, the rabbit sera immunoprecipitated a protein whose migration rate corresponded to 15.5 kd (Figure 2), which corresponds to the size expected for an unmodified product that was translated from the first AUG of the E6 ORF. Cells transformed by the full-length BPV genome also contain this protein, but at significantly lower levels (not shown). That this protein is encoded by E6 was confirmed by its absence in cells transformed by other genes (Figure 2) and the ability of bacterial extracts containing the E6 fusion protein to specifically inhibit its immunoprecipitation from E6-transformed mouse cells (not shown). Sub-cellular fractionation studies indicate that the protein is divided about equally between the nucleus and non-nuclear membranes. It is not clear how the protein associates with the membrane fraction, since its predicted amino acid sequence does not have a large hydrophobic domain or other sequences that might help direct the protein to membranes.

TRANSFORMATION BY E5

Prior to identifying E6 as a transforming gene located at the 5′ end of 69T, an LTR-activated sub-genomic BPV fragment containing the 3′ end of 69T (pHLB1 in Figure 1 and Figure 3) had been shown to be able to transform both NIH 3T3 and C127 cells (Nakabayashi *et al.* 1983). Although this fragment was only 2.3 kb long, it contained 4 intact ORF (E2, E3, E4, and E5) as well as the 3′ end of E1. In order to determine which BPV sequences were required for transform-

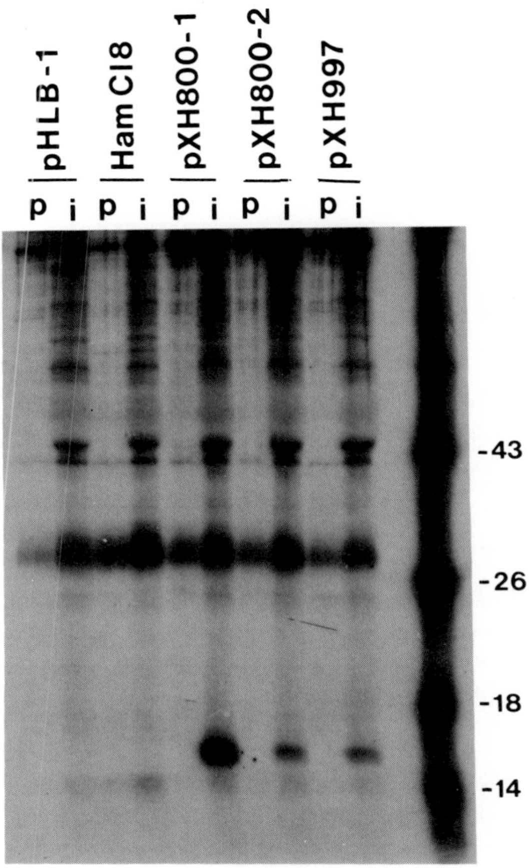

Fig. 2. Immunoprecipitation of E6 from transformed cell lines. p = pre-immune sera, i = immune sera. 25×10^6 CPM of ^{35}S-cysteine labeled cells were used per lane. Cell lines are designated by the plasmids (Figure 1) that were used to generate them. Ham C18 is a *ras*^{H} transformed C127 cell line. (From Androphy *et al.* 1985).

ation by pHLB1, a series of XhoI linker insertion/deletion mutants were generated and analyzed for their transforming activity (Figure 3) (Schiller *et al.* 1986).

Prior to undertaking this analysis, it was believed that E2 would be the transforming gene, since it was the most highly conserved of the 4 complete ORFs in pHLB1. However, the results indicated that sequence in E5 were required for transformation by pHLB1 and that E2, E3, and E4 were not required for this activity (Figure 3). The introduction of an XhoI linker at multiple locations

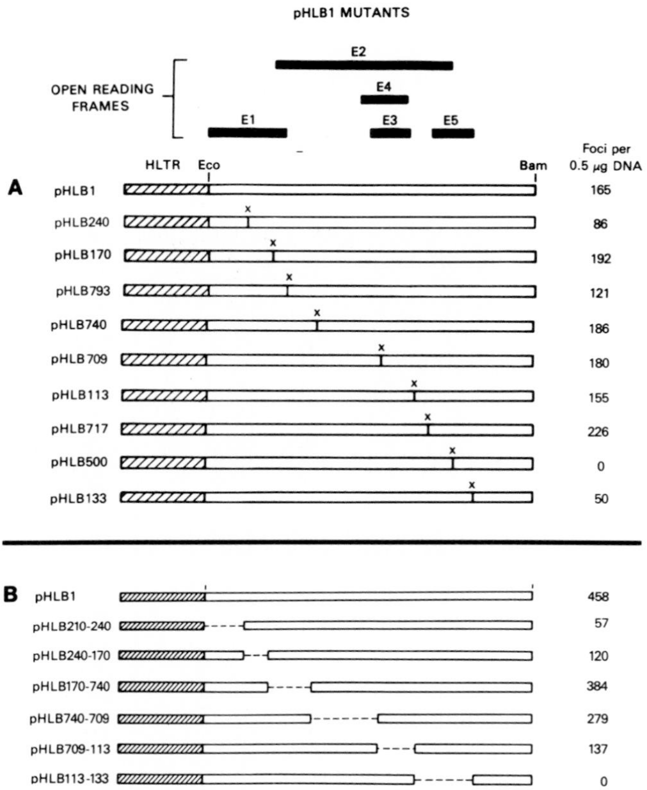

Fig. 3. Mutational analysis of the 3′ transforming region. Designations are the same as in Figure 1. Only the carboxy-terminus of E1 is present in the 2.3 kb EcoRI to BamHI fragment of pHLB1. DNAs were transfected onto NIH3T3 cells in both experiments A and B. (From Schiller *et al.* 1986).

within E2 did not significantly reduce the transforming activity of pHLB1, although one of these XhoI linkers with intact transforming activity simultaneously introduced in a frame shift in E2, E3, and E4. On the other hand, a frame shift mutation limited to E5 (clone pHLB500) was transformation defective. Results obtained with deletion mutants strongly suggested that coding sequences outside E5 are not required for transformation by pHLB1. ORFs that by sequence analysis may be analogous to BPV E5 have been noted only in some HPVs (Schwarz *et al.* 1983), but they may be common to all animal PVs that induce fibropapillomas (Groff & Lancaster 1985).

The BPV E5 ORF by itself would be predicted to encode a very small (44 amino acid) product that is extremely hydrophobic. Using antibodies generated from a synthetic peptide encoding the C-terminal amino acids of E5, Schlegel *et al.* (1986) have detected this predicted (7 kd) product in cells transformed by BPV E5. As might be expected for a peptide that is composed principally of hydrophobic residues, the E5 protein has been found in the membrane fraction of the transformed cells.

E1 INHIBITS CELL TRANSFORMATION

In addition to these 2 transforming genes, other BPV products can influence the transforming activity of the BPV genome positively and negatively. The positive effects on transformation would appear to be indirect, since our analysis of LTR-activated BPV DNA fragments suggest that only E5 and E6 can directly transform mouse C127 cells. Furthermore, a double mutant of the full length BPV DNA carrying frame shift mutations limited to E5 and E6 is transformation defective.

One inhibitory effect we have noted involves a product of the E1 ORF. Sequences within E1 have previously been shown to be required for maintenance of the BPV DNA as an extrachromosomal element (Lusky & Botchan 1985). We have found that each of a series of full length BPV DNA mutants with mutations in E1 (created via the insertion of single XhoI linkers) transform C127 cells more efficiently than does the wild type BPV DNA (Schiller *et al.* manuscript in preparation). This increased efficiency is seen both with respect to the number of foci of transformed cells per microgram of DNA, as well as in the ability of the E1 mutants to produce much larger colonies in soft agar.

Two different experiments suggest that the increased transforming activity of the mutants results from loss of an E1-encoded function. First, cotransfection of wild type BPV and E1 mutants resulted in suppression of the high transformation phenotype of the E1 mutants. The second experiment tested the ability of a plasmid carrying an LTR-activated E1 ORF (pE1) to suppress the high transformation phenotype. In cotransfection experiments, pE1 was able to suppress the high transformation phenotype of a full length BPV DNA plasmid carrying an E1 mutation, while an isogenic construction with a frame shift mutation in E1 (pE1-760) had no effect (Figure 4). These results strongly suggest that E1 encodes a *trans* acting (protein) product that inhibits BPV transformation.

The HPV genomes associated with human genital tumors are usually integrated, and the E1 ORF is often interrupted or deleted (Schwarz *et al.* 1985). When considered in conjunction with our BPV E1 data, it may be postulated that the E1 product of HPVs may inhibit tumor progression.

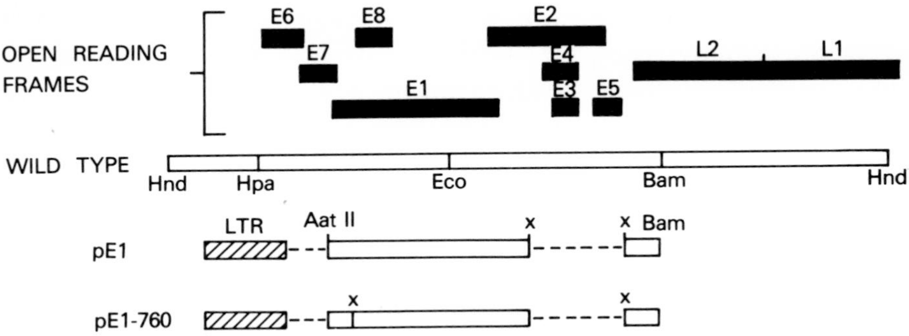

DNA clones Transfected	Foci per 0.5 μg DNA
pE1	0
pE1-760	0
FL-WT	22
FL-WT+pE1	30
FL-WT+pE1-760	12
FL-760	330
FL-760+pE1	63
FL-760+pE1-760	366

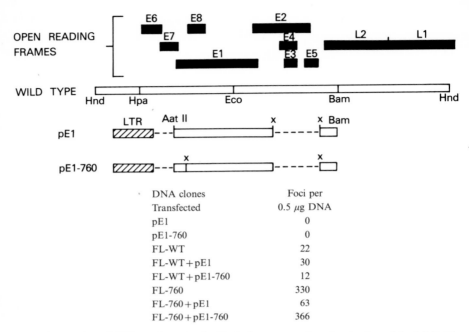

Fig. 4. Suppression of BPV transformation by E1. In the upper portion, the structure of the full length BPV genome, E1 expression vector (pE1), and an E1 frame shift mutant (pE1-760) are shown. In the experiment (lower portion), pE1 and pE1-760 were cotransfected onto C127 cells either with the full-length wild type BPV DNA (FL-WT) or with a full-length BPV carrying the same E1 frame shift mutation as in pE1-760 (FL-760).

E2 ENHANCES TRANSFORMATION AND BINDS AN E2 PEPTIDE TO BPV DNA

Others have shown that mutations in E2 significantly decrease the transforming activity of full length BPV DNA (Sarver *et al.* 1984, de Maio 1985). In contrast to E5 and E6, however, E2 does not appear to encode a protein that directly transforms cells. Rather, E2 probably encodes a product that enhances the transcriptional activity of BPV genes (such as E6) whose expression depends on the upstream regulatory region (URR) of the BPV genome (located between 2.5 and 3.5 [HindIII/HpaI], see Figure 1). This hypothesis is based on the experimental observation that the BPV URR can function as an enhancer element in the presence of a *trans* acting factor encoded by E2 (Spalholz *et al.* 1985).

We have begun to dissect the mechanism by which E2 enhances URR-mediated transcription (Androphy *et al.,* in press). Some virally encoded activators of transcription, such as the E1A product of adenovirus, act indirectly via cellular factors that enhance viral RNA transcription. By contrast, our biochemical experiments suggest that E2 enhancement results from the direct interaction of an E2-encoded peptide and the BPV URR.

If the E2 product is derived from the E2 ORF, it would be a 410 amino acid product, assuming the first AUG in the ORF is utilized. In our experiments, we have expressed the C-terminal 297 amino acids as a bacterial fusion protein that also includes a small number of N-terminal residues from the lambda CII gene. Rabbits immunized with this peptide developed antibodies that reacted specifically with the E2 determinants in the fusion protein.

We assessed the ability of the E2 fusion protein to bind to restriction endonuclease-digested BPV DNA fragments in an immunoprecipitation assay (McKay 1981). Under these conditions, the protein bound only to a small subset of the BPV DNA fragments (Figure 5, lane 2). The 2 fragments to which the fusion protein bound contain sequences from the URR. Specific binding occurred only with double-stranded fragments: single-stranded DNA was bound non-specifically (lane 3). Additional experiments indicated that BPV contains at least 5 binding sites that vary in their affinity for E2. Four of them are located in the URR; the other, which is among the weaker binding sites, is located in E2. Three of the binding sites in the URR have been localized to restriction endonuclease fragments that are 54–70 nucleotides in length. Inspection of the DNA sequences in the fragments to which the fusion protein binds reveals a common motif whose prototype may be ACC(G)XXXPyCGGT(G)(C). This motif has been noted to be conserved among the URRs of all PV that have been sequenced (Dartmann *et al.* 1986). Each fragment to which the fusion protein binds contains at least one copy of this motif with a match for at least 9 of the 11 presumably important nucleotides. Consistent with this hypothesis, the BPV E2 fusion protein also binds specifically to an HPV-16 restriction endonuclease fragment. Not only is this fragment located in the HPV-16 URR, it also contains a segment that matches 9 nucleotides in the 11 nucleotide BPV motif.

These results strongly imply that E2 functions by binding specifically to multiple sites in the URR. Before this hypothesis can be accepted, genetic studies must validate the functional significance of these binding sites for an authentic E2 product.

CONCLUSIONS

In summary, our studies on BPV indicate that the transforming activity of the viral genome results from the complex interaction between genes that can directly

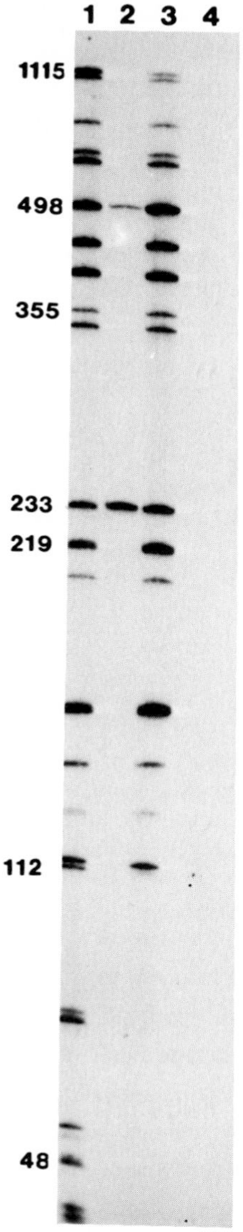

Fig. 5. Protein-DNA complex immunoprecipitation assay. 50 ng of partially purified E2 was immunoprecipitated with E2-specific antisera. The complexes were then incubated with full-length BPV DNA that had been digested with Sau96I, BamHI, and HindIII and end-labeled with ^{32}P-dNTPs. After dissociation and denaturation, DNA was analyzed on an acrylamide sequencing gel. Lane 1, end-labeled marker DNA; lane 2, immunoprecipitation with E2 protein and E2 antisera; lane 3, as lane 2, except that DNA was denatured prior to binding; lane 4, E2 protein and pre-immune sera. (From Androphy *et al.,* in press.)

transform cells (E5 and E6) and those that can influence transformation indirectly in either a negative (E1) or positive (E2) manner. It is likely that additional viral and cellular factors that significantly influence transformation will be identified as greater understanding of BPV genetics develops. Identification of the mechanisms by which these products function should provide important insights into BPV biology and may also have relevance to the involvement of HPV in benign and malignant disease.

REFERENCES

Androphy, E. J., Lowy, D. R. & Schiller, J. T. (1987) A peptide encoded by the bovine papillomavirus E2 *trans*-activating gene binds to specific sites in papillomavirus DNA. *Nature (London)* (in press).

Androphy, E. J., Schiller, J. T. & Lowy, D. R. (1985) Identification of the protein encoded by the E6 transforming gene of bovine papillomavirus. *Science 230*, 442–5.

Danos, O., Engel, L. W., Chen, E. Y., Yaniv, M. & Howley, P. M. (1983) Comparative analysis of the human type 1a and bovine type 1 papillomavirus genomes. *J. Virol. 46*, 557–66.

Dartmenn, K., Schwarz, E., Gissmann, L. & zur Hausen, H. (1986) The nucleotide sequence and genome organization of human papilloma virus type 11. *Virology 151*, 124–30.

DiMaio, D. (1986) Nonsense mutation in open reading frame E2 of bovine papillomavirus DNA. *J. Virol. 57*, 475–80.

Dvoretzky, I., Shober, R., Chattopadhyay, S. K. & Lowy, D. R. (1980) A quantitative in vitro focus assay in mouse cells for bovine papillomavirus. *Virology 103*, 369–75.

Engel, L. W., Heilman, C. A. & Howley, P. M. (1983) Transcriptional organization of the bovine papillomavirus type 1. *J. Virol. 47*, 516–28.

Gissmann, L. (1984) Papillomaviruses and their association to cancer in animals and in man. *Cancer Surv. 3*, 161–81.

Groff, D. E. & Lancaster, W. D. (1985) Molecular cloning and nucleotide sequence of deer papillomavirus. *J. Virol. 56*, 85–91.

Jablonska, S. & Orth, G. (1985) Epidermodysplasia Verruciformis. In: *Clinics in Dermatology: Warts/Human Papillomaviruses*, eds. Jablonska, S. & Orth, G., Vol. 3, No. 4, pp. 83–96. J. B. Lippincott, Philadelphia.

Lancaster, S. E. & Olson, C. (1982) Animal papillomaviruses. *Microbiol. Rev. 46*, 191–207.

Law, M.-F., Lowy, D. R., Dvoretzky, I. & Howley, P. M. (1981) Mouse cells transformed by bovine papillomavirus contain only extrachromosomal viral DNA sequences. *Proc. Natl. Acad. Sci. U.S.A. 78*, 2727–31.

Lowy, D. R., Dvoretsky, I., Shober, R., Law, M.-F., Engel, L. & Howley, P. M. (1980) *In vitro* transformation by a defined subgenomic fragment of bovine papillomavirus DNA. *Nature (London) 287*, 72–4.

Lusky, M. & Botchan, M. R. (1985) Genetic analysis of bovine papillomavirus type 1 *trans*-acting replication factors. *J. Virol. 53*, 955–65.

McKay, R. D. G. (1981) Binding of a simian virus 40 T antigen-related protein to DNA. *J. Mol. Biol. 145*, 471–88.

Nakabayashi, Y., Chattopadhyay, S. K. & Lowy, D. R. (1983) Organization of the transforming function of bovine papillomavirus DNA. *Proc. Natl. Acad. Sci. U.S.A. 80*, 5832–6.

Pfister, H. (1984) Biology and biochemistry of papillomaviruses. *Rev. Physiol. Biochem. Pharmacol. 99*, 112–68.

Sarver, N., Byrne, J. C. & Howley, P. M. (1982) Transformation and replication in mouse cells of a bovine papillomavirus-pML2 plasmid vector that can be rescued in bacteria. *Proc. Natl. Acad. Sci. USA 79*, 7147–51.

Sarver, N., Rabson, M. S., Yang, Y.-C., Byrne, J. C. & Howley, P. M. (1984) Localization and analysis of bovine papillomavirus type 1 transforming functions. *J. Virol. 52*, 377–88.

Schiller, J. T., Vass, W. C. & Lowy, D. R. (1984) Identification of a second transforming region in bovine papillomavirus DNA. *Proc. Natl. Acad. Sci. USA 81*, 7880–4.

Schiller, J. T., Vass, W. C., Vousden, K. H. & Lowy, D. R. (1986) E5 open reading frame of bovine papillomavirus type 1 encodes a transforming gene. *J. Virol. 57*, 1–6.

Schlegel, R., Wade-Glass, M., Rabson, M. S. & Yang, Y.-C. (1986) The E5 transforming gene of bovine papillomavirus encodes a small, hydrophobic polypeptide. *Science 233*, 464–7.

Schwarz, E., Durst, M., Demankowski, C., Lattermann, O., Zech, R. Wolfsperger, E., Suhai, S. & zur Hausen, H. (1983) DNA sequence and genome organization of genital human papillomavirus type 6b. *EMBO J. 2*, 2341–8.

Schwarz, E., Freese, U. K., Gissmann, L., Mayer, W., Roggenbuck, B. Stremlau, A. & zur Hausen, H. (1985) Structure and transcription of human papillomavirus sequences in cervical carcinoma cells. *Nature 314*, 111–4.

Spalholz, B. A., Yang, Y.-C. & Howley, P. M. (1985) Transactivation of a bovine papilloma virus transcriptional regulatory element by the E2 gene product. *Cell 42*, 183–91.

Stenlund, A., Zabielski, J., Ahola, H., Moreno-Lopez, H. & Pettersson, U. (1985) Messenger RNAs from the transforming region of bovine papilloma virus type I. *J. Mol. Biol. 182*, 541–54.

Watts, S. L., Phelps, W. C., Ostrow, R. S., Zachow, K. R. & Faras, A. J. (1984) Cellular transformation by human papillomavirus DNA *in vitro*. *Science 225*, 634–6.

Yang, Y.-C., Okayama, H. & Howley, P. M. (1985) Bovine papillomavirus contains multiple transforming genes. *Proc. Natl. Acad. Sci. USA 82*, 1030–4.

Yasumoto, S., Burkhardt, A. L., Doniger, J. & DiPaolo, J. A. (1986) Human papillomavirus type 16 DNA-induced malignant transformation of NIH 3T3 cells. *J. Virol. 57*, 572–7.

DISCUSSION

WONG-STAAL: Have you demonstrated that the sequences that bind the E2 protein are indeed the target sequences for transactivation?

LOWY: No.

ZUR HAUSEN: Are you postulating that the E2 equivalent exists exclusively in the malignant tumors or will be repressed in malignant tumors?

LOWY: I do not have a preconceived notion about this speculation.

HANAHAN: I have two questions: The first is in reference to the mutation that Mike Botchan has produced in this region, which results in an overreplication of episomal BPV. I was wondering whether in your supertransformed lines you have looked at the copy number of BPV.

LOWY: It seems to be integrated.

HANAHAN: The second question is in terms of E5 being deleted in human papillomas. Is there any sense that the two oncogenes could each be more effective in the different cell types. Is there any indication *in vivo* in bovine tumors where the E5 and E6 oncogenes are? Is one predominantly expressed in epidermal cells, and the other preferentially in dermal fibroblasts.

LOWY: I do not think that there is any direct evidence using subgenomic DNAs or deleted DNAs that one is more potent than the other. But I think that all of the papillomaviruses that cause fibropapillomas have E5 open reading frames, and one would presume that E5 was important for fibroblast transformation. I should point out, however, that HPV6 also has a similar E5 open reading frame, and it does not cause fibropapillomas.

BALTIMORE: Do you have any thoughts about how a short peptide like that, and one that is that so hydrophobic, could transform? How small is it?

LOWY: 44 amino acids.

BERNS: Have you an idea if E1 or E2 is dominant, one dominant over the other, if you express both to high levels, what are the effects in transformation?

LOWY: No, I really do not.

YANIV: In some cases E2 does not transactivate and even represses as in the case of HPV18 promoter.

LOWY: It may be very complex. Obviously, Dr. Roizman can talk about particular transacting factors that were simultaneously either inhibited or enhanced depending on the situation.

WONG-STAAL: Have you looked at the distribution of E6 in BPV-infected cells that are not morphologically transformed?

LOWY: The level of protein is too low for us to detect it after fractionation.

VII. Cytoskeleton and Activation of C-ONC

Interaction between Herpes Simplex Virus and the Cellular Cytoskeleton Structures

B. Norrild[1], L. N. Nielsen[1] & J. Forchhammer[2]

Three fibrillar structures of the cytoskeleton of tissue culture cells have been well characterized, the actin-containing microfilaments, the microtubules and the intermediate filaments. The actins and the tubulins are present in all cells, but the intermediate filament proteins are specific for each category of cells as demonstrated by the production of antibodies specific for the various proteins (for reviews see Weber & Osborn 1981, Sun & Green 1978, Dustin 1978, Osborn & Weber 1983, Moll *et al.* 1982).

The main function of the microfilaments is apparently maintenance of cell morphology (Tilney 1983), while microtubules are important for intracellular transport and motility (Schliwa 1984, Pollard *et al.* 1976, Dustin 1978), but the function of the intermediate filaments is largely unknown. However, in fibroblasts the vimentin-containing intermediate filaments might anchor the nuclei (Lehto *et al.* 1978, Menko *et al.* 1983), and in epithelial cells the keratin protein expression might program the maturation of the keratinocytes (Sun *et al.* 1983, 1984, Fuchs & Green 1980). The functions of the cytoskeleton fibrillar structures might be further elucidated by analysis of the changes induced in the microfilaments, microtubules and intermediate filaments in cells infected with either RNA or DNA virus.

The present review will describe the interactions between the cell cytoskeleton and viral proteins or nucleic acid. The main emphasis will be on the possible involvement of the cytoskeleton elements of human fibroblasts and rat epithelial cells in the intracellular transport and sorting of herpes simplex virus particles and proteins.

---

[1]Institute of Medical Microbiology, Juliane Maries Vej 22, DK-2100 Copenhagen and [2]The Fibiger Institute, Danish Cancer Society, Copenhagen, Denmark.

VIRAL CARCINOGENESIS, Alfred Benzon Symposium 24,
Editors: N. O. Kjeldgaard, J. Forchhammer, Munksgaard, Copenhagen 1987.

TABLE I

RNA Virus	Micro-filaments	Micro-tubules	Intermediate filaments	References
Reo	–	–	Disruption/CPE	Sharpe *et al.* *Virology 120*, 399 (1982)
VSV	Reduction	–	–	Rutter *et al.* *J. Gen. Virol. 37*, 233 (1977)
NDV/Measles/ Mumps	Increase	–	–	Fagraeus *et al.* *Arch. Virol. 57*, 291 (1978)
VSV		m-RNA attachment/ Translation		Cervera *et al.* *Cell 23*, 113 (1981)
Polio	–	–	RNA attachment/ Replication	Lenk *et al.* *Cell 16*, 289 (1979)

A. MORPHOLOGIC CHANGES OF MICROFILAMENTS, MICROTUBULES AND INTERMEDIATE FILAMENTS IN VIRUS-INFECTED CELLS

The changed organization of the microfilaments, microtubules and intermediate filaments in virus-infected cells is summarized in Tables I and II. Among the RNA, viruses, only reovirus induces a disruption of the intermediate filaments (Sharpe *et al.* 1982), but vesicular stomatitis virus (VSV), measles and mumps infections all change the number of microfilaments in the cells (Rutter & Mannweiler 1977). In poliovirus-infected cells, RNA is apparently associated with the intermediate filaments during replication and the function of the filaments seems to be that of a structural support matrix (Lenk *et al.* 1979). The cytoskeleton is also involved in the association of RNA during translation, as observed in VSV-infected cells, but it is not known to which of the fibrillar structures the RNA is attached (Cervera *et al.* 1981).

DNA virus interferes more generally with the various filament structures of the cytoskeleton. The actin-containing filaments are changed/disrupted by most of the viruses studied. For poxvirus, the actin filaments are reorganized and accumulated around the virus factory. They might function as an anchor of the replication complexes (Hiller *et al.* 1981), and for Frog-3 virus the microfilaments are reassembled late in the infection and might thus be involved in the proper release of virus particles (Murti & Goorha 1983). The rearrangement of the actin-containing filaments observed in HSV-infected cells (Figure 1) has at present not been related to any specific function in virus morphogenesis or in intracellular transport.

The microtubules are affected by the infection with both Frog-3 virus and HSV

TABLE II

DNA Virus	Micro-filaments	Micro-tubules	Intermediate filaments	References
Frog-3	Disruption/ Reformed-virus release	Disruption	Disruption	Murti *et al.* *J. Cell Biol. 96*, 1248 (1983)
Pox	?/Anchor for factory + virus transport	–	–	Hiller *et al.* *Exp. Cell Res. 132*, 81 (1981)
Adeno	No change	No change/ Virus attachment	No change	Luftig *et al.* *J. Virol. 16*, 696 (1975) Dales *et al.* *Virology 56*, 465 (1973)
HSV	Disruption Degradation	–	–	Heeg *et al.* *Arch. Virol. 70*, 233 (1981) Winkler *et al.* *Arch. Virol 72*, 95 (1982) Bedows *et al.* *Mol. Cell Biol. 3*, 712 (1983)
HSV (fibro-blasts)	Reorganiza-tion	Reorganiza-tion/gp transport	No change	Norrild *et al.* *J. Gen. Virol. 67*, 97 (1986)
HSV (epi-thelial cells)	Disruption	Disruption	Keratin cleavage	Nielsen *et al.* (submitted)
HPV (warts)			Keratin change	Staquet *et al.* *Arch. Dermatol. Res. 271*, 83 (1981)
SV-40 (transformation)	–	–	Keratin change	Hronis *et al.* *Cancer Res. 44*, 5797 (1984)

(Figure 1) (Murti & Goorha 1983, Norrild *et al.* 1986). Intact microtubules are necessary for normal intracellular transport of the HSV glycoproteins gB, gC and gD and for their insertion into the plasma membrane (for nomenclature of the proteins see section C). Studies done in HSV-infected fibroblasts showed that reorganization of microtubules with the alkaloid Taxol being present before and during infection allowed normal synthesis of the viral glycoproteins, but these were not transported to and inserted into the plasma membrane. Disintegration of microtubules with demecolcine also allowed synthesis of glycoproteins, but they remained in small vesicles which were distributed to the periphery of the cell, and again there was no integration of glycoproteins into the plasma membrane (Norrild *et al.* 1986).

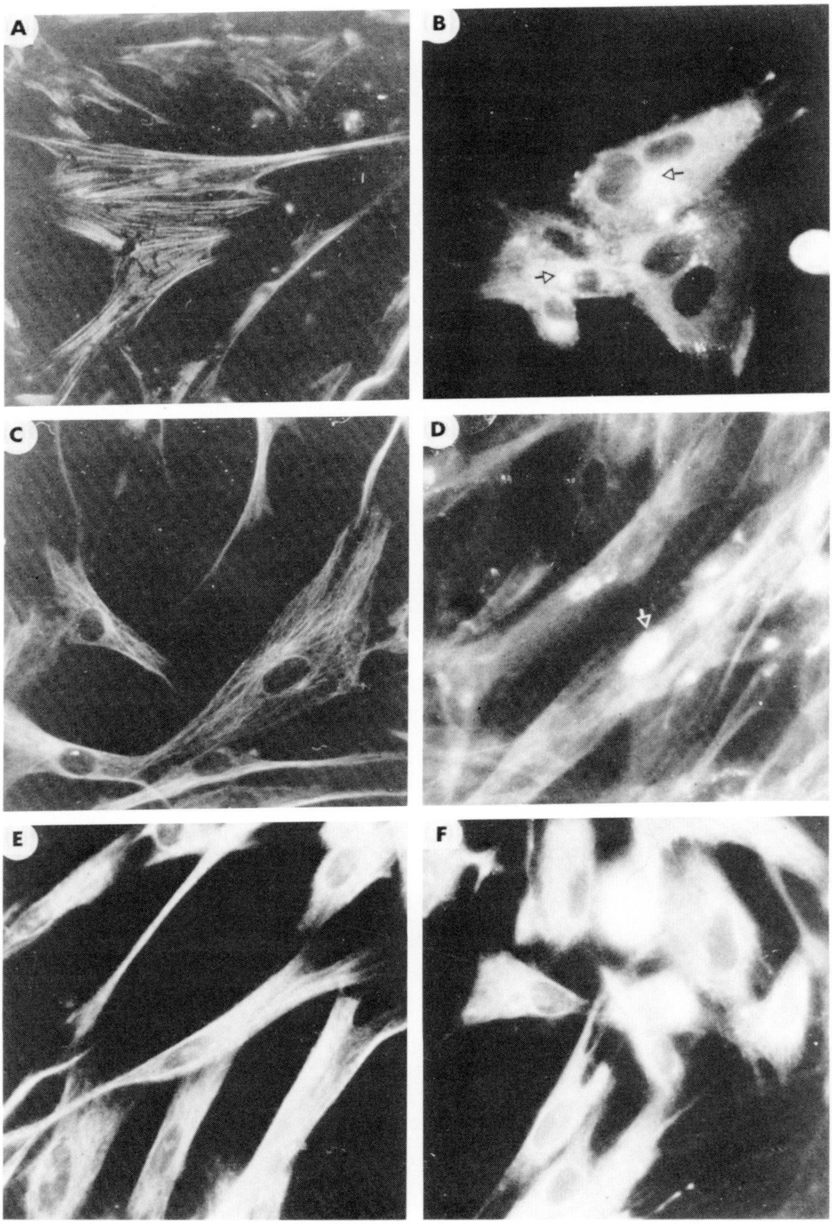
A
B
C
D
E
F

The intermediate filaments are disrupted in Frog-3-infected cells but they are apparently more resistant to infection with HSV (Figure 1) (Murti & Goorha 1983, Norrild *et al.* 1986). No function can at present be ascribed to the intermediate filaments in the infected cells.

B. BIOCHEMICAL CHANGES IN MICROFILAMENT, MICROTUBULE AND INTERMEDIATE FILAMENT PROTEINS OF INFECTED EPITHELIAL CELLS

The DNA viruses which have been reported to lead to changes in cytoskeleton proteins are HSV and human papilloma virus (HPV). The natural host cells for HSV are epithelial cells, which are not easily established in culture. In an epithelial cell line established from rat palate, HSV infection leaves the actin and tubulin proteins unchanged, but virus induces a change in the keratin proteins (Table II) (Nielsen *et al.* submitted). Analysis of the cytoskeleton proteins from MOCK- and HSV-infected cells by 2D-gel electrophoresis shows that actin (AC), alpha (t_1)- and beta (t_2)-tubulin have an unchanged electrophoretic mobility (Figure 2). Two keratin proteins focusing at pI 5.3–5.5 and with apparent molecular weights of 44K and 48K are identified in extracts from MOCK-infected cells, whereas an additional 46K keratin protein is present in HSV-infected cells (Figures 2, 3). The 46K keratin of infected cells appeared after 8 h of infection, and when HSV-infected cells were labeled in a 10 min pulse 12 h postinfection and chased for an additional 8 h it was shown that the 46K protein occurred as a processed form of the 48K keratin (Figure 3). The 2 acidic keratins in uninfected rat cells were correlated to the better characterized human keratins by reaction with a monoclonal antibody, AE1, described by Sun *et al.* 1984. The 44K and 48K keratins correspond to the human 48K and 50K keratins, respectively (Nielsen *et al.* submitted).

HPV-infected skin also has a changed keratin pattern compared to normal skin (Table II). In certain warts, a "new" 72K keratin was identified, and the quantity of the normal 67K keratin was drastically decreased (Staquet *et al.* 1981).

Fig. 1. Uninfected human fibroblasts (A, C, E) and HSV-infected cells (B, D, F) are fixed in methanol 9 h postinfection. The cells are stained by indirect immunofluorescence technique with rabbit antibodies to myosin (A, B), tubulin (C, D) and vimentine (E, F). FITC-coupled swine anti-rabbit immunoglobulin is added as the second antibody. The myosin and tubulin aggregates formed in infected cells are indicated by arrows (B, D). Magnification ×320.

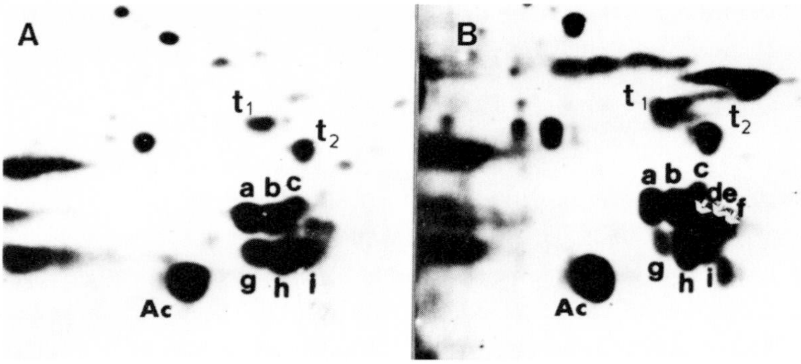

Fig. 2. Autoradiogram of keratins from 35-S-methionine-labeled MOCK- and HSV-infected cells separated by 2D-gel electrophoresis. Mock-infected cells A, HSV-infected cells B. The arrowheads in B mark the 46K series of keratins.

The abnormal keratin pattern obtained from infected tissue/cells might lead to speculation about the influence of virus on cell differentiation. The high molecular weight keratins are present in final differentiated cells, whereas more low molecular weight keratins are present in basal cells and in less differentiated tissue (Fuchs & Green 1980, Sun *et al.* 1985).

Infection of human keratinocytes with simian virus-40 (SV-40) changes the keratin expression drastically, the high molecular weight keratins disappearing and only the low molecular weight keratins being present in the cells (Table II) (Hronis *et al.* 1984).

The observations made in infected rat keratinocytes, where the 46K keratin accumulates, do not seem to support the hypothesis of a dedifferentiation of cells during the infection, as the processing from the 48K to the 46K keratin is most likely a proteolytic degradation induced by the virus.

C. HERPES SIMPLEX VIRUS PROTEIN SYNTHESIS

The detailed knowledge of the synthesis of herpes simplex virus proteins in infected tissue culture cells will not be reviewed here, but information pertinent to the present paper will be described in brief.

Viral proteins are synthesized in a sequential order early after infection. The alpha-proteins are synthesized before the beta-1 and beta-2 proteins, which again are necessary for the synthesis of the gamma-1 and gamma-2 proteins (Honess & Roizman 1973, Roizman & Batterson 1985). The alpha-proteins accumulate in

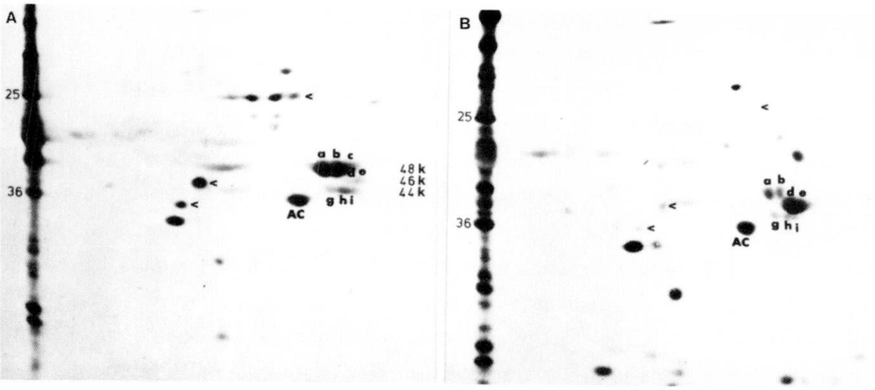

Fig. 3. Autoradiogram of keratins from HSV-1-infected cells analyzed by 2-D gel electrophoresis. A: Infected cells labeled 12 h postinfection in a 10 min 35-S-methionine pulse and harvested immediately. B: Infected cells labeled as in A, but chased in excess of unlabeled methionine for an additional 8 h before harvest. The lanes in the left side of both panel A and panel B represent HSV-1-infected cell protein (ICP) extracts added as molecular weight markers. The numbers refer to ICP number 25 (mol.wt. 63K) and number 36 (mol.wt. 42K). AC is actin, and the arrows in A and B illustrate 3 HSV-1 proteins which are unstable during the chase period.

the nucleus (Fenwick & Roizman 1977, Fenwick *et al.* 1978, Pereira *et al.* 1977), but of the 5 proteins the function is only well known for ICP-4 (174K) (Wilcox *et al.* 1980, Preston 1979).

The beta-1 group of proteins includes, among others, a DNA binding protein (ICP-8) (Knipe & Spang 1982), and the thymidine kinase enzyme belongs to the beta-2 group of proteins (Morse *et al.* 1978). The structural proteins are gamma proteins, and the glycoproteins named gB, gC, gD and gE of HSV-1 belong to this group (for reviews see Norrild 1980, Spear 1984). gH, which was identified recently, has not yet been characterized as beta or gamma (Buchmaster *et al.* 1984). The synthesis, processing and transport of the glycoproteins have been extensively studied (for reviews see Norrild 1980, Spear 1984). It is well known that the glycosylation occurs in descrete steps where the precursor molecules in most cases are high mannose type glycoproteins which are trimmed as the proteins are translocated from one cellular compartment to the other (Johnson & Spear 1982, Cerafini-Cessi & Campadelli-Fiume 1981, Compton & Courtney 1984). The glycoproteins are eventually transported to and inserted into the plasma membrane of the infected cells and into the envelope of the virions (for review see Norrild 1980).

D. ASSOCIATION OF VIRAL PROTEINS/PARTICLES TO CYTOSKELETON PROTEINS

The role of the cytoskeleton elements in transport of viral proteins has been most convincingly demonstrated for the transport of the major DNA binding protein (ICP-8) and of the capsid protein (ICP-5) of HSV (Quinlan & Knipe 1983, Ben-Ze'ev *et al.* 1983). Both proteins are associated with the cytoplasmic fibrillar structures early after their synthesis. It is presently unknown if the binding of ICP-8 and ICP-5 is to microfilaments, microtubules or intermediate filaments. However, ICP-8 remains associated with the cytoskeleton proteins also after extraction under conditions which disrupt microtubules (Quinlan & Knipe 1983), but the binding of ICP-8 to the cytoskeleton is sensitive to mechanical shear forces. ICP-8 is known to be translocated into the cell nuclei in a multistep process. The protein is first present on the outside of the nuclei and is then transported into the nuclei, where it associates with the chromatin fraction, whereas ICP-5 apparently binds to the nuclear matrix (Knipe & Spang 1982, Ben Ze'ev *et al.* 1983).

The hypothesis for transport of the 2 proteins into the nucleus is that the proteins are guided by the cytoskeleton to the nuclear pore areas where intermediate filaments might be involved in the transfer of the proteins to the fibrillar structures of the nuclear matrix. These structures might be contiguous with the filaments on the cytoplasmic site of the nuclear membrane (Quinlan & Knipe 1983). Among other viruses, it has been shown that the membrane proteins of the paramyxoviruses, Newcastle disease virus and Sendai virus bind firmly to actin in reconstitution experiments. This is a more indirect piece of evidence for the function of microfilaments in the intracellular transport of the so-called M-glycoprotein (Giuffre *et al.* 1982). Few studies were designed to look for a direct binding of virus particles to the cytoskeleton. One of the most convincing studies shows binding between adenovirus and microtubules. The conclusions are based on reconstitution experiments where purified virus and microtubules are mixed and prepared for electromicroscopy (Luftig & Weihing 1975). Although the data are elegant it cannot be excluded that the binding observed only occurs under the experimental conditions used and that the binding is not representative for the *in vivo* situation.

CONCLUSION

The detailed understanding of the possible participation of the cellular cytoskeletal elements in virus morphogenesis awaits more experimental evidence. However, the association of adenovirus with purified microtubules supports the hypothesis of a direct involvement of microtubules in transport of the virus particles. That the microtubules are also important for the correct transport of HSV glycoproteins has been experimentally confirmed, but it is still unknown whether the microtubules function as a kind of "railway track" for transport vesicles and virus particles. The hypothesis is attractive, however, especially when the elegant experiments on intraaxonal transport by Allen *et al.* (1985) are considered. The microtubules have been shown to support bidirectional transport of organelles, and a translocator protein "kinesin" has been identified in squid axoplasm, a protein which might coat the organelles and make the transport machinery work (Vale *et al.* 1985). It still remains to be elucidated whether the cytoskeletal fibrillar structures are directly participating in intracellular transport and sorting, where certain signals call for transport along one or the other type of filaments, or whether the functions of the cytoskeleton fibrillar structures are highly specialized and are communicated among the compartments in a very specific way.

REFERENCES

Allen, R. D., Weiss, D. G., Hayden, J. H., Brown, D. T., Fujiwake, H. & Simpson, M. (1985) Gliding movement of and bidirectional organelle transport along single native microtubules from squid axoplasm: Evidence for an active role of microtubules in cytoplasmic transport. *J. Cell Biol. 100*, 1736–52.

Ben-Ze'ev, A., Abulafia, R. & Bratosin, S. (1983) Herpes simplex virus and protein transport are associated with the cytoskeletal framework and the nuclear matrix in infected BSC-1 cells. *Virology 129*, 501–7.

Buckmaster, E. A., Gompels, V. & Minson, T. (1984) Characterization and physical mapping of an HSV-1 glycoprotein of approximately 115×10^3 molecular weight. *Virology 139*, 408–13.

Cervera, M., Dreyfuss, G. & Penman, S. (1981) Messenger RNA is translated when associated with the cytoskeletal framework in normal and VSV-infected HeLa cells. *Cell 23*, 113–20.

Compton, T. & Courtney, R. J. (1984) Evidence for post-translational glycosylation of a nonglycosylated precursor protein of herpes simplex virus type 1. *J. Virol. 52*, 630–7.

Dales, S. & Chardonnet, Y. (1983) Early events in the interaction of adenoviruses with HeLa cells. IV. Association with microtubules and the nuclear pore complex during vectorial movement of the inoculum. *Virology 56*, 465–83.

Dustin, P. (1978) *Microtubules*. Springer Verlag, Berlin-Heidelberg-N.Y.

Fenwick, M. & Roizman, B. (1977) Regulation of herpes virus macromolecular synthesis. VI. Synthesis and modification of viral polypeptides in enucleated cells. *J. Virol. 22*, 720–5.

Fenwick, M. L., Walker, M. J. & Petkevich, J. M. (1978) On the association of virus proteins with the nuclei of cells infected with herpes simplex virus. *J. Gen. Virol. 39*, 519–29.

Fuchs, E. & Green, H. (1980) Changes in keratin gene expression during terminal differentiation of the keratinocyte. *Cell 19*, 1033–42.

Giuffre, R. M., Tovell, D. R., Kay, C. M. & Tyrell, D. J. J. (1982) Evidence for an interaction between the membrane protein of a paramyxovirus and actin. *J. Virol. 42*, 963–8.

Hiller, G., Jungwirth, C. & Weber, K. (1981) Fluorescence microscopical analysis of the life cycle of vaccinia virus in chick embryo fibroblasts. *Exp. Cell Res. 132*, 81–7.

Honess, R. W. & Roizman, B. (1973) Proteins specified by herpes simplex virus. XI. Identification and relative molar rates of synthesis of structural and nonstructural herpes virus polypeptides in the infected cell. *J. Virol. 12*, 1347–65.

Hronis, T. S., Steinberg, M. L., Defendi, V. & Sun, T.-T. (1984) Simple epithelial nature of some simian virus-40-transformed human epidermal keratinocytes. *Cancer Res. 44*, 5797–804.

Johnson, D. C. & Spear, P. C. (1982) Monensin inhibitis the processing of herpes simplex virus glycoproteins, their transport to the cell surface and the egress of virions from infected cells. *J. Virol. 43*, 1102–12.

Knipe, D. M. & Spang, A. E. (1982) Definition of a series of stages in the association of two herpesviral proteins with the cell nucleus. *J. Virol. 43*, 314–24.

Lehto, V.-P., Virtanen, I. & Kurki, P. (1978) Intermediate filaments anchor the nuclei in nuclear monolayers of cultured fibroblasts. *Nature, London 272*, 175–7.

Lenk, R. & Penman, S. (1979) The cytoskeletal framework and poliovirus metabolism. *Cell 16*, 289–301.

Luftig, R. B. & Weihing, R. R. (1975) Adenovirus binds to rat brain microtubules *in vitro*. *J. Virol. 16*, 696–706.

Menko, A. S., Toyama, Y., Boettiger,D. & Holzer, H. (1983) Altered cell spreading in cytochalasin B: A possible role for intermediate filaments. *Mol. Cell. Biol. 3*, 113–25.

Moll, R., Franke, W. W., Schiller, D. L., Geiger, B. & Krepler, R. (1982) The catalogue of human cytokeratin polypeptides: Patterns of expression of cytokeratins in normal epithelia, tumors and cultured cells. *Cell 31*, 11–24.

Morse, L. S., Pereira, L., Roizman, B. & Schaffer, P. A. (1978) Anatomy of herpes simplex virus (HSV) DNA. X. Mapping of viral genes by analysis of polypeptides and functions specified by HSV-1 × HSV-2 recombinants. *J. Virol. 26*, 389–410.

Murti, K. G. & Goorha, R. (1983) Interaction of Frog Virus-3 with the cytoskeleton. I. Altered organization of microtubules, intermediate filaments, and microfilaments. *J. Cell Biol. 96*, 1248–57.

Nielsen, L. N., Forchhammer, J., Dabelsteen, E., Jepsen, A., Teglbjærg, C. S. & Norrild, B. (1986) Herpes simplex virus induced changes of the keratin type intermediate filament in rat epithelial cells. (Submitted).

Norrild, B. (1980) Immunochemistry of herpes simplex virus glycoproteins. *Curr. Top. Microbiol. Immunol. 90*, 67–106.

Norrild, B., Lehto, V.-P. & Virtanen, I. (1986) Organization of cytoskeleton elements during herpes simplex virus type 1 infection of human fibroblasts: An immunofluorescence study. *J. Gen. Virol. 67*, 97–105.

Osborn, M. & Weber, K. (1983) Biology of disease. Tumor diagnosis by intermediate filament typing: A novel tool for surgical pathology. *Lab. Invest. 4*, 372–94.

Pereira, L., Wolff, M., Fenwick, M. & Roizman, B. (1977) Regulation of herpesvirus macromolecular synthesis. V. Properties of *α* polypeptides made in HSV-1 and HSV-2 cells. *Virology 77*, 733–49.

Pollard, T. D., Fujiwara, K., Niedermann, R. & Maupin-Szamier, P. (1976) In: *Cell Motility. Cold Spring Harbor Conference on Cell Proliferation*, vol. 3, eds. Goldman, R. D., Pollard, T. D. & Rosenbaum, J. L., pp. 689–724. Cold Spring Harbor Laboratory, New York.

Preston, C. M. (1979) Control of herpes simplex virus type 1 mRNA synthesis in cells infected with wild-type virus or the temperature-sensitive mutant tsK. *J. Virol. 29*, 275–84.

Quinlan, M. P. & Knipe, D. M. (1983) Nuclear localization of herpes-virusproteins: Potential role for the cellular framework. *Mol. Cell. Biol. 3*, 315–24.

Roizman, B. & Batterson, W. (1985) Herpesviruses and their replication. In: *Virology*, ed. Fields, B. N., pp. 497–526. Raven Press, New York.

Rutter, G. Mannweiler, K. (1977) Alterations of actin-containing structures in BHK 21 cells infected with Newcastle Disease virus and vesicular stomatitis virus. *J. Gen. Virol. 37*, 233–42.

Schliwa, M. (1984) Mechanisms of intracellular organelle transport. In: *Cell and Muscle Motility*, vol. 5, pp. 1–81. Plenum Publishing Corporation, New York.

Serafini-Cessi, F. & Campadelli-Fiume, G. (1981) Studies on benzhydrazone, a specific inhibitor of herpesvirus glycoprotein synthesis. Size distribution of glycopeptides and endo-β-N-acetylglucosidase H treatment. *Arch. Virol. 70*, 331–43.

Sharpe, A. H., Chen, L. B. & Fields, B. N. (1982) The interaction of mammalian reoviruses with the cytoskeleton of monkey kidney CV-1 cells. *Virology 120*, 399–411.

Spear, P. G. (1984) Glycoproteins specified by herpes simplex viruses. In: *The Herpesviruses*, vol. 3, ed. Roizman, B., pp. 315–56. Plenum Publishing Corporation, New York.

Staquet, M. J., Viac, J. & Thivolet, J. (1981) Keratin polypeptide modifications induced by human papilloma viruses (HPV). *Arch. Dermatol. Res. 271*, 83–90.

Sun, T.-T., Eichner, R., Nelson, W. G., Tseng, S. C. G., Weiss, R. A., Jarvinen, M. & Woodcock-Mitchell, J. (1983) Keratin classes: Molecular markers for different types of epithelial differentiation. *J. Invest. Dermatol. 81*, suppl. 1, 1095–155.

Sun, T.-T., Eichner, R., Schermer, A., Cooper, D., Nelson, W. G. & Weiss, R. A. (1984) Classification, expression, and possible mechanisms of evolution of mammalian epithelial keratins: A unifying model. *Cancer Cells 1*, 169–76.

Sun, T.-T. & Green, H. (1978) Keratin filaments of cultured human epidermal cells. *J. Biol. Chem. 253*, 2053–60.

Sun, T.-T., Tseng, S. C. G., Huang, A. J.-W., Cooper, D., Schermer, A., Lynch, M. H., Weiss, R. & Eichner, R. (1985) Monoclonal antibody studies of mammalian epithelial keratins: A review. *Ann. N.Y. Acad. Sci. 455*, 307–29.

Tilney, L. G. (1983) Interaction between actin filaments and membranes give spatial organization to cells. *Modern Cell Biol. 2*, 163–99.

Vale, R. D., Schnapp, B. J., Mitchison, T., Steuer, E., Reese, T. S. & Scheetz, M. P. (1985) Different axoplasmic proteins generate movement in opposite directions along microtubules *in vitro*. *Cell 101*, 623–32.

Weber, K. & Osborn, M. (1981) Microtubule and intermediate filament networks in cells viewed by immunofluorescence microscopy. In: *Cytoskeletal Elements and Plasma Membrane Organization*, eds. Poste, G. & Nicolson, G. L., pp. 1–53. Elsevier/North-Holland Biomedical Press, Amsterdam.

Wilcox, K. W., Kohn, A., Sklyanskaya, E. & Roizman, B. (1980) Herpes simplex virus phosphoproteins. I. Phosphate cycles on and off some viral polypeptides and can alter their affinity for DNA. *J. Virol. 33*, 167–82.

DISCUSSION

ROIZMAN: Richard Courtney recently reported that transport of glycoproteins is polar in the sense that HSV glycoproteins go to one surface or another.

NORRILD: That could still be a possibility, I agree. But then we have to use the polar cells and use dog kidney cells. As far as I recall, these cells are not very permissive. It is a little hard to find polar cells where you can still study the HSV infection and follow the transport to the apical or to the basal-lateral part of the cell. But it should be done.

ROIZMAN: Would there be different elements involved?

NORRILD: What I think is the case and what has been published about the influenza- and the VSV-infected cells is that in double infected cells you can demonstrate a trans Golgi sorting. One virus is released from the upper site, the other one from the basal site of the cell. I think that we are also going to find that some of the HSV glycoproteins go to the apical and others to the basal-lateral part of the cell. So the sorting might tell us something about function.

ROIZMAN: When you disrupt the microtubules, is there a change in the glycosylation of the glycoproteins?

NORRILD: No, it is not changed. They are still fully glycosylated which means they at least go to the Golgi.

WEINBERG: Does the glycoprotein D go directly to the plasma membrane of the cell surface?

NORRILD: All of the glycoproteins go first to the ER, then to the Golgi, and finally to the plasma membrane where they are inserted. But what makes the difference is that what we observe with the antibodies we have used is that you had this aggregation of gD at the adhesion plaques. We showed that it was the adhesion plaques by double labelling the cells with glycoprotein and vinculin antibodies. It seems to be only glycoprotein D that specifically accumulates in adhesion areas. We still do not know too much about the function of gD. You heard yesterday from Dr. Roizman that you can delete at least some of the glycoproteins and still have a functional virus. So, we do not have too much information about the real function of the glycoproteins, but only know that they have to be present in the virus envelope in order to get infectious virus particles.

WEINBERG: Is glycoprotein D one of the glycoproteins that is dispensable for infectivity?

ROIZMAN: We have not tried to delete it yet, but there are no ts mutations in the gene, so I suspect it could be deleted.

BALTIMORE: Is that the gene whose product you are trying to make the target for vaccine?

ROIZMAN: Yes.

BALTIMORE: Can you make a vaccine against something that is dispensable?

ROIZMAN: They are dispensable only in the sense that the virus lacking the gene will grow in cell culture. Most of the viruses with gene deletions do not grow in animals, or grow very poorly. Most of the "dispensable" genes are required for growth in animals and presumably in people.

NORRILD: We should stress at this point that human serologic studies indicate the presence of high amounts of antibodies to glycoproteins D in human sera. It seems to be one of the most immunogenic in humans and there is no reason to expect that humans would be infected with deleted viruses. That has not been seen so far.

YANIV: There were old reports that SV40 assembled along fibres in the nuclei. Do you know anything about such nuclear viruses?

NORRILD: I can only say that if we look at the EM studies done 10 years ago, it looks convincing, but the pictures do not prove any association. I must admit that even though we have taken the approach to look with gold-staining for the possible association between microtubule and glycoproteins, we will not be able to prove function of the cytoskeleton. With EM you might see structures in a close association, but it does not mean they are necessarily functional units. That is why I wanted to include the studies that were done from extracted cells without Icp5 that showed that if you had a real attachment you could not dissociate Icp5 and insoluble cytoskeleton protein. But again, to that you can argue that because they were insoluble they will stay in the same fraction. So, it is very hard to prove that we are looking at a real association.

WEINBERG: When one talks about adhesion plaques on cells growing in monolayer, what is the *in vivo* correlate to that? Has one ever made such a correlate?

NORRILD: Not to my knowledge. That is not really my special field so I would not be able to say that there is an analogue.

Changes in the Levels of Expression of Human Tropomyosins IEF 52, 55 and 56 in Normal and SV40-Transformed MRC-5 Fibroblasts

Julio E. Celis[1], Borbala Gesser[1], J. Victor Small[2], Søren Nielsen[1] & Ariana Celis[1]

Malignant transformation of cultured cells is often characterized by changes in cell morphology. While the patterns of microtubules and intermediate filaments appear unchanged in transformed cells, the organization of the actin containing microfilaments is significantly altered (McNutt *et al.* 1973, Pollack & Rifkin 1975, Pollack *et al.* 1975, Goldman *et al.* 1976). Often, the actin cables observed in well-spread normal cells are either lost or present in reduced numbers in transformed cells and, instead, diffuse meshworks are observed.

To date little is known concerning the molecular mechanisms underlying the changes in the actin cytoskeleton that accompany transformation. Several proteins such as myosin (Weber & Groeschel-Stewart 1974, Fujiwara & Pollard 1976), tropomyosin (Lazarides 1975), alpha-actinin (Lazarides & Burridge 1975, Fujiwara *et al.* 1978), filamin (Wang *et al.* 1975, Heggeness *et al.* 1977) and vinculin (Geiger 1979), have been shown to be components of the actin microfilaments, but it is not clear whether alterations in the relative proportion of these proteins play a role in the control of actin microfilament organization. We (Bravo *et al.* 1981a, Bravo & Celis 1982b, Celis *et al.* 1984a), as well as others (Paulin *et al.* 1979, Linder *et al.* 1981, Hendricks & Weintraub 1981, 1984, Forchhammer 1982, Matsumara *et al.* 1983a, Lin *et al.* 1984, 1985, Franza & Garrels 1984) have identified a few non-muscle tropomyosins whose rate of synthesis is sensitive to changes in growth rate and transformation, and the possibility has been raised

[1]Department of Medical Biochemistry, Ole Worms Alle, Building 170, University Park, Aarhus University, DK-8000 Aarhus C, Denmark, and [2]Institute of Molecular Biology, Austrian Academy of Sciences, Salzburg, Austria.

VIRAL CARCINOGENESIS, Alfred Benzon Symposium 24,
Editors: N. O. Kjeldgaard, J. Forchhammer, Munksgaard, Copenhagen 1987.

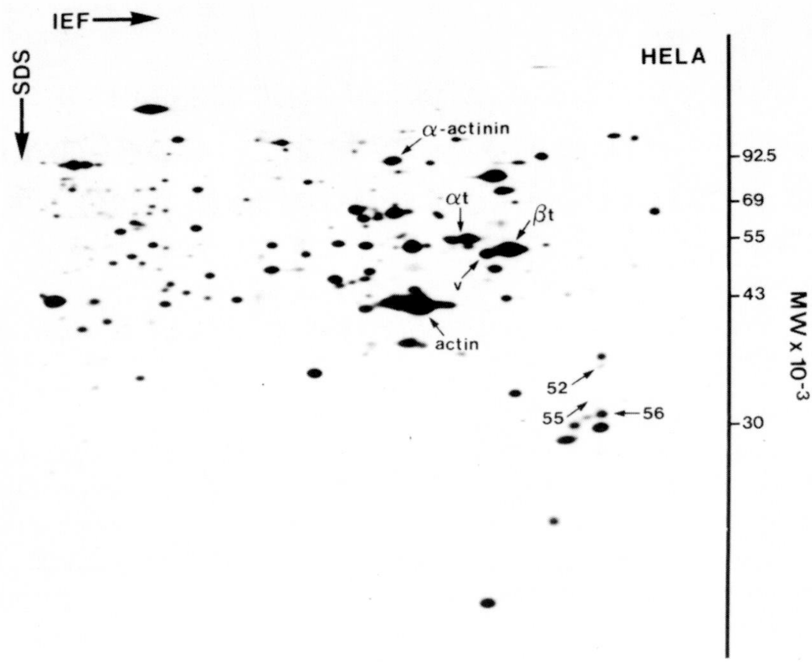

Fig. 1. Position of tropomyosins IEF's 52, 55 and 56 in a two dimensional gel map (IEF) of (^{35}S)-methionine-labelled proteins from HeLa cells. Cells were labelled as previously described (Bravo *et al.* 1981b, Celis & Bravo 1981). The positions of actin, alpha-actinin, alpha-tubulin (alpha-t), beta-tubulin (beta-t) and vimentin (v) are indicated for reference.

that these changes may be responsible, in part, for the rearrangement of actin cables into meshworks in the neoplastic cells (Hendricks & Weintraub 1981, 1984, Leonardi *et al.* 1982, Matsumara *et al.* 1983a, Lin *et al.* 1984, 1985).

In an effort to determine to what extent changes in the expression of tropomyosins relate to differences in the actin cytoskeleton of normal and SV40-transformed human MRC-5 fibroblasts, we present studies in which we have determined the steady state levels of 3 tropomyosins recognized by a mouse monoclonal antibody raised against purified human tropomyosin IEF 52 (Mr= 35 kd, Figure 1; HeLa protein catalogue number, Bravo & Celis 1982a, 1984, Celis *et al.* 1986). These, as well as other results reported here suggest that changes in the levels of these 3 tropomyosins are not enough to account for the magnitude of the loss of actin cables observed in the transformed cells.

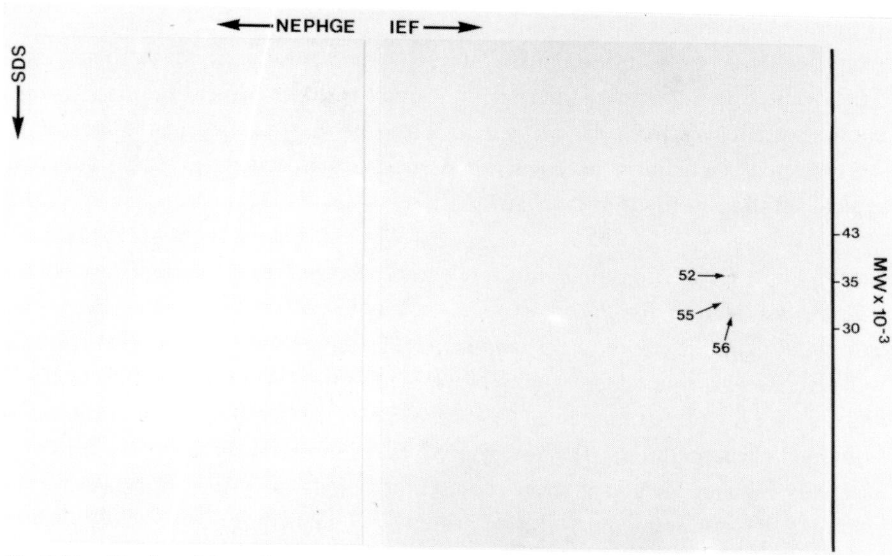

Fig. 2. Identification of the antigens reacting with mAb 1D122G9 by two dimensional gel electrophoresis and immunoblotting. HeLa cell proteins were separated by two dimensional gel electrophoresis (NEPGHE, IEF) and transferred to nitrocellulose sheets. Antigens were identified by superposition of immunoblots with their corresponding (identical) autoradiograms.

CHARACTERIZATION OF A MOUSE MONOCLONAL ANTIBODY RAISED AGAINST TROPOMYOSIN IEF52

Identification of human proteins IEF's 55 and 56 as tropomyosins

The specificity of a mouse monoclonal antibody (mAb 1D122G9) raised against purified tropomyosin IEF 52 (Celis *et al.* 1986) was determined by two dimensional gel immunoblotting of HeLa cell proteins. As shown in Figure 2, the antibody reacts with tropomyosin IEF 52 (Mr=35 kd; Bravo *et al.* 1981a, Bravo & Celis 1982ab, 1984) and 2 acidic proteins previously identified as tropomyosin-like proteins (Bravo *et al.* 1981a, Bravo & Celis 1982ab, 1984). These proteins, which lack proline and tryptophan, correspond to IEF 55 (Mr=31.8 kd) and 56 (Mr=31 kd) in the HeLa protein catalogue (Bravo & Celis 1982a, 1984). The 3 HeLa tropomyosins correspond most likely to TM1,3 and 4 in the nomenclature of Matsumara *et al.* (1983ab) (see also Franza & Garrels 1984).

Immunofluorescence microscopy of goat synovial cultured cells treated with 1% Triton X-100 in Pipes buffer (Small & Celis 1978), fixed with 3.6% formaldehyde and reacted with mAb 1D122G9 revealed an actin-like staining pattern

(Figure 3A; compare with actin staining, Figure 3B) that was clearly different from that obtained with tubulin (Figure 3C) or vimentin antibodies (Figure 3D). The striated or interrupted pattern on the actin cables typical of tropomyosin staining was clearly seen in well-spread areas of the cells (see also Figure 6C). As expected, incubation of pig striated muscle with mAb 1D122G9 gave the typical staining of myofibrils (Figure 3E).

Steady state levels of tropomyosins IEF's 52, 55 and 56 in normal and SV40-transformed MRC-5 fibroblasts

The steady state levels of the 3 tropomyosins recognized by mAb 1D122G9 as well as of total actin and alpha-actinin was determined in normal (slowly cycling) and SV40-transformed human MRC-5 fibroblasts using quantitative two dimensional gel electrophoresis (IEF) (O'Farrell 1975, O'Farrell *et al.* 1977, Bravo *et al.* 1982). Figures 4A and B show representative gels (only the appropriate area is shown) of (^{35}S)-methionine-labelled proteins (18 h labelling) from slowly cycling and SV40-transformed MRC-5 fibroblasts, respectively. Individual spots were excised from the gels, counted, and their radioactivity normalized against polypeptide IEF 35 (mitochondrial) as the level of this protein remains constant upon transformation or following changes in growth rate (Bravo & Celis 1982b). As shown in Table I, the levels of tropomyosins 52 (ratio transformed vs normal cells = 0.53) and 56 (ratio = 0.83), as well as of actin (ratio = 0.73), were significantly lower in the SV40-transformed human fibroblasts compared to their normal counterparts. The level of tropomyosin IEF 55 (ratio = 1.6) on the other hand was substantially higher in the transformed cells, while that of alpha-actinin remained constant. The ratio of actin to total tropomyosin was found to be unchanged on transformation (Table I). A similar decrease in the rate of synthesis of tropomyosins IEF 52 and 56 has been observed in other pairs of normal and transformed cells such as human amnion (Figure 5A)/transformed amnion cells (AMA, Figure 5B) (Bravo & Celis 1982b), human WI-38 fibroblasts/WI-38 SV40 (Bravo & Celis 1982b), human epidermal basal cells/SV40-transformed basal cells (Celis *et al.* 1984b) and mouse 3T3 fibroblasts/3T3 SV40 (Celis *et al.* 1984a).

Levels of tropomyosins IEF's 52, 55 and 56 in Triton cytoskeletons from normal and SV40-transformed MRC-5 fibroblasts

The levels of actin, alpha-actin and tropomyosins IEF's 52, 55 and 56 in Triton cytoskeletons from normal and SV40-transformed human MRC-5 fibroblasts was determined as described above. Radioactivity in each spot was normalized

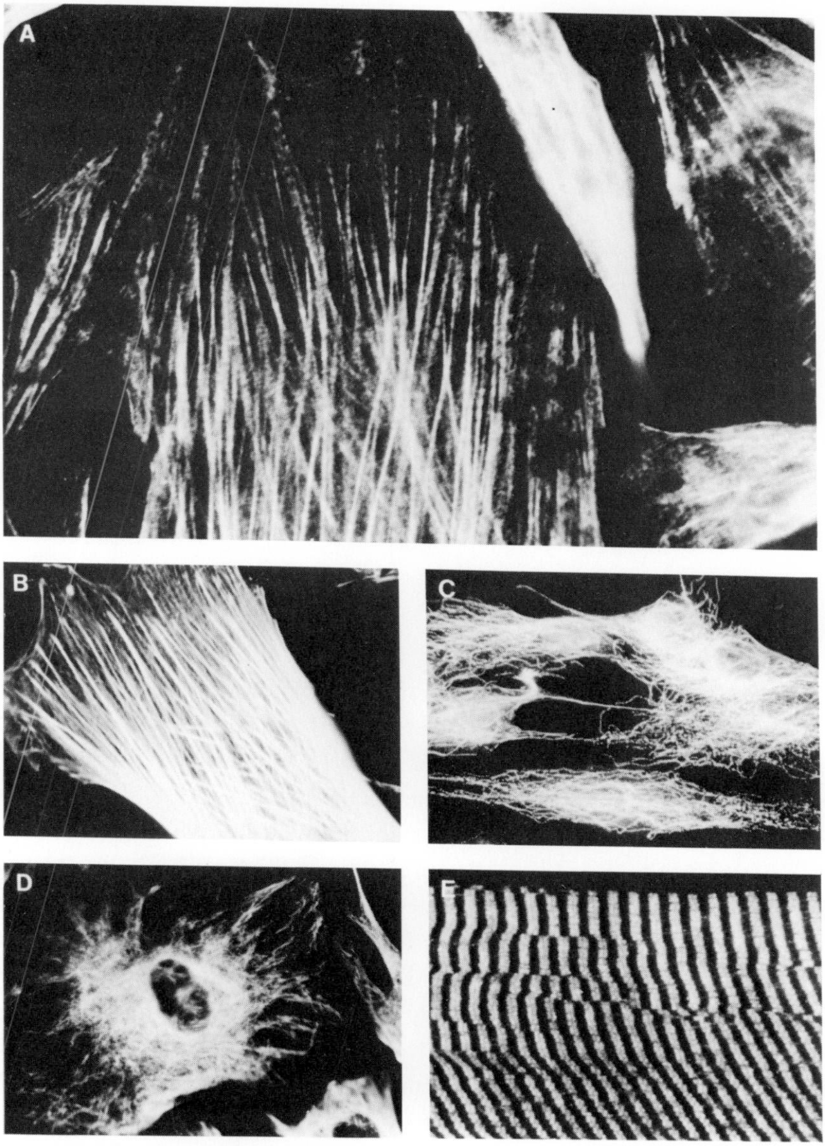

Fig. 3. Indirect immunofluorescence of cytoskeletal proteins in goat synovial cells and pig striated muscle. (A-D) goat synovial Triton cytoskeletons reacted with tropomyoxin (mAb 1D122G9) (A), actin (B), tubulin (C) and vimentin (D) antibodies. (E) pig striated muscle reacted with mAb 1D122G9. (A) ×838, (B–D) ×405 and (E) ×507.

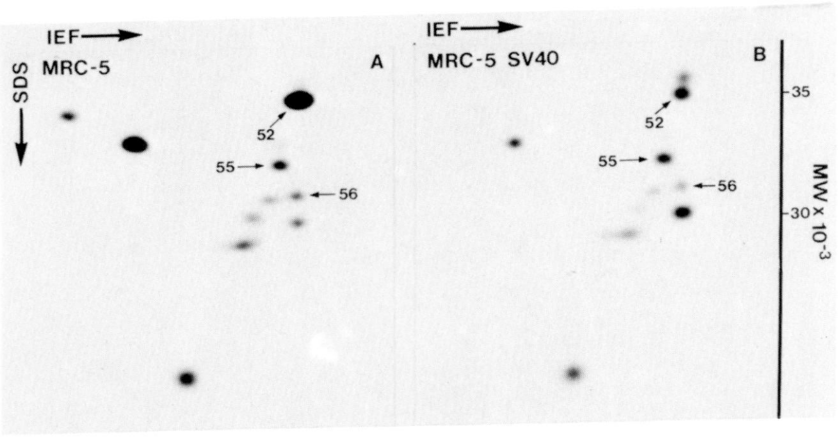

Fig. 4. Two-dimensional gel electrophoresis (IEF) of (^{35}S)-methionine-labelled proteins from normal (A) and SV40-transformed (B) human MRC-5 fibroblasts. Only the appropriate region of the gel is shown.

against IEF 35, a mitochondrial protein that is present mainly in Triton cytoskeletons (Bravo & Celis 1982a, 1984). Furthermore, it was corrected to allow a direct comparison with the data presented in Table I. As shown in Table II, Triton cytoskeletons from transformed cells contain about 52 and 62% of the actin and alpha-actinin present in similar skeletons of normal cells. A comparison of this data with the steady state levels of both proteins in whole cells (Table I) showed that more of these 2 proteins is extracted by Triton in the transformed cells (52.3% actin, 53.3% alpha-actinin) as compared to their normal counterparts (32.5% actin, 18.7% alpha-actinin). As a result, the ratio actin/alpha-actinin observed in both cytoskeletons was very similar (Table II).

Analysis of the cpm recovered in the tropomyosin spots showed that the Triton cytoskeletons from the SV40-transformed cells contained 52 and 72% of the tropomyosins IEF's 52 and 56 present in their normal counterparts (Table II). Their ratio to actin, however, was very similar in both types of cytoskeleton (Table II). This was not the case for tropomyosin IEF 55, which was present in nearly twice the amount in Triton cytoskeletons from transformed cells (Table II). The ratio of actin to total tropomyosin was found to be 21.3 for normal cytoskeletons and 14.7 for their transformed counterparts.

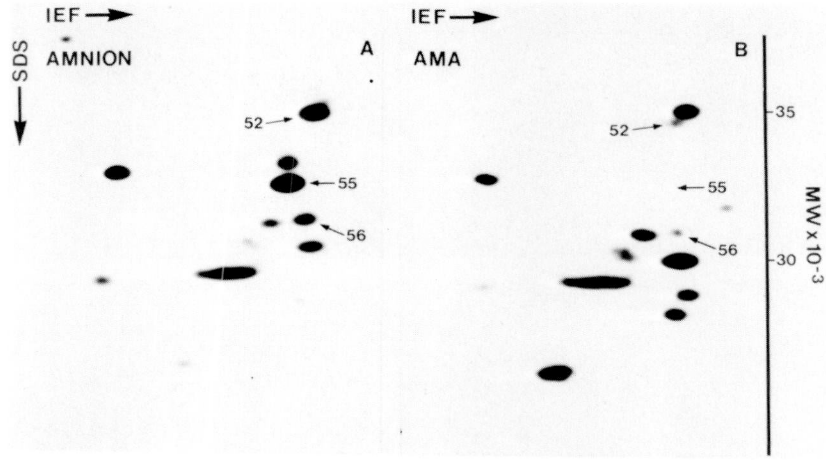

Fig. 5. Two-dimensional gel electrophoresis (IEF) of (^{35}S)-methionine-labelled proteins from normal (A) and transformed (B) human amnion cells (AMA). Only the pertinent area of the gel is shown.

SV40-transformed MRC-5 fibroblasts contain fewer actin cables than their normal counterparts

The presence of actin cables in normal and SV40-transformed human MRC-5 fibroblasts was assessed by indirect immunofluorescence. Figure 6 shows representative photomicrographs of these cells permeated with 0.1% Triton, fixed with 3.6% formaldehyde and reacted with actin antibodies (Figures 6A and B) and mAb 1D122G9 (Figures 6C and D). Few actin cables were observed in the transformed cells and in many cases they delineated the edges of the cells. In contrast, the slowly growing normal cells exhibited abundant actin cables and were much larger and flatter in morphology.

DISCUSSION

Several lines of experimentation have suggested that non-muscle tropomyosins may play a role in stabilizing actin microfilaments (Kawamura & Muruyama 1970, Fujime & Ishiwata 1971, Takebayashi *et al.* 1977, Maupin-Szamier & Pollard 1978, Lehrer 1981, Bernstein & Bamburg 1982, Fattoum *et al.* 1983, Lin *et al.* 1985). Accordingly, there has been mounting interest in determining to what extent changes in the expression of various non-muscle tropomyosins relate to differences in the organization of the actin cytoskeleton of normal and transformed cells (Paulin *et al.* 1979, Linder *et al.* 1981, Hendricks & Weintraub 1981,

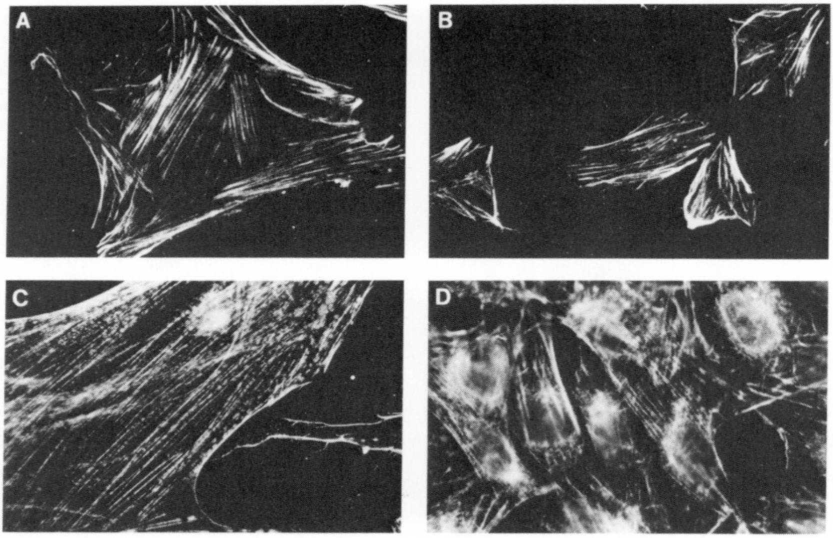

Fig. 6. Actin cables in normal (A) and SV40-transformed human MRC-5 fibroblasts (B). Triton cytoskeletons from normal and transformed cells reacted with actin antibodies (A,B) and with mAb 1D122G9 (C,D). ×405.

TABLE I

Quantitation of actin, alpha-actinin and tropomyosins in normal and SV40-transformed human MRC-5 fibroblasts

Cell type	Actin[a]	Alpha-actinin[a]	Tropomyosins[a] IEF 52	Tropomyosins[a] IEF 55	Tropomyosins[a] IEF 56	Ratio Actin/total tropomyosin
MRC-5	46381	3122	951	245	497	27.4
MRC-5 SV40	33 985 (0.73)[b]	3283 (1.05)[b]	502 (0.53)[b]	392 (1.6)[b]	414 (0.83)[b]	26

[a] cpm for each protein were normalized against IEF 35.

[b] Ratio of incorporation of transformed vs normal cells is given in brackets.

Forchhammer 1982, Matsumara *et al.* 1983ab, Lin *et al.* 1984, 1985, Franza & Garrells 1984, Celis *et al.* 1984a).

The results reported in this article showed significant changes in the levels of expression of 3 tropomyosins (IEF 52, 55 and 56; HeLa protein catalogue number, Bravo & Celis 1982a, 1984) between normal MRC-5 human fibroblasts and their SV40-transformed counterparts. Changes in the levels of expression of these tropomyosins have also been observed in other pairs of normal and transformed cells such as human amnion (Figure 5A)/transformed amnion cells (AMA, Figure 5B) (Bravo & Celis 1982b), human W1-38 fibroblasts/Wl-38 SV40 (Bravo & Celis 1984a), human epidermal basal cells/SV40-transformed basal cells (Celis *et al.* 1984b) and mouse 3T3 fibroblasts/3T3 SV40 (Celis *et al.* 1984a).

Similar variations in the levels of tropomyosins IEF 52 (TM-1, Matsumara *et al.* 1983ab) have been reported by other laboratories in SV40, Kirsten murine sarcoma virus and adenovirus-transformed REF 52 rat cultured cells (Matsumara *et al.* 1983a, Lin *et al.* 1984, Franza & Garrells 1984) and in Rous sarcoma virus-transformed chicken embryo fibroblasts (Hendricks & Weintraub, 1981, 1984, Lin *et al.* 1985), as well as in human HUT-11, EJ, CRL-1420 and MCF-7 cells (Lin *et al.* 1984). Moreover, an increase in the levels of tropomyosin 55 (TM-3, Matsumara *et al.* 1983b) has been observed in SV40 (Lin *et al.* 1984), and some adenovirus-transformed REF 52 rat cells (Franza & Garrells 1984), but was not observed in Kirsten murine virus-transformed cells (Lin *et al.* 1984, Franza & Garrells 1984).

Taken together, our results and those of others have shown that a decrease in the steady state levels of tropomyosin IEF 52 (TM1) is one of the most reproducible and ubiquitous changes in tropomyosin expression accompanying spontaneous and viral transformation (Bravo *et al.* 1981a, Hendricks & Weintraub 1981, 1984, Forchhammer 1982, Bravo & Celis 1982b, Matsumura *et al.* 1983a, Lin *et al.* 1984, Franza & Garrells 1984).

Contrary to earlier reports (Hendricks & Weintraub 1981), we found that the ratio of total tropomyosin to actin is the same in both normal and transformed cells (Table I). Taking into account the approximately three-fold higher methionine content in actin, as compared to tropomyosin (Nakamura *et al.* 1979, Matsumara & Matsumara 1985) and the dimeric structure of tropomyosin, we estimated a molar ratio of actin to tropomyosin of 1 to 15–20 in whole cells. If we assume that this tropomyosin is complexed with actin in a 1:6 or 1:7 ratio, depending on molecular length (see Coté 1983), then around one third of the cells' actin will be occupied by tropomyosin molecules.

Recent data on the relative affinities of the different non-muscle tropomyosins for striated muscle actin suggest that factors may influence the relative stabilizing effect of the different tropomyosin variants on actin filaments (Lin *et al.* 1985). From comparative sequence analyses and functional studies on platelet and muscle tropomyosins (reviewed in Coté 1983) we would expect the higher molecular weight and therefore longer tropomyosins to enhance actin stability via end-

TABLE II

Quantitation of actin, alpha-actinin and tropomyosins in Triton cytoskeletons from normal and SV40-transformed MRC-5 fibroblasts

	cpm in[a]					Ratio Actin/			
			Tropomyosin				Tropomyosin		
Triton cytoskeleton[b]	Actin	Alpha-actinin	IEF 52	IEF 55	IEF 56	Alpha-actinin	IEF 52	IEF 55	IEF 56
MRC-5	31292	2538	932	214	326	12.3	33.6	146.2	96
MRC-5 SV40	16 209 (0.52)[c]	1565 (0.62)[c]	485 (0.52)[c]	378 (1.77)[c]	236 (0.72)[c]	10.4	33.4	42.9	68.7

[a] cpm have been normalized against IEF 35, a mitochondrial protein that is mainly present in the cytoskeleton (Bravo & Celis 1982a, 1984). Furthermore, the cmp have been corrected to allow a direct comparison with the data presented in Table I.

[b] Cells labelled with (^{35}S)-methionine were extracted with 0.1% Triton X-100 in Pipes buffer as described in Material and Methods.

[c] Ratio of cpm in transformed vs normal cytoskeletons is given in brackets.

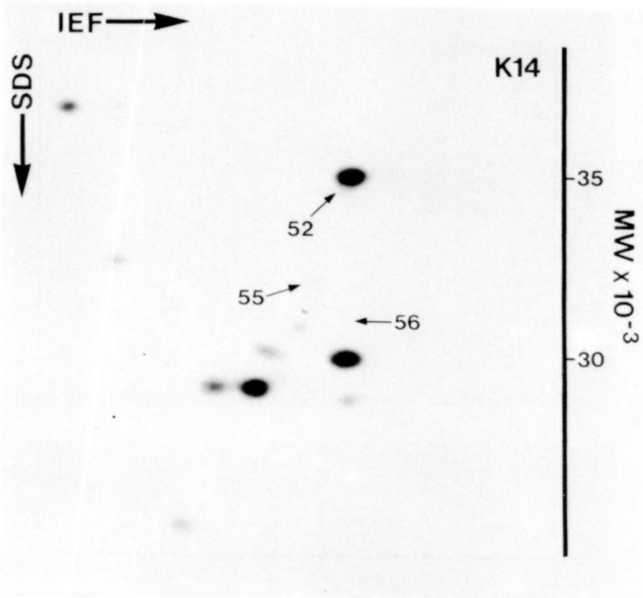

Fig. 7. Two dimensional gel electrophoresis (IEF) of (^{35}S)-methionine-labelled proteins from SV40-transformed human keratinocytes (K14). Only the appropriate region of the gel is shown.

to-end associations. The results of Lin and coworkers suggest that this may be so. But whether or not such longitudinal coupling of tropomyosin on actin leads to more or less stability of the cytoplasmic stress fibre bundles that contain a whole spectrum of actin-associated proteins remains to be established.

The present data, taken together with earlier studies, indicate that there is no simple correlation between the synthesis of tropomyosin variants and the expression of stress fibre bundles. In particular, the following observations may be noted: First, Triton cytoskeletons from SV40-transformed MRC-5 fibroblasts contain, relative to untransformed cells, about 52 and 72% of tropomyosins IEF 52 and 56 and yet exhibit much less than half of the actin cables seen in their normal counterparts. Second, SV40-transformed human epidermal keratinocytes (K14) contain less IEF's 52 and 56 (Figure 7) than AMA cells (Figure 5B) but exhibit higher numbers of actin cables (Figure 8A and B); and third, quantitative two-dimensional gel electrophoretic analysis (IEF) of these 2 tropomyosins throughout the cell cycle of AMA cells have shown that the relative levels of these proteins remain more or less constant during the division cycle (Table III,

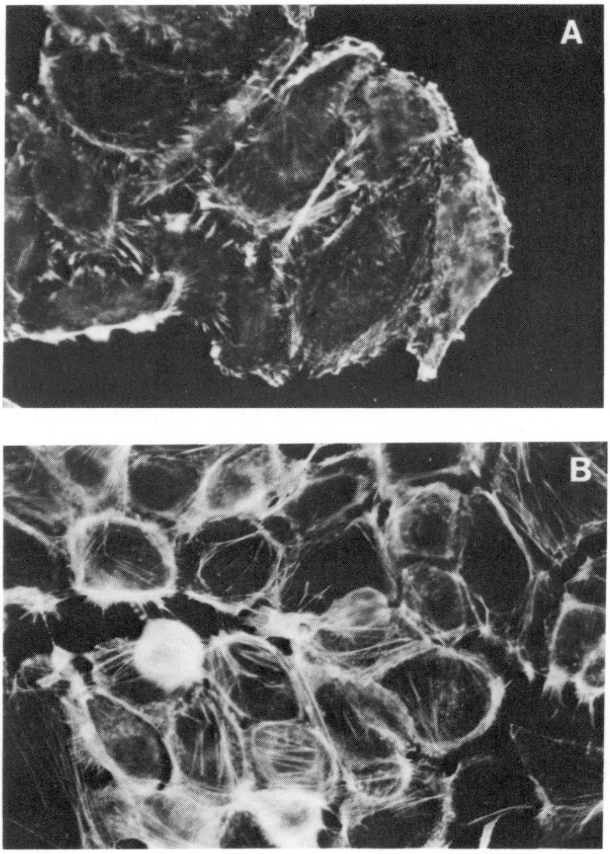

Fig. 8. Actin cables in human AMA (A) and SV40-transformed keratinocytes (K14) (B).

Figures 9A and B, only gels of mitotic and G_1 cells are shown) despite the fact that mitotic cells do not exhibit any actin cables.

Whether or not changes in tropomyosin levels effect stress fibre assembly in interphase or are a secondary consequence of the transformed state still remains to be elucidated. In this respect the microinjection of distinct tropomyosin variants, in excess, into cultured cells may be informative.

ACKNOWLEDGMENTS

We would like to thank Susan Himmelstrup Jørgensen for typing the manuscript and Orla Jensen for photography. We also thank Peder Madsen and Gitte

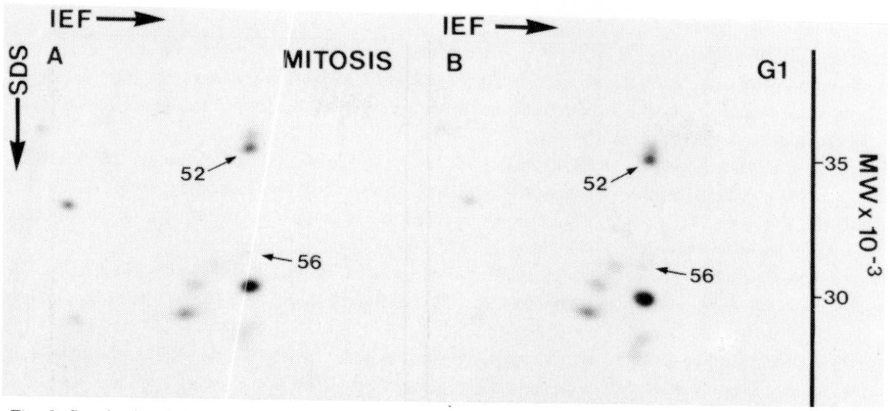

Fig. 9. Synthesis of (^{35}S)-methionine labelled tropomyosins IEF's 52 and 56 during the mitotic (A) and G_1 phases (B) of the cell cycle of AMA cells. Mitotic AMA cells obtained by mechanical detachment were labelled for 45 min while G_1 were labelled for 2 h, 3 h after plating mitotic cells. Only the appropriate region of the gel is shown.

TABLE III

Relative rate of synthesis of tropomyosins IEF 52 and 56 throughout the cell cycle of AMA cells

Cell cycle stage[a]	Relative rate of synthesis of tropomyosins[b] IEF 52	IEF 56
Mitosis	154	108
G_1	170	112
S	185	90
G_2	200	116

[a] Mitotic AMA cells were obtained by mechanical detachment. Two 250 ml flasks containing $1–2 \times 10^6$ cells per flask were used. Synchronized interphase cells were labelled with (^{35}S)-methionine for 2 h throughout interphase while mitotic cells were labelled for 45 min. Entrance of the cells into S-phase was monitored by indirect immunofluorescence using PCNA antibodies specific for cyclin (Miyachi *et al.* 1978, Celis & Celis 1985). The radioactivity in both tropomyosins was normalized against actin, a protein that is synthesized at a constant rate throughout the cell cycle (Bravo & Celis 1980).

Petersen Ratz for help and discussion. Søren Nielsen is a recipient of a fellowship from the Danish Medical Research Councils. This work was supported by grants from the Danish Medical and Natural Science Research Councils, the Danish Cancer Foundation, NOVO (to JEC) and from the Austrian Science Research Council (to JVS).

REFERENCES

Bernstein, B. W. & Bamburg, J. R. (1982) Tropomyosin binding to F-actin protects the F-actin from disassembly by brain actin-depolymerizing factor (ADF). *Cell Motility 2,* 1–8.

Bravo, R. & Celis, J. E. (1980) A search for differential polypeptide synthesis throughout the cell cycle of HeLa cells. *J. Cell Biol. 84,* 795–802.

Bravo, R. & Celis, J. E. (1982a) Updated catalogue of HeLa cell proteins: percentages and characteristics of the major cell polypeptides labelled with a mixture of 16 (^{14}C) aminoacids. *Clin. Chem. 28,* 766–81.

Bravo, R. & Celis, J. E. (1982b) Human proteins sensitive to neoplastic transformation in cultured epithelial and fibroblast cells. *Clin. Chem. 28,* 949–54.

Bravo, R. & Celis, J. E. (1984) Catalogue of HeLa cell proteins. In: *Two-dimensional gel electrophoresis of proteins: Methods and applications,* eds. Celis, J. E. & Bravo, R., pp. 445–76. Academic Press, New York.

Bravo, R., Fey, S. J., Bellatin, J., Mose Larsen, P., Arevalo, J & Celis, J. E. (1981a) Identification of a nuclear and of a cytoplasmic polypeptide whose relative proportions are sensitive to changes in the rate of cell proliferation. *Exp. Cell Res. 136,* 311–19.

Bravo, R., Fey, S. J., Small, J. V., Mose Larsen, P. & Celis, J. E. (1981b) Coexistence of three major isoactins in a single Sarcoma 180 cell. *Cell 25,* 195–202.

Bravo, R., Small, J. V., Fey, S. J., Mose Larsen, P. & Celis, J. E. (1982) Architecture and polypeptide composition of HeLa cell cytoskeletons. Modification of cytoarchitectural proteins during mitosis. *J. Mol. Biol. 159,* 121–43.

Celis, J. E. & Bravo, R. (1981) Cataloguing human and mouse proteins. *Trends Biochem. Sci. 6,* 197–201.

Celis, J. E., Bravo, R., Mose Larsen, P., Fey, S. J., Bellatin, J. & Celis, A. (1984a) Expression of cellular proteins in normal and transformed human cultured cells and tumors. In: *Two-dimensional gel electrophoresis of prteins: Methods and applications,* eds. Celis, J. E. & Bravo, R., pp. 307–62. Academic Press, New York.

Celis, J. E. & Celis, A. (1985) Cell cycle-dependent variations in the distribution of the nuclear protein cyclin (proliferating cell nuclear antigen) in cultured cells: Subdivision of S-phase. *Proc. Natl. Acad. Sci. USA 82,* 3262–66.

Celis, J. E., Fey, S. J., Mose Larsen, P. & Celis, A. (1984b) Expression of the transformation-sensitive protein "cyclin" in normal human epidermal basal cells and simian virus 40-transformed keratinocytes. *Proc. Natl. Acad. Sci. USA 81,* 3128–32.

Celis, J. E., Gesser, B., Small, J. V., Nielsen, S. & Celis, A. (1986) Changes in the levels of human tropomyosins IEF 52, 55 and 56 do not correlate with the loss of actin cables observed in SV40 transformed MRC-5 fibroblasts. *Protoplasma* (in press).

Coté, G. P. (1983) Structural and functional properties of the non-muscle tropomyosins. *Mol. Cell. Biochem. 57,* 127–46.

Fattoum, A., Hartwig, J. H. & Stossel, T. P. (1983) Isolation and some structural and functional properties of macrophage tropomyosin. *Biochemistry 22,* 1187–93.

Forchhammer, J. (1982) Quantitative changes of some cellular polypeptides in C3H mouse following transformation by Moloney-Sarcoma virus. In: *Biological markers of neoplastic transformation,* ed. Chandra, P., pp. 445–55. Plenum Press, New York.

Franza, B. R. & Garrells, J. I. (1984) Transformation sensitive proteins of REF 52 cells detected by computer-analyzed to dimensional gel electrophoresis. *Cancer Cells 1,* 137–46.

Fujime, S. & Ishiwata, S. (1971) Dynamic study of F-actin by quasielastic scattering of laser light. *J. Mol. Biol. 62,* 251–65.

Fujiwara, K. & Pollard, T. (1976) Fluorescent antibody localization of myosin in the cytoplasm cleavage furrow and mitotic spindle of humans. *J. Cell Biol. 71,* 848–75.

Fujiwara, K., Porter, M. E. & Pollard, T. D. (1978) Alpha-actinin localization in the cleavage furrow during cytokinesis. *J. Cell Biol. 79,* 268–75.

Goldman, R. D., Yerna, M. J. & Schloss, J. A. (1976) Localization and organization of microfilaments and related proteins in normal and virus-transformed cells. *J. Supramol. Struct. 5,* 155–83.

Heggeness, M. H., Wang, K. & Singer, S. J. (1977) Intracellular distributions of mechanochemical proteins in cultured fibroblasts. *Proc. Natl. Acad. Sci. USA 74,* 3883–87.

Hendricks, M. & Weintraub, H. (1981) Tropomyosin is decreased in transformed cells. *Proc. Natl. Acad. Sci. USA 78,* 5633–37.

Hendricks, M. & Weintraub, H. (1984) Multiple tropomyosin polypeptides in chicken embryo fibroblasts: Differential repression of transcription by Rous sarcoma virus transformation. *Mol. Cell Biol. 4,* 1823–33.

Kawamura, M. & Maruyama, K. (1970) Electron microscopic particle lengths of F-actin polymerized *in vitro. J. Biochem. 67,* 437–57.

Lazarides, E. (1975) Tropomyosin antibody: the specific localization of tropomyosin in nonmuscle cells. *J. Cell Biol. 65,* 549–61.

Lazarides, E. & Burridge, K. (1975) Alpha-actinin immunofluorescent localization of a muscle structural protein in non-muscle cells. *Cell 6,* 289–98.

Lehrer, S. S. (1981) Damage to actin filaments by glutaraldehyde: Protection by tropomyosin. *J. Cell Biol. 90,* 459–66.

Leonardi, C. L., Warren, R. H. & Rubin, R. W. (1982) Lack of tropomyosin correlated with the absence of stress fibres in transformed rat kidney cells. *Biochim. Biophys. Acta 720,* 154–62.

Lin, J. J.-C., Helfman, D. M., Hughes, S. H. & Chou, C.-S. (1985) Tropomyosin isoforms in chicken embryo fibroblasts: Purification, characterization and changes in Rous sarcoma virus transformed cells. *J. Cell Biol. 100,* 692–703.

Lin, J. J.-C., Yamashiro-Matsumara, S. & Matsumara, F. (1984) Microfilaments in normal and transformed cells: changes in the multiple forms of tropomyosin. *Cancer Cells 1,* 57–65.

Linder, S., Krondahl, U., Sennerstam, R. & Ringertz, N. (1981) Retinoic acid-induced differentiation of F9 embryonal carcinoma cells. *Exp. Cell Res. 132,* 453–60.

Matsumara, F., Lin, J. J.-C., Yamashiro-Matsumara, S., Thomas, G. P & Topp, W. C. (1983a) Differential expression of tropomyosin forms in the microfilaments isolated from normal and transformed rat cultured cells. *J. Biol. Chem. 258,* 13954–64.

Matsumara, F., Yamashiro-Matsumara, S. & Lin, J. J.-C. (1983b) Isolation and characterization of tropomyosin-containing microfilaments from cultured cells. *J. Biol. Chem. 258,* 6636–44.

Maupin-Szamier, P. & Pollard, T. D. (1978) Actin filament destruction by osmium tetroxide. *J. Cell Biol. 77,* 837–52.

McNutt, N. S., Culp, L. A. & Black, P. H. (1973) Contact-inhibited revertant cell lines isolated from SV40-transformed cells. *J. Cell Biol. 56,* 412–28.

Miachi, K., Fritzler, M. J. & Tan, E. (1978) Autoantibody to a nuclear antigen in proliferating cells. *J. Immunol. 121,* 2228–34.

Nakamura, T., Yamaguchi, M. & Yanagisawa, T. (1979) Comparative studies of actins from various sources. *J. Biochem. (Tokyo) 85,* 627–31.

O'Farrell, P. H. (1975) High-resolution two dimensional electrophoresis of proteins. *J. Biol. Chem. 250,* 4007–21.

O'Farrell, P. Z., Goodmann, H. M. & O'Farrell, P. H. (1977) High resolution two-dimensional electrophoresis of basic as well as acidic proteins. *Cell 12,* 1131–42.

Paulin, D., Perreu, J., Jakob, F. & Yaniv, M. (1979) Tropomyosin synthesis accompanies formation of actin filaments in embryonal carcinoma cells induced to differentiate by hexamethylene bisacetamide. *Proc. Natl. Acad. Sci. USA 76,* 1891–95.

Pollack, R., Osborn, M. & Weber, K. (1975) Patterns of organization of actin and myosin in normal and transformed non-muscle cells. *Proc. Natl. Acad. Aci. USA 72,* 994–98.

Pollack, R. & Rifkin, D. (1975) Actin-containing cables within anchorage-dependent rat embryo cells are dissociated by plasmin and trypsin. *Cell 6,* 495–506.

Small, J. V. & Celis, J. E. (1978) Filament arrangements in negatively stained cultured cells: The organization of actin. *Cytobiology 16,* 308–25.

Takebayashi, T., Morita, Y. & Oosawa, F. (1977) Electron microscopic investigation of the flexibility of F-actin. *Biochim. Biophys. Acta 492,* 357–63.

Wang, K., Ash, J. F. & Singer, S. J. (1975) Filamin, a new high-molecular-weight protein found in smooth muscle and non muscle cells. *Proc. Natl. Acad. Sci. USA 72,* 4483–86.

Weber, K. & Groeschel-Stewart, U. (1974) Antibody to myosin: the specific visualization of myosin-containing filaments in non muscle cells. *Proc. Natl. Acad. Sci. USA 71,* 4561–64.

DISCUSSION

RIGBY: Do you have any clue as to the mechanism by which the tropomyosin synthesis is regulated? Have you looked at mRNA levels?

CELIS: Yes, there is a paper by Weintraub's group (Hendricks & Weintraub 1984 *Mol. Cell. Biol 4,* 1823) showing that synthesis of these proteins is transcriptionally regulated. Transformation of chicken embryo fibroblasts with Rous sarcoma virus leads to a complete repression of transcription.

BRUGGE: Were the labelling conditions for all the 2D-gels such that you would only be measuring the rate of synthesis and not the steady state levels of these proteins?

CELIS: We have done both, and the results are very similar. Most of the results that I have shown here correspond to steady state levels. We cannot find much difference.

BRUGGE: Have you looked at any cells transformed by temperature-sensitive mutant viruses?

CELIS: No, we have not done that, but Weintraub and colleagues have done it. They have used temperature-sensitive mutants of RSV, and in general their results are similar to ours. They differ, however, in the fact that we have found that the ratio of total tropomyosin to actin is the same in both normal and transformed human fibroblasts. This difference may be related to the fact that various cell types exhibit different combinations of tropomyosins. In any case, I feel that the data currently available is not sufficient to explain the loss of cables observed in transformed cells. Also, we have to take into account that, with few exceptions, cables have not been shown in tissues. So, we do not know to what extent this sort of structures are artefacts and due to culturing of cells.

FORCHHAMMER: I have done similar experiments to those by Weintraub with Moloney sarcoma virus. You did report some experiments where the actin tropomyosin ratio was constant in transformed and untransformed with SV40. Weintraub finds an increase in tropomyosin with the Rous sarcoma virus. I found a decrease with the MoMSV, and furthermore pulse chase experiments indicated that one of the tropomyosins was labile in the transformed cells. This was reduced 3-fold in contrast to the other two which remained constant during a 24-h chase.

CELIS: One of the main problems has been the lack of standardization of these proteins. The tropomyosins are a family of proteins that may play different roles in various cells, and that are differently regulated.

FORCHHAMMER: I agree with your conclusion that there was not sufficient explanation for the degradation of the microfilaments, because there are these inconclusive results. But the tropomyosin I studied was defined by immunoprecipitation.

CELIS: Yes. I think this just points to the fact that we are dealing with very complex phenomena.

WEINBERG: As we have some time left I am wondering whether there are issues that people would like to discuss from this morning or this afternoon or other issues?

KJELDGAARD: During the first day, we have been talking a lot about the changes accompanying transformation, we have looked at activation of genes, and we have heard about various ways this can occur. This morning we have heard about transgenic mice and how activation of the *myc* gene by by translocation in many cases involved the immunoglobulin genes. This is one group of genes which are rearranged, a rearrangement which must involve very specific mechanisms. By this rearrangement one could imagine changes in the gross chromosome structure, which would transfer genes from inactive regions of the chromosome to active regions. We could thus envisage an activation of genes which would not involve any of the mechanisms which we have talked about up to now, enhancers and promoters and insertion models. I wonder what is known about the biology and biochemistry which is involved in the gene rearrangement. Is this something which is entirely restricted to the immunoglobin genes or could it also involve other genes which we have not recognized yet?

BALTIMORE: There is really very little known about the mechanisms involved in rearrangements of immunoglobulin genes. There are clearly two systems of rearrangement: there are the joining rearrangements, that bring together the variable regions, and then there are the switch recombinations, which are particular to heavy chain, and allow multiple heavy chain isotypes to be made. But, although the sequences involved in these sometimes were defined many years ago, with joining sequences involving the characteristic heptamer/nonomer sequences and the switch regions involving the repeated sequences that were referred to this morning, the actual mechanisms or enzymes involved are not really known. A number of laboratories, including my own, identified an endonuclease a number of years ago, that might play a role or that might not. It was never terribly clear. But a good *in vitro* system that carries out either switch rearrangements or joining rearrangements has not been identified. So, I think, the only thing we really have to go on is the sequence that allowed the work that Suzanne Cory and colleagues and a number of other laboratories have done. Looking at the sequences before and after rearrangements gives some idea that the kinds of sequences involved in switch rearrangements or in joining rearrangements may lead to translocations. There are six places known on various chromosomes where there are rearranging families of immunoglobulin or T-cell receptor genes. Aside from that there really is no evidence that I know of of other rearrangements that occur at high frequencies and are involved in differentiation, although people keep expecting to find other gene systems of this sort. The classical translocation of the Philadelphia chromosome that involves the *bcr* region on one chromosome with the *abl* region on another, which is as characteristic as any of the immunoglobulin-related ones, does not seem to involve a sequence which has a propensity to rearrange at any other time.

YANIV: In the rearrangement that involves the Abelson virus, do you activate *abl* transcription?

BALTIMORE: No, you change the nature of the protein that is made. The actual amount of Abelson protein made is increased, but it is not clear whether that is important or not.

YANIV: You change the primary structure?

BALTIMORE: Yes, because you change the 5′ end of the protein. That is probably the critical event.

YANIV: One can maybe rephrase the question and ask whether activation of oncogenes can be restricted only to the immunoglobulin region because it has some mechanisms for DNA rearrangement or alternatively any active gene when it is fused at a certain distance to an oncogene can activate the oncogene by transmitting long distance changes in chromatin structure that are inheritable from mother to daughter cell.

BALTIMORE: I was struck this morning when there was the discussion about the rat plasmacytomas. I do not know what the number was, but some large fraction involved well-recognized genes and then there was one that was not part of the immunoglobulin gene series, that was in fact a LINE sequence that was rearranged, at a ratio of 1:10 or whatever the number was.

SÜMEGI: It is very low. We have found only one tumor where the c-*myc* gene was rearranged with a LINE sequence. As LINE sequences have been found inside the immunoglobulin heavy chain region, we do not know the significance of this rearrangement. We do not know either whether the LINE activates the c-*myc* gene or the c-*myc* is activated because it has been juxtaposed to the immunoglobulin heavy chain region.

YANIV: So, you would bring it as an argument to say that it is only the immunoglobulin region that can easily move next to *myc*.

BALTIMORE: Yes, I think easily is the right word.

YANIV: Because it is apt to translocate or recombine?

BALTIMORE: Because recombination is a common act, and therefore the states of recombination are more likely there. But clearly, it can happen elsewhere, and without the immunoglobulin rate of recombination we might be focusing on other ones.

WEINBERG: The number of documented instances of other gene fusions, unnatural fusions, is rather small. The only one that comes to my mind, but there must be others known, is the recent work of Barbacid in which he had found a *trk* gene, which is some sort of kinase, becomes fused to one of the tropomyosin genes. I believe that this oncogene already pre-existed in the tumor cells prior to being manipulated by transfection. That represents an example where there is another activating gene besides immunoglobulin. But they certainly are not a large catalogue for such examples that are well known.

YANIV: So the immunoglobulin domain should have some attraction for recombination enzymes that make mistakes sometimes?

CORY: If you look at the sequences around many of the plasmacytoma c-*myc*-Ig switch region junctions there is enough sequence ressemblance to make you suspect that the enzymes catalyzing Ig switch recombination also sometimes cleave the *myc* locus. But the sequence homology is not great.

VAN DER EB: I would like to ask Suzanne Cory, do you think there is a chance that it is *myc* that translocates to the immunoglobulins?

CORY: I have a bias, but I think that *myc* deregulation may be particularly significant for lymphoid cells.

WEINBERG: The paradox here is that there is a number of other *myc*-like genes in the cell which are physiologically very similar to one another in their mode of action. I could cite the N-*myc* gene or L-*myc* gene, both of which would seem to have very similar biological effects. Thus, it is not at all obvious why c-*myc* is the preferred target for joining.

SÜMEGI: We screened 26 multiple myelomas for the rearrangement of the N-*myc* or L-*myc* genes, and we did not find any. We did not find proviral DNA integration in the vicinity of the N-*myc* or the L-*myc* gene in murine T-cell lymphomas, either. Perhaps the N-*myc* and the L-*myc* genes cannot be activated by enhancing their transcription by viral LTR or immunoglobulin enhancers. Their activation may require some other mechanisms than rearrangements.

BERNS: There could also be another reason for that, because proviral integration might just go to those chromosomal sites which are already active. So, if N-*myc* and L-*myc* are inactive in lymphoid cells that might not be a very preferred site to go to, so there could be an alternative mechanism.

BALTIMORE: Yes, I think it is important to remember, as has been known for many years, that there are levels of regulation of genes, and that a gene to be activated by translocation might already have to be in a partially active state, i.e. at the initial state there is a determination, to use an old embryogical word.

CORY: We also have to remember the existence of fragile sites. Maybe there are topographical strains in certain regions that renders them much more fragile. If these happen to coincide with a gene involved in growth control then you will find these genes being altered with increased frequency.

Induction of c-*fos* and c-*myc* by Growth Factors: Role in Growth Control

Rodrigo Bravo & Rolf Müller

Several proto-oncogenes have recently been shown to be part of the mechanisms involved in the control of cell proliferation. Among them are the c-*sis* gene which encodes for the B subunit of the platelet-derived growth factor (PDGF) (Doolittle *et al.* 1983, Waterfield *et al.* 1983), and the c-*erb*B and c-*fms* genes that encode for the epidermal growth factor (EGF) (Downward *et al.* 1984) and the macrophage colony-stimulating factor-1 (CSF-1) receptor, respectively (Sherr *et al.* 1985). The possible role of proto-oncogenes as transducers of the signals from the activated receptor has been recently demonstrated by the observation that c-*src* is specifically phosphorylated in a serine residue shortly after PDGF treatment of fibroblasts (Gould *et al.* 1985). Other observations indicate that proto-oncogenes of the *ras* family can be implicated in the cAMP metabolism (Kataoka *et al.* 1985, Koda *et al.* 1985). Several studies suggest that proto-oncogenes encoding for nuclear proteins are also involved in the control of cell proliferation. Stimulation of fibroblasts with mitogenic agents leads to a dramatic increase in c-*fos* expression, followed by accumulation of c-*myc* RNA (Kelly *et al.* 1983, Greenberg & Ziff 1984, Cochran *et al.* 1984, Kruijer *et al.* 1984, Müller *et al.* 1984). The results of these studies, together with subsequent investigations analyzing the expression of c-*fos* and c-*myc* during the cell cycle and the role of protein kinase C as a signal transducer for the induction of the proto-oncogenes, will be discussed.

INDUCTION OF c-*fos* AND c-*myc* RNAs BY GROWTH FACTORS

Stimulation of quiescent (i.e., serum-deprived) NIH3T3 cells with fetal calf serum (FCS) leads to the synchronous entry of the cell population into the S-phase of the cell cycle within 10–12 h (Müller *et al.* 1984). The earliest known change in gene expression following stimulation of fibroblasts with FCS is the induction

---

European Molecular Biology Laboratory, Postfach 10.2209, D-6900 Heidelberg, F.R.G.

VIRAL CARCINOGENESIS, Alfred Benzon Symposium 24,
Editors: N. O. Kjeldgaard, J. Forchhammer, Munksgaard, Copenhagen 1987.

of c-*fos* transcription (Cochran *et al.* 1984, Greenberg & Ziff 1984, Kruijer *et al.* 1984, Müller *et al.* 1984). c-*fos* mRNA levels increase at least 50-fold within 30 min and decrease rapidly to basal levels within the following 90 min. On the other hand, c-*myc* RNA shows an approximately 20-fold increase followed by a slow decrease, reaching basal levels of expression at approximately 18 h. Purified growth factors, such as PDGF and fibroblast growth factor (FGF), are equally efficient at inducing c-*fos* and c-*myc* expression as is 10% fetal calf serum (FCS). Although the increased levels of c-*fos* mRNA have been shown to be largely due to transcriptional activation (Greenberg & Ziff 1984), the levels of c-*myc* mRNA are controlled by post-transcriptional mechanisms (i.e., by modulation of the stability of c-*myc* mRNA; Blanchard *et al.* 1985). While the induction of c-*fos* precedes the accumulation of c-*myc* mRNA it is unlikely that the c-*fos* gene product is involved in the post-transcriptional control of c-*myc,* as both genes can be induced in the presence of protein synthesis inhibitors (Cochran *et al.* 1984, Greenberg & Ziff 1984, Müller *et al.* 1984).

INDUCTION OF c-*fos* PROTEIN BY SERUM AND GROWTH FACTORS

Immunoprecipitation experiments demonstrated that the induction of c-*fos* mRNA is rapidly followed by the synthesis of c-*fos* protein. Maximum synthesis is observed approximately 1 h after stimulation, and c-*fos* protein synthesis is greatly reduced at 2 h, correlating with the observed levels of c-*fos* mRNA. The growth factor-induced c-*fos* protein is post-transcriptionally modified, as judged from its increased molecular mass. While unmodified c-*fos* protein migrates as a 54 KD protein on reducing polyacrylamide gels, several forms in the range of 55–65 KD are observed in stimulated 3T3 cells (Müller *et al.* 1984, Kruijer *et al.* 1984). Two-dimensional gel analyses of c-*fos* protein have demonstrated that the changes in molecular weight of the protein are associated with a decrease in the isoelectric point (pI). These changes in pI are possibly due to phosphorylations of the protein.

As determined by immunofluorescence, the levels of c-*fos* protein are maximal 1–2 h after stimulation by either serum or purified growth factors. c-*fos* protein is barely detectable at 4 h. An increase in c-*fos* protein can be observed after 15 min of stimulation. The typical granular staining pattern of c-*fos* throughout the nucleus, with the exception of the nucleoli, presents no changes in distribution during the whole period of induction (Müller *et al.* 1984).

EXPRESSION OF c-*fos* AND c-*myc* DURING THE CELL CYCLE

Recent studies have demonstrated that c-*fos* mRNA levels are extremely low or even undetectable throughout the cell cycle in NIH3T3 cells (Bravo *et al.* 1986), suggesting that a high expression of c-*fos* is not required for the continuous cycling of NIH3T3 cells. A role of c-*fos* in the normal proliferation of cells, however, cannot be completely ruled out since c-*fos* expression in growing cells has been shown to be slightly elevated compared to quiescent cells (Müller *et al.* 1984).

Analyses of the expression of c-*myc* mRNA and protein have shown that, in contrast to c-*fos*, c-*myc* is constantly expressed throughout the cell cycle in several cell types, indicating that the c-*myc* product plays a role during normal cell proliferation (Han *et al.* 1985, Rabbitts *et al.* 1985, Thompson *et al.* 1985). However, the level of expression during the cell cycle is approximately 10-fold lower than in growth factor-stimulated cells. An interesting observation is that the expression of c-*myc* in asynchronous cultures requires the constant presence of external growth factors in the medium. A short serum deprivation of 2 h leads to a dramatic decrease in the level of c-*myc* RNA (Bravo *et al.* 1986).

Another set of results shown in these studies indicates that cells at any phase of the cell cycle other than mitosis are able to respond to growth factor by inducing c-*fos* and c-*myc*, indicating that this effect is not restricted to only quiescent cells.

There is some evidence suggesting that c-*fos* and c-*myc* must act in concert to efficiently promote cell division. First, cells transformed by *fos* oncogenes require PDGF to respond with growth to progression factors (Bravo and Müller, unpublished observations). Second, cells transfected with c-*myc* become only partially independent of PDGF for responding to progression factors (Armelin *et al.* 1984).

The observation that several other unidentified genes are induced by growth factors or serum in quiescent cells (Cochran *et al.* 1983, Lau & Nathans 1985) suggests that the transition from G0 to G1 is possibly the result of a complex interaction of several genes, including c-*fos* and c-*myc*.

ROLE OF PHOSPHOLIPID DEGRADATION, PROTEIN KINASE C AND Ca^{2+} IN THE INDUCTION OF c-*fos* AND c-*myc* EXPRESSION

Binding of a growth factor to its receptor leads to the immediate phosphorylation of the receptor and the activation of its intrinsic tyrosine kinase activity. This is followed by stimulation of inositol phospholipid turnover, activation of the

TABLE I
*Induction of c-*fos *and c-*myc *by different agonists*

Agonist	Quiescent		TPA-treated (48 h)	
	c-*fos*[a]	c-*myc*[b]	c-*fos*[a]	c-*myc*[b]
PDGF	++++	++++	++	++
Diacylglycerol (OAG)	++	++	−	−
TPA	+	+	−	−
A23187	++	++	++	++
Phospholipase C	++++	++++	−	−
TPA+A23187	++++	++++	+	+

Concentrations of the different agonists were: PDGF, 25 ng/ml; OAG, 150 μg/ml; TPA, 50 ng/ml; A23187, 3 μM; Phospholipase C, 5 units/ml. When TPA and A23187 were added together, the concentrations used were: TPA, 10 ng/ml; 1 μM.

[a] Analyzed after 1-h stimulation.

[b] Analyzed after 2-h stimulation. The intensity of the induction is expressed as: (++++), 75–100%; (+++), 50–75%; (++), 25–50%; (+), 10–25%, compared to maximal PDGF induction.

Na^+/H^+ antiporter system, changes in nucleotide metabolism and phosphorylation of specific cellular proteins. A central role to the signal pathway that uses phosphoinositol for generating second messengers such as inositol 1,4,5-triphosphate (Ins-P3) and diacylglycerol (DG) to transmit information into the cell from the growth factor/receptor interaction has been suggested (for reviews see Nishizuka 1984, Berridge & Irvine 1984, Ashendel 1985). These 2 second messengers regulate different pathways, possibly essential for the induction of specific genes required for the onset of DNA synthesis. Diacylglycerol stimulates protein kinase C (PKC) which then activates the Na^+/H^+ antiporter system, raising the intracellular pH, while Ins-P3 triggers the release of Ca^{2+} from the endoplasmic reticulum (Nishizuka 1984, Berridge & Irvine 1984). To elucidate the importance of these pathways in the transduction of growth factor signals to the c-*fos* and c-*myc* genes, we used several compounds known to affect phospholipid degradation, activation of PKC and release of intracellular Ca^{2+}. We also made use of the observation that the constant exposure (for several days) of 3T3 cells to the tumor promoter 12-O-tetradecanoyl-13-acetate (TPA), a phorbol ester able to stimulate PKC, leads to a dramatic decrease of PKC activity in the cell (Rodriguez-Pena & Rozengurt 1984). The results of these studies showed that the degradation of PtdIns-P2 and the subsequent activation of PKC and release of Ca^{2+} seem to play an important role in c-*fos* and c-*myc* induction. The observations that led us to this conclusion are the following: (i) phospholipase C, an enzyme that specifically cleaves inositol phospholipids to generate diacylglycerol and Ins-P3, therefore activating protein kinase C and Ca^{2+} release, is a potent inducer of c-*fos* and c-*myc* expression. (ii) Activators of protein kinase C, such as TPA and OAG, a diacylglycerol analogue, induce c-*fos* and c-*myc* expression, albeit at lower levels than those observed with PDGF. This would suggest that protein kinase C is an important, but not the only element involved in the signal transduction to the c-*fos* and c-*myc* genes. (iii) PDGF, which stimulates protein kinase C through the production of DG, and increases intracellular Ca^{2+} by Ins-P3 production, is a strong inducer of c-*fos* and c-*myc,* similarly to phospholipase C. In cells pretreated with TPA for 3 d (protein kinase C-depleted), PDGF induced the proto-oncogenes, but at lower levels, suggesting that both pathways – protein kinase C- and Ca^{2+}-dependent – are important in c-*fos* and c-*myc* induction. This is further supported by: (iv) treatment of the cells with the calcium ionophore A23187 induces the expression of c-*fos* and c-*myc* to significant levels; and (v) a synergistic effect was observed in the induction of c-*fos* when cells were treated simultaneously with TPA and A23187. A summary of these results is shown in Table I.

REFERENCES

Armelin, H. A., Armelin, M. C. S., Kelly, K., Stewart, T., Leder, P., Cochran, B. H. & Stiles, C. D. (1984) Functional role for c-*myc* in mitogenic response to platelet-drived growth factor. *Nature 310,* 655–60.

Ashendel, C. L. (1985) The phorbol ester receptor: a phospholipid regulated protein kinase. *Biochem. Biophys. Acta 822,* 219–42.

Berridge, M. J. & Irvine, R. F. (1984) Inositol trisphosphate, a novel second messenger in cellular signal transduction. *Nature 312,* 315–21.

Blanchard, J. M., Piechazyk, M., Dani, C., Chambard, J. C., Franchi, A., Poyssegur, J. & Jeanteur, P. (1985) C-*myc* gene is transcribed at high rate in Go-arrested fibroblasts and is post-transcriptionally regulated in response to growth factors. *Nature 317,* 443–45.

Bravo, R., Burckhardt, J., Curran, T. & Müller, R. (1986) Expression of c'*fos* in NIH3T3 cells is very low but inducible throughout the cell cycle. *EMBO J. 5,* 695–700.

Cochran, B. H., Reffel, A. C. & Stiles, C. D. (1983) Molecular cloning of gene sequences regulated by platelet-derived growth factor. *Cell 33,* 939–47.

Cochran, B. H., Zullo, J., Verma, I. M. & Stiles, C. D. (1984) Expression of the c-*fos* gene and of a *fos*-related gene is stimulated by platelet-derived growth factor. *Science 226,* 1080–82.

Doolittle, R. F., Hunkapillar, M. W., Hood, L. E., Deware, S. G., Robbins, K. C., Aaronson, S. A. & Antoniades, H. N. (1983) Simian sarcoma virus *onc* gene v-*sis,* is derived from the gene (or genes) encoding a platelet-derived growth factor. *Science 221,* 275–77.

Downward, J., Yarden, Y., Mayes, E., Scrace, G., Totty, N., Stockwell, P., Ullrich, A., Schlessinger, J. & Waterfield, M. D. (1984) Close similarity of epidermal growth factor receptor and v-*erb-B* oncogene protein sequences. *Nature 307,* 521–27.

Gould, K. L., Woodgett, J. R., Cooper, J. A., Buss, J. E., Shalloway, D. & Hunter, T. (1985) Protein kinase C phosphorylates, $pp60^{src}$ at a novel site. *Cell 42,* 849–57.

Greenberg, M. E. & Ziff, E. M. (1984) Stimulation of 3T3 induces transcription of the c-*fos* proto-oncogene. *Nature 311,* 433–38.

Hann, B. R., Thompson, C. B. & Eisenmann, R. E. (1985) C-*myc* oncogene protein synthesis is independent of the cell cycle in human and avian cells. *Nature 314,* 366–69.

Katoaka, T., Powers, S., Cameron, S., Fasano, O., Goldfarb, M., Broach, J. & Wigler, M. (1985) Functional homology of mammalian and yeast RAS genes. *Cell 40,* 19–26.

Kelly, K., Cochran, B. H., Stiles, C. D. & Leder, P. (1983) Cell-specific regulation of the c-*myc* gene by lymphocyte mitogens and platelet-derived growth factors. *Cell 35,* 603–10.

Kruijer, W., Cooper, J. A., Hunter, T. & Verma, I. M. (1984) Platelet-derived growth factor induces rapid but transient expression of the c-*fos* gene and protein. *Nature 312,* 711–16.

Lau, L. F. & Nathans, D. (1985) Identification of a set of genes expressed during the Go/G1 transition of cultured mouse cells. *EMBO J. 4,* 3145–51.

Müller, R., Bravo, R., Burckhardt, J. & Curran, T. (1984) Induction of c-*fos* gene and protein by growth factor precedes activation of c-*myc*. *Nature 312,* 716–20.

Nishizuka, Y. (1984) The role of protein kinase C in cell surface signal transduction and tumour promotion. *Nature 308,* 693–98.

Rabbitts, P. H., Watson, J. V., Lamond, A., Forster, A., Stinson, M. A., Evan, G., Fisher, W., Atherton, E., Sheppard, R. & Rabbitts, T. H. (1985) Metabolism of c-*myc* gene products: c-*myc* mRNA and protein expression in the cell cycle. *EMBO J. 4,* 2009–15.

Rodriguez-Pena, A. & Rozengurt, E. (1984) Disappearance of Ca^{2+} sensitive, phospholipid dependent protein kinase activity in phosbol ester-treated 3T3 cells. *Biochem. Biophys. Res. Commun. 120,* 1053–59.

Sherr, C. J., Rettenmier, C. W., Sacca, R., Roussel, M. F., Look, A. T. & Stanley, E. R. (1985) The c-*fms* proto-oncogene product is related to the receptor for the mononuclear phagocyte growth factor, CSF-1. *Cell 41,* 665–76.

Thompson, C. B., Challoner, P. B., Neiman, P. E. & Groudine, M. (1985) Levels of c-*myc* oncogene mRNA are invariant throughout the cell cycle. *Nature 314,* 363–66.

Toda, T., Uno, I. Iskikawa, T., Powers, S., Kataoka, T., Broek, D., Cameron, S., Broach, J. Matsumoto, K. & Wigler, M. (1985) In yeast, RAS proteins are controlling elements of adenylate cyclase. *Cell 40,* 27–36.

Waterfield, M. D., Scrace, G. T., Whittle, N., Stroobant, P., Johnsson, A., Westeson, A., Westermark, B., Heldin, C.-H., Huang, J. S. & Deuel, T. F. (1983) Platelet-derived growth factor is structurally related to the putative transforming protein $p28^{sis}$ of simian sarcoma virus. *Nature 304,* 35.

DISCUSSION

YANIV: You showed that calcium ionophore in the presence of TPA present an exponential and not a linear increase in the response of *fos*?

BRAVO: Yes. At the highest concentrations of calcium ionophore used (.1 to 1 pM) the expression of *fos* jumped around three or four times. I do not have a clear explanation for these results. TPA could need a minimum dose of $Ca^{2}+$ ionophore for the synergistic effect in *fos* induction.

YANIV: The conclusion you draw is that *myc* and *fos* synthesis is a direct response of the cell to growth factors, and they are not related to the cell cycle.

BRAVO: No, they are not cell cycle-regulated. That is very clear.

YANIV: What do they do?

BRAVO: I do not know. Possibly, they need to be expressed to keep the cell going. That is why I am saying that the easiest explanation is that they are only required for taking cells from the G_0 state. That is the easiest explanation.

YANIV: Are they expressed also in all other phases of cell growth?

BRAVO: *Myc*, but not *fos* is expressed at detectable levels during the cell cycle.

YANIV: No, but if you did add growth factors?

BRAVO: *Myc* is induced, and then goes down, but never disappears. It is always expressed at a certain level in growing cells. That is completely different to *fos*. *Fos* can be induced in growing cells, but the expression is switched off immediately after as in quiescent cells.

YANIV: But they are induced in a situation where you do not need them.

WEINBERG: Yes, I think he is saying that the expression of *fos* and *myc* may be observed throughout the cell cycle, but it is fortuitous for successful progression of the cell cycle.

SHERR: I would just like to make a comment about the notion that these pathways are really triggered independently. It is probably premature to consider them in this way, and there are at least two examples that I am aware of where there is "cross talk" between the two signal pathways. It is becoming apparent, for example, that there is more than one phospholipase C activity. There is a GTP-driven membrane-bound enzyme which is thought to be the hormone-responsive enzyme, but there is also cytosolic enzyme which has been already purified. The latter enzyme is calcium-responsive, and in the presence of higher concentrations of calcium will degrade both PI and PIP to generate diacylglycerol. So, one might at least keep in mind the possibility that after a calcium signal, the cell might produce diacylglycerol in the absence of a strict inducer of the protein kinase C pathway. Secondly, Phil Majerus has shown that the 5′ phosphatase that degrades inositol triphosphate to IP2 is a preferred substrate for protein kinase C. In platelets, the phosphatase is a 40 K protein substrate recognized years ago, and the enzyme is apparently activated by phosphorylation. So, by triggering protein kinase C, for example, the cell can degrade IP3 and alter its calcium response. Thus, there are examples of positive and negative controls whereby signals in one arm of the pathway influence the outcome of the other.

BRAVO: There are experiments showing that, in fibroblast, TPA does not increase the concentration of internal calcium in the cells. That is very clear. So, if TPA is able in some way to activate calcium-related bands, it has not been detected at least in fibroblast. That is the only thing that I can say. And also, the observation that the calcium ionophore can work in cells that have been treated with TPA. It is an evidence that maybe they are not using the protein kinase C pathway. And a third observation is that EGF does not induce phosphoslipid breakdown, and does not activate protein kinase C, and still can do it. Another observation that has been done by Axel Ullrich and Peter Parker is that there are several possible protein kinase C-related activities, and they have cloned a protein kinase C and then found more than one sequence that is very related to protein kinase C. I agree that one can be oversimplifying the pathways at present, but it seems that there are two pathways at least, a protein kinase C-dependent and a calcium-dependent pathway.

WEINBERG: Walther Moolenaar from the Netherlands says that EGF does indeed induce phosphoinositol degradation resulting in a different kind of breakdown product than does PDGF.

BRAVO: In A431 cells it is known that EGF triggers phospholipid degradation, in fibroblast it does not.

WEINBERG: Perhaps Dr. Schlessinger could comment on this?

SCHLESSINGER: Well, to make the story more complicated, indeed, EGF does stimulate kinase C in A431 cells. By an unclear mechanism it stimulates the entry of calcium into these cells, so it stimulates kinase C and probably regulates its own receptor. But I would like to add another complexity to the problem. It appears that all these receptors communicate by an unclear mechanism, for example PDGF and bombesin decrease the affinity of EGF to the EGF-receptor and also reduce its tyrosine kinase activity. However, these growth factors act in synergy for the stimulation of mitogenesis. So, there is a "cross-talk" at the level of the receptor, which by an unknown mechanism leads to enhanced mitogenesis.

BALTIMORE: In one of your slides you showed that *myc* levels fell. I am not particularly sure of what you did, but you took serum out of the medium?

BRAVO: That is right.

BALTIMORE: They were just growing cells?

BRAVO: Asynchronous cultures.

BALTIMORE: O.K. And you changed the medium to a medium lacking serum.

BRAVO: That is right. Two hours after serum deprivation you do not find *myc* expression.

BALTIMORE: Well, in two hours the *myc* goes down to the basal level. That suggests the hypothesis which you referred to that the low basal level of *myc* is to tell the cell it ought to progress through the next cell cycle because everything is alright. Historically, there are a whole lot of ways of putting cells into G_0, for instance by removing isoleucine. Have you seen whether those treatments also reduce *myc* expression.

BRAVO: Yes, they do not. We have done those experiments with isoleucin deprivation. It is something in serum, definitely. The interesting thing is, you can add FCS, leave it for 30 minutes, enough time to have a good induction. Then take it out for another hour, and put it back again, and you can re-induce *fos* and *myc*. This can be repeated several times.

BALTIMORE: I always thought that the system was reset immediately in the ...

BRAVO: The cells reset extremely fast, including after PDGF, EGF, FGF, or FCS treatment. You can switch on and off. So, the cell is extremely responsive.

RAPP: You had with EGF a fast response of *fos* and *myc*, and with PDGF a slower response for *myc*. If you mix the two growth factors, which one prevails?

BRAVO: We have mixed EGF with bombesin. The cells respond initially like EGF-treated. Then they behave like bombesin-treated cells, in the case of *myc* expression remaining high for several hours.

RAPP: How do you imagine that *fos* virus transformation works? Does it block those cells from entering a G_0 phase?

BRAVO: That is a very simple explanation. In collaboration with Rolf Müller we have taken primary fibroblasts and made them express high levels of v-*fos*, and then tried to grow the cells in a medium that does not have PDGF. The cells do not grow. One could argue that maybe *fos* expression is not enough, and I would agree with that. There are at least 50–100 independent genes induced by growth factors, so it must be a complement between several genes to keep all the cells from G_0.

LOWY: I was wondering if there is additional evidence for the necessary involvement of *fos* in the progression from G_0 to G_1 other than that it is part of a response that occurs invariably.

BRAVO: There is an interesting observation by R. Müller. If you take a monolayer of cells which has been deprived of growth factor for at least 48 hours, and then wound the monolayer, you can detect *fos* induction by immunofluorescence or *in situ* hybridization. That is a very clear observation. How to interpret it is not easy, because there is no growth factor in the medium. No analysis has been done for *myc* in this experiment.

BALTIMORE: Wounding is generally done with serum in a medium. The cells have become quiescent, because the serum has been depleted of a factor. But serum is still there?

BRAVO: That is right, but the serum depletion means that the growth factor has disappeared.

BALTIMORE: I believe that the only useful interpretation of the wounding experiment is that when you wound the cells you lower the threshold for growth stimulation. So the amount of residual growth factor in the medium is still important. Because there is residual growth factor. You can show that by taking sparse cells and placing them in the medium, and they will grow.

BRAVO: That is right, in the case there is remaining growth factor. However, in the experiments that I mentioned the cells were deprived of serum for 48 hours before wounding. Therefore, no growth factors were present.

BALTIMORE: What you are saying is really that wounding is like adding serum and not really making a difference?

VAN DER EB: I was wondering, does EGF cause activation of the sodium proton pump in fibroblasts?

BRAVO: Yes, EGF also produces changes in pH in fibroblasts, and it has not been possible to explain how it does it.

WEINBERG: I would just mention that Paolo Dotto at the Whitehead Institute has made a line of keratinocytes which can no longer be induced to differentiate by adding either calcium or TPA, and when you add calcium and TPA to those keratinocytes, unlike normal keratinocytes, *fos* is not turned on. However, if you add PDGF to those cells, *fos* is turned on perfectly normally. That suggests that PDGF is working in those cells via a mechanism which is neither TPA-responsive nor calcium-responsive. This suggests that there are more than just those two second two messengers able to turn on *fos*.

BRAVO: I want to add something on top of that. If you expose quiescent cells to TPA for a long time so that they become quiescent again, and then you treat them with different growth factors, PDGF and EGF induce the cells to DNA synthesis, but bombesin, that in principle works very similar to PDGF, does not. That is another argument for other mechanisms working in parallel.

V-*myc* Regulation of c-*myc* Expression

John L. Cleveland[1], *Mahmoud Huleihel*[1], *Robert Eisenman*[2], *Ulrich Siebenlist*[3], *James N. Ihle*[4] & *Ulf R. Rapp*[1]

Infection of mouse cells from a variety of lineages with retroviruses expressing high levels of avian v-*myc* was found to be invariably associated with a lack of c-*myc* expression. In order to distinguish between v-*myc*-induced shut-down versus cell-programmed downregulation of c-*myc* expression, we have determined steady state levels of c-*myc* mRNA in 3 different cell lines in culture which express various levels of c-*myc* prior to infection. Extreme levels of v-*myc* expression (10- to 100-fold excess over c-*myc*) were achieved in a myeloid (FDC-P1) and a T lymphoid (CTB-6) cell line. In both lines c-*myc* expression was absent in the infected cells and, in the case of FDC-P1, could not be induced by growth factor (IL-3) or inhibitors of protein synthesis (to remove a labile repressor). Moreover, DNAase I hypersensitive sites typical for active c-*myc* alleles were absent in FDC-P1-infected cells. In NIH 3T3 fibroblast cells, v-*myc* was expressed at levels at least 5 times those of c-*myc* present in uninfected cells. Again, v-*myc* expression was associated with down-regulation of c-*myc*. However, in contrast to infected FDC-P1 cells, downregulated NIH 3T3 cells could be induced to express c-*myc* by treatment with anisomycin for periods corresponding to ≥ 6 times the half-life of v-*myc* protein in these cells. A deletion analysis of v-*myc* revealed that removal of 109 amino acids from the middle of the c-*myc* coding exons was sufficient to abolish its ability to downregulate c-*myc* expression.

LACK OF c-*myc* EXPRESSION IN v-*myc*-INDUCED TUMORS

The effect of v-*myc* on c-*myc* expression was examined by use of a series of v-*myc*-carrying viruses (Rapp *et al.* 1985a, b, c) which shared common transcription

---

[1]Laboratory of Viral Carcinogenesis, National Cancer Institute, Frederick, Maryland 21701, [2]Frederick Hutchinson Cancer Research Center, Seattle, Washington, [3]Laboratory of Immunoregulation, National Institute of Allergy & Infectious Diseases, Bethesda, Maryland 20205, [4]LBI-Basic Research Program, NCI-Frederick Cancer Research, Post Office Box B, Frederick, Maryland 21701, U.S.A.

VIRAL CARCINOGENESIS, Alfred Benzon Symposium 24,
Editors: N. O. Kjeldgaard, J. Forchhammer, Munksgaard, Copenhagen 1987.

regulatory elements (LTRs) (Figure 1). Briefly, 3 of the viruses, J-2, J-3 and J-5, contain a complete v-*myc* gene. In the case of J-2 and J-3 this is a composite of MH2 (5′ half) and MC29 (3′ half) v-*myc* which is expressed from a subgenomic mRNA. In J-5, v-*myc* is derived entirely from MC29 and expressed as part of gag-v-*myc* fusion protein. A similar construct, HF, was obtained from Hung Fan and included in some of the experiments described below. The 4 other viruses contain deletions in v-*myc:* J-1 and J-1A lack all sequences derived from the 2nd c-*myc* coding exon; J-1 Bal lacks the first 123 amino acids from the 2nd c-*myc* coding exon but has the 54 C-terminal amino acids of v-*myc;* and J-5Q contains a deletion of 109 amino acids in the middle of the c-*myc* coding exons.

The J-2 virus, which contains 2 intact oncogenes, v-*raf* and v-*myc,* induces rapid hemopoietic and epithelial neoplasms after a latency period that is much shorter than that of viruses carrying only the v-*raf* (3611 MSV) or v-*myc* (J-3, J-5) oncogenes (Rapp *et al.* 1985b). J-3 and J-5 viruses predominantly induce lymphomas (T and B lineage cells) followed by pancreatic adenocarcinoma and other less frequent tumors, including mammary adenocarcinoma, after a latency period which depends on the replication efficiency of the pseudotyping helper MuLV (Morse *et al.* 1986). Examination of a large number of tumors induced by the J-2, J-3 and J-5 viruses for expression of v-*myc* and c-*myc* polyA RNA showed that high levels of v-*myc* were invariably associated with lack of expression of c-*myc* (Morse *et al.* 1986). A representative Northern analysis of five J-5 virus-induced tumors, a T cell lymphoma, a B cell lymphoma, two pancreatic and one mammary adenocarcinoma is shown in Figure 2. All tumors express high levels of v-*myc* hybridizing transcripts typical for the J-5 virus, whereas c-*myc* transcription was undetectable.

DOWNREGULATION OF c-*myc* BY v-*myc* IN MYELOID/LYMPHOID CELL LINES

A similar observation, that high levels of c-*myc* from an altered allele were often associated with absence of expression from the normal allele, had been made previously, mainly with Burkitt's lymphoma and mouse plasmacytoma cell lines, and led to the proposal of a feedback mechanism for regulation of c-*myc* expression (Leder *et al.* 1983, Stanton *et al.* 1983, Siebenlist *et al.* 1984). We have extended the original observation to include a variety of other cell lineages. However, in the case of Burkitt's lymphoma and mouse plasmacytomas, as well as in our v-*myc* virus-induced tumors, one can not distinguish between cell-regulated versus *myc* -induced downregulation of c-*myc*. In order to address this question, we have infected 3 different cell lines, which normally express c-*myc,*

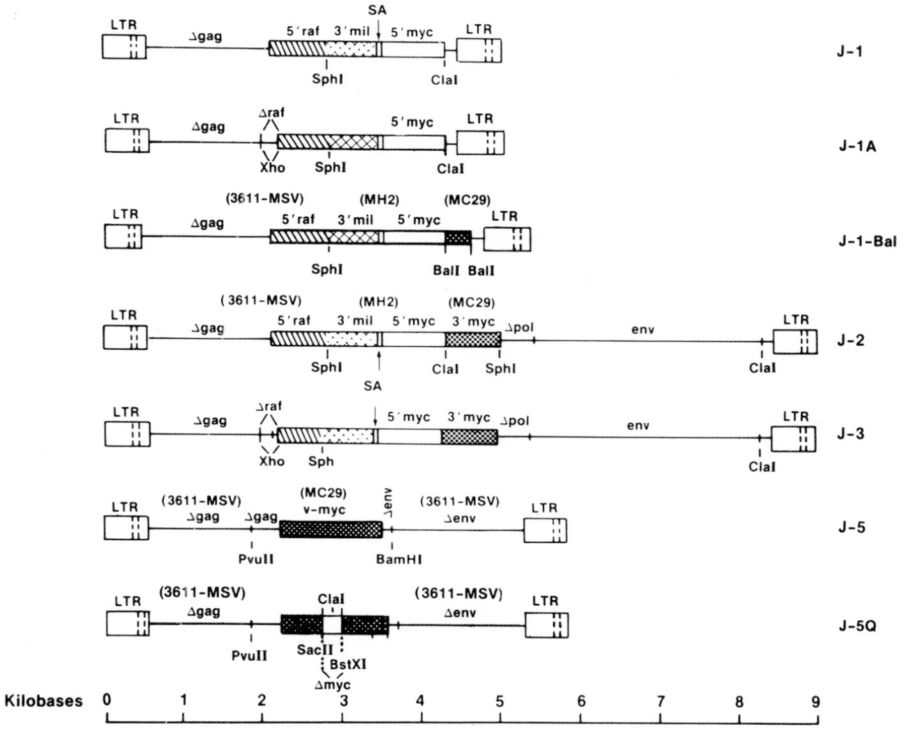

Fig. 1. Genomic organization of the construct retroviruses J-1, J-2, J-3, and J-5 and derivatives J-1A, J-1-Bal, and J-5Q. Important restriction enzyme sites are indicated and the sizes of the various viruses are shown in kilobases. The origin of specific v-*onc* sequences in the constructs is indicated by specific shading: , 3611 MSV v-*raf;* , MH2 v-*mil;* MH2 v-*myc;* , MC29 v-*myc*. SA indicates location of splice acceptor sequence present in MH2 v-*myc* which allows expression of v-*myc* protein from a subgenomic mRNA (Rapp *et al.* 1986b). The construction of J-1 through J-5 recombinant retroviruses has been previously described (Rapp *et al.* 1985b, c). The J-1A virus was constructed by deletion of a 257 bp *Xho*I fragment which spans the gag-*raf* border in J-1; this deletion causes an out-of-frame fusion of *gag* and *raf*. The J-1-Bal virus was constructed by an in-frame fusion of J-1 v-*myc* at the *Cla*I site with a 450 bp *Bal*I fragment from the C-terminus of MC29 v-*myc* (163 nucleotides of coding and 287 nucleotides of noncoding sequence). The J-5Q virus was derived from J-5 by deletion of a 328 bp *Sac*II-*Bst*XI fragment from the 1st and 2nd coding exon of v-*myc*, followed by an in-frame ligation.

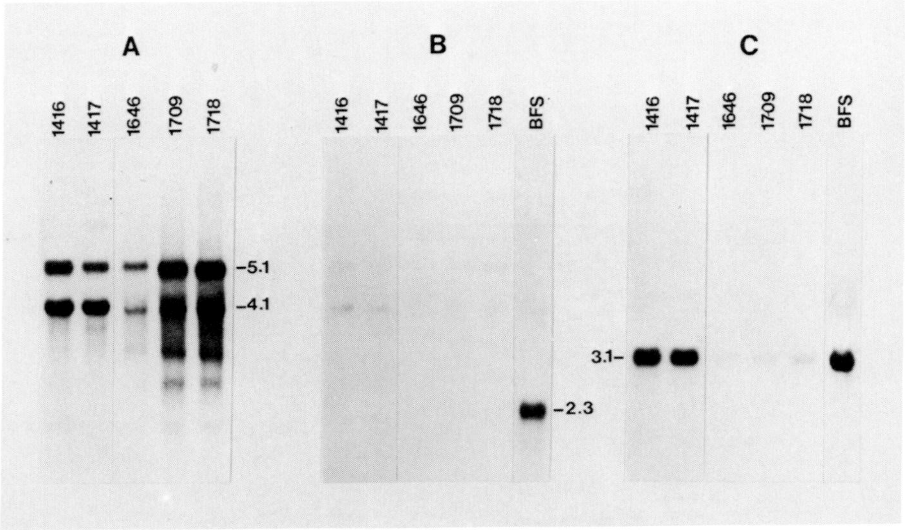

Fig. 2. Northern hybridization analysis of v-*myc,* c-*myc,* and c-*raf*-1 expression in J-5 virus-induced lymphomas (1416, 1417), and pancreatic (1646, 1709) and mammary (1718) adenocarcinoma. Total polyA mRNA was prepared from tumors or from the murine T cell BFS as previously described (Rapp *et al.* 1985c), blotted (5 μg), and hybridized under stringent conditions with v-*myc* (A), c-*myc* (B), or c-*raf*-1 (C) probes. Sizes of hybridizing mRNA are given in kb. Blots were exposed for 20 h.

with v-*myc*-expressing viruses and determined the effect of v-*myc* on c-*myc* expression. One of the cell lines was NIH 3T3 fibroblasts, whereas the other 2 were growth factor-dependent lines: a myeloid line, FDC-P1, which requires IL-3 for growth, and a T lymphoid line, CTB-6, which normally requires IL-2 for growth.

Both FDC-P1 and CTB-6 cells are fairly resistant to virus infection. However, due to the fact that high level expression of v-*myc* in these cells abrogates their respective growth factor requirements. (Rapp *et al.* 1985a), we were able to generate homogenously-infected cells by selection for growth factor independence. The phenotype of the factor-abrogated lines was indistinguishable in terms of a variety of differentiation markers from that of the parental factor-dependent lines. The expression of v-*myc* retroviruses in infected fibroblasts, FDC-P1, and CTB-6 cells is shown in Figure 3. The Northern blot in panel A was hybridized with v-*myc* under conditions where cross-hybridization with c-*myc* does not occur. Both FDC-P1 and CTB-6 express high levels of v-*myc* hybridizing transcripts typical of the infecting virus. In fact, the levels of retrovirus v-*myc* mRNA in the myeloid and lymphoid cells appeared markedly higher than in the control fibro-

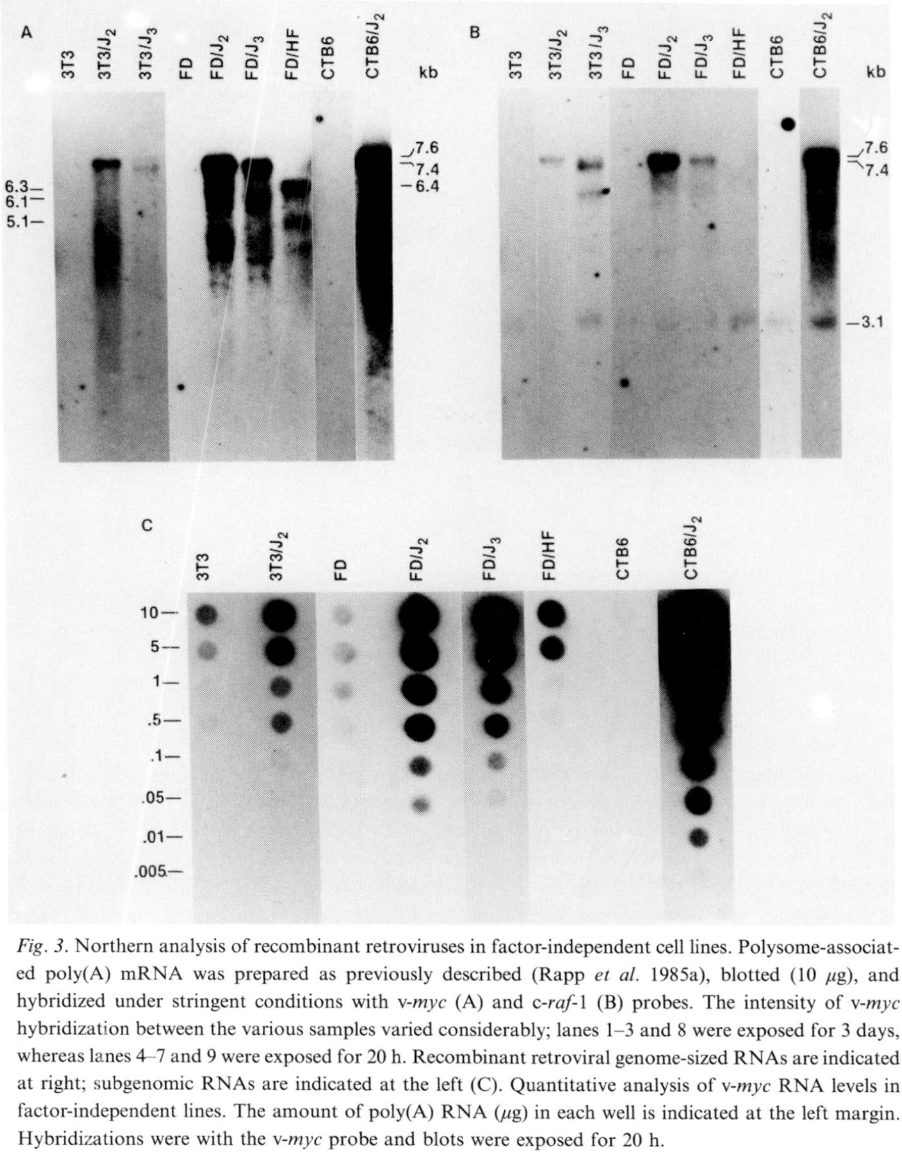

Fig. 3. Northern analysis of recombinant retroviruses in factor-independent cell lines. Polysome-associated poly(A) mRNA was prepared as previously described (Rapp *et al.* 1985a), blotted (10 μg), and hybridized under stringent conditions with v-*myc* (A) and c-*raf*-1 (B) probes. The intensity of v-*myc* hybridization between the various samples varied considerably; lanes 1–3 and 8 were exposed for 3 days, whereas lanes 4–7 and 9 were exposed for 20 h. Recombinant retroviral genome-sized RNAs are indicated at right; subgenomic RNAs are indicated at the left (C). Quantitative analysis of v-*myc* RNA levels in factor-independent lines. The amount of poly(A) RNA (μg) in each well is indicated at the left margin. Hybridizations were with the v-*myc* probe and blots were exposed for 20 h.

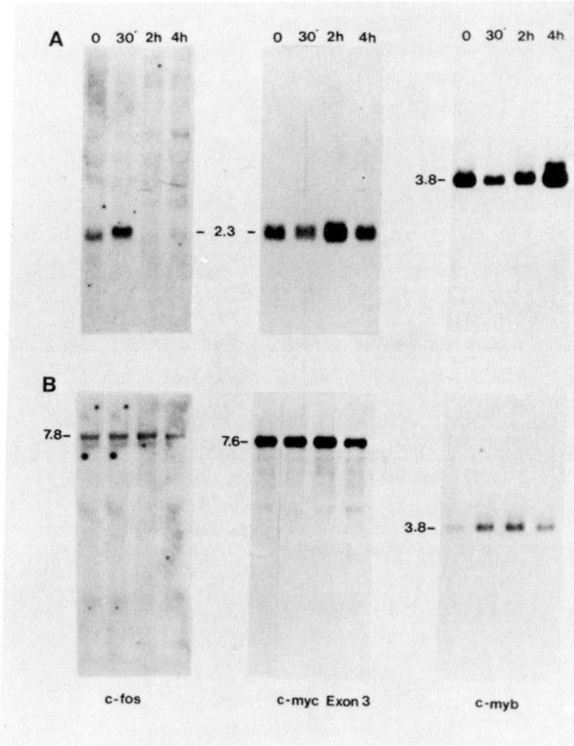

Fig. 4. Interleukin-3 (IL-3)-stimulated proto-oncogene expression in FDC-P1 (A) and FD/J-2 (B) cells. The cells were washed twice in RPMI-1640 and then stimulated with recombinant IL-3 for the times indicated and total poly(A) RNA prepared as previously described. Purified RNA (5 μg) was analyzed for c-*fos*, c-*myc*, and c-*myb* RNA by blotting and hybridization with mouse c-*fos*, c-*myc* exon 3, and c-*myb* DNA probes. Blots of c-*myc* and c-*myb* were exposed for 20 h; blots of c-*fos* were exposed for 4 d.

blast lines. To evaluate this difference further, quantitative dot blot analysis was performed (panel C in Figure 3) which revealed that levels of v-*myc* RNA were approximately 10-fold higher in FDC-P1 and 100-fold higher in CTB-6 cells than in the corresponding fibroblast lines. When tested for c-*myc* expression in these cells, c-*myc* RNA was not detectable in any of the infected cell lines (Rapp *et al.* 1985a, Cleveland *et al.* 1986). This is shown for FDC-P1 infected with the J-2 virus in Figure 4. In this case hybridization with a murine c-*myc* exon 3 probe was done under conditions which also detect v-*myc*. c-*myc* RNA was not detectable in

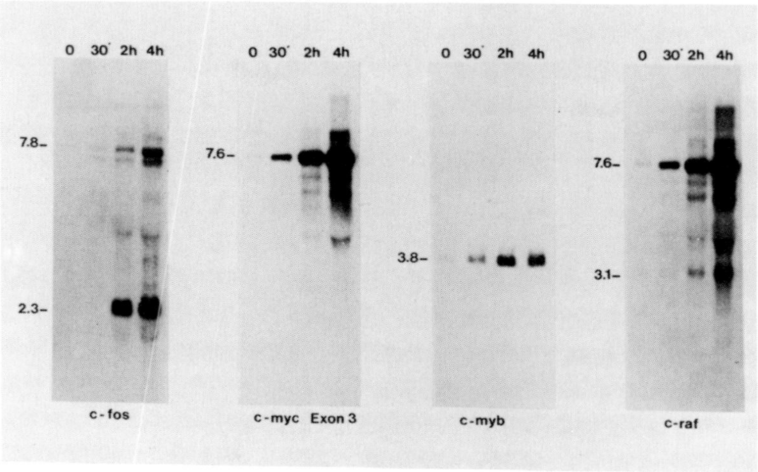

Fig. 5. Expression of proto-oncogenes in FD/J-2 cells treated with anisomycin. Total poly(A) RNA was prepared from FD/J-2 cells treated with 10 μM anisoymcin at the times indicated, blotted (5 μg/lane), and hybridized with c-*fos*, c-*myc*, c-*myb*, and c-*raf* probes. Blots were exposed for 20 h.

the FD/J-2 cell line but was readily apparent in the uninfected FDC-P1 cell (Figure 4).

Since FD/J-2 cells retained functional receptors for IL-3, we were in a position to test whether c-*myc* could be re-induced in these cells by treatment with ligand. IL-3 increased the levels of c-*myc*, c-*fos* and c-*myb* RNA in control FDC-P1, but not in FD/J-2 cells (Figure 4). The fact that not only c-*myc*, but also c-*fos* and c-*myb* were refractory to induction suggested that the IL-3 receptor, while perfectly capable of binding ligand, may be mute in terms of signalling to the nucleus. To further test the reversibility of c-*myc* expression we therefore turned to another approach, treatment with the protein synthesis inhibitor anisomycin to remove a labile repressor that might be involved in c-*myc* downregulation. Anisomycin rapidly increased steady state levels of c-*fos*, c-*myb* and c-*raf* RNA but c-*myc* RNA remained undetectable (Figure 5). Moreover, the conformation of the c-*myc* gene has changed in FD/J-2 to that of the inactive allele. DNA from normal FDC-P1 cells grown in IL-3 has a series of DNAase I-hypersensitive sites typical for the active gene (Siebenlist *et al.* 1984). In contrast, FD/J-2 cells lack the two promoter-associated hypersensitive sites III1 and III2, while site I, a hypothetical repressor binding site (Siebenlist *et al.* 1984) is still present (data not shown).

Thus, in myeloid FDC-P1 cells the generation of homogenously-infected cul-

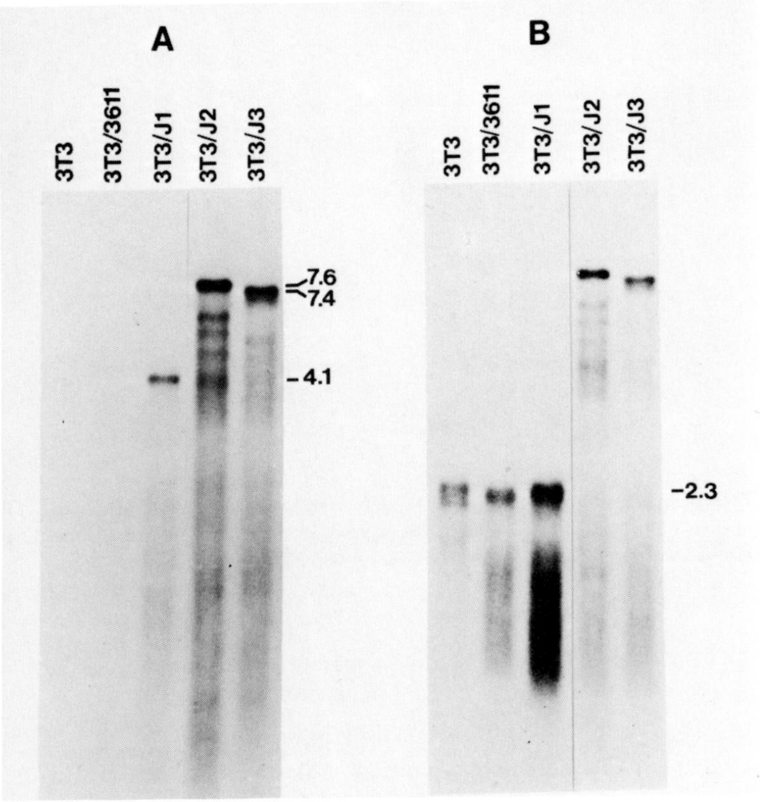

Fig. 6. C-*myc* expression in NIH 3T3 fibroblast cells infected with *raf*- and *myc*-containing retrovirus. Total poly(A) RNA was prepared from the indicated cells (at 80% confluency), blotted (5 μg) and hybridized with either v-*myc* (A) or c-*myc* (B) probes. Blots were exposed for 20 h (v-*myc*) or 3 d (c-*myc*).

tures by selection for growth factor independence yielded cells in which v-*myc* was expressed at ~10 times the level of c-*myc* expression before infection. C-*myc* expression was invariably downregulated in infected cells and the gene took on the conformation of an inactive allele. In the 3rd cell line that was tested for *in vitro* infection, NIH 3T3 fibroblasts, the scenario of v-*myc*-regulated suppression of c-*myc* appeared somewhat different. As is apparent from the Northern blot in Figure 3, the ratio of v-*myc* in infected cells to c-*myc* in uninfected fibroblasts before infection was not as extreme as in the FDC-P1 (Figures 3 and 4) or CTB-6 (Figure 3, and data not shown) cells and if dosage of v-*myc* is an important

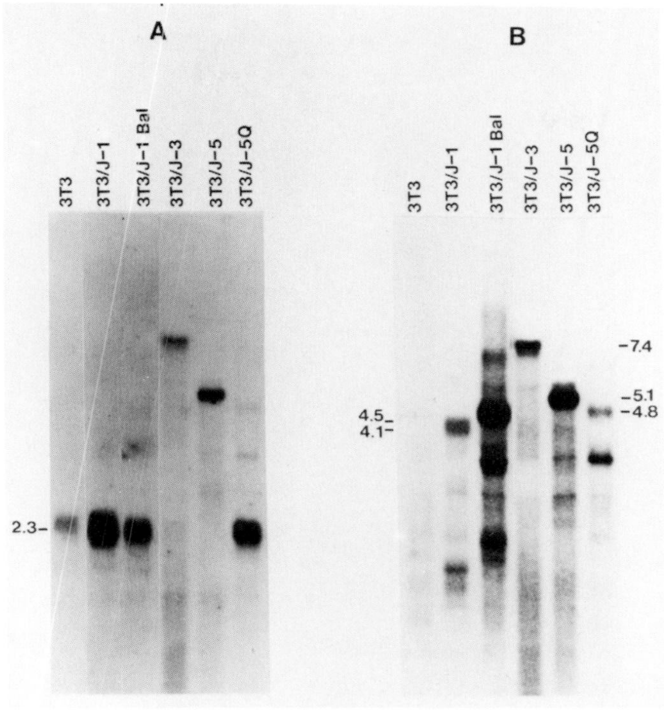

Fig. 7. C-*myc* expression in NIH 3T3 fibroblast cells infected with derivatives of J-1 and J-5. Total poly(A) RNA was prepared from the indicated cells (80% confluency) blotted (5 μg) and hybridized with either c-*myc* (A) or v-*myc* (B) probes. Blots were exposed for 20 h (v-*myc*) or 3 d (c-*myc*).

parameter for c-*myc* regulation one would predict that it may have less drastic effects on c-*myc* expression.

DOWNREGULATION OF c-*myc* BY v-*myc* IN NIH 3T3 FIBROBLASTS

C-*myc* RNA became undetectable in NIH 3T3 cells upon infection with either J-2, J-3 (Figure 6) or J-5 (Figure 7) viruses. In contrast, the v-*raf*-carrying virus 3611 MSV did not appreciably affect c-*myc* expression and infection with the J-1 virus which, in addition to an active *raf* gene, expresses high levels of a C-terminally truncated v-*myc*, lead to increased levels of c-*myc* RNA (Figure 6). Whereas the shut down of c-*myc* was uniformly observed upon analysis of 10 independent v-*myc* virus-positive cell clones infected with any of the *myc* viruses, c-*myc* induction by J-1 virus, although the predominant outcome, was not a

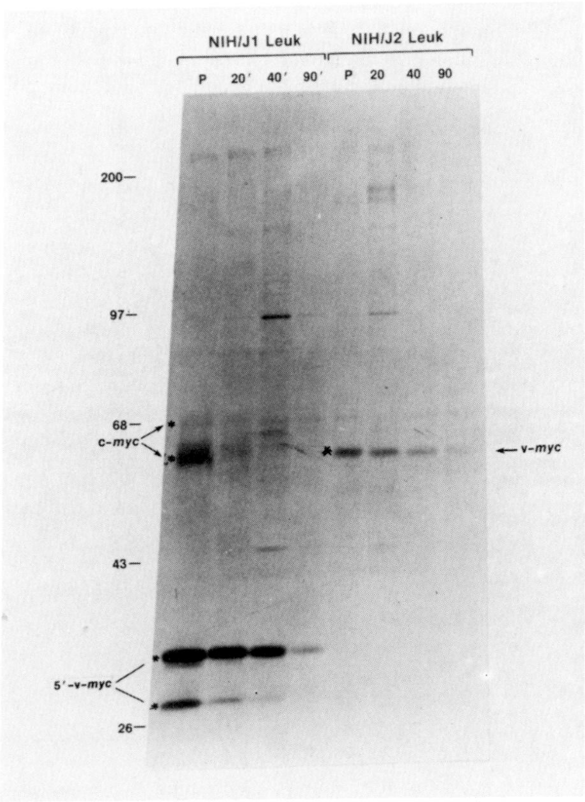

Fig. 8. Half-lives of c-*myc* and v-*myc* protein in J-1- and J-2-infected fibroblasts. ^{35}S-methionine pulse J-1 and J-2 virus-infected cultures of NIH 3T3 cells were chased and, at the times indicated, the lysates precipitated with antisera specific for the C-terminus of v-*myc* (J-2) or with antisera raised against bacterially-expressed c-*myc* (J-1). Gels were exposed for 2 d.

uniform property of infected single cell clones. Analysis of the effects of deletion mutants of v-*myc* on c-*myc* expression showed that neither J-1-Bal nor J-5Q were able to downregulate c-*myc*. This is apparent from Northern analysis (Figure 7) as well as from c-*myc* and v-*myc* protein determination (data not shown). Is the downregulation of c-*myc* expression by v-*myc* in fibroblasts reversible? Anisomycin treatment for periods up to 6 times the half-life of v-*myc* protein in these cells (Figure 8) demonstrated low, but detectable, steady state levels of c-*myc* RNA (Figure 9), although this effect was less impressive than the effect of the inhibitor on increasing steady state levels of c-*fos* transcripts (Figure 9).

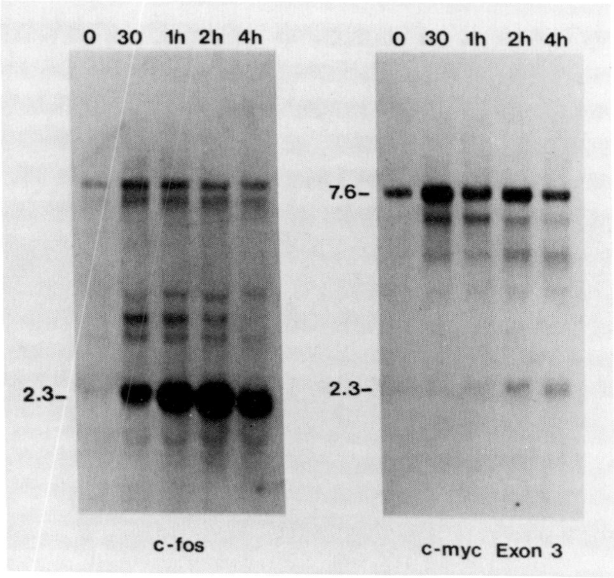

Fig. 9. C-*myc* expression in 3T3/J-2 cells treated with anisomycin. Total poly(A) RNA was prepared from 3T3/J-2 cells treated with (10 μM) anisomycin at the times indicated, blotted (5 μg), and hybridized with c-*fos* and c-*myc* exon 3 probes. Blots were exposed for 20 h. Sizes of RNAs are given in kb.

CONCLUSIONS

We have shown that high levels of v-*myc* expression *in vivo* are generally associated with absence of expression of c-*myc* in all cell lineages tested. While we cannot exclude the possibility that a cellular differentiation program mediated or contributed to downregulation *in vivo,* we have established that c-*myc* shut-down is induced by v-*myc* for 3 different cell lines *in vitro*. In the infected myeloid FDC-P1 cells, which had been selected for abrogation of growth factor-dependence by the virus, downregulation of c-*myc* did not appear to be due to a reversible, repressor mediated regulation, since the IL-3-mediated signal transduction pathway appears altered and the c-*myc* gene took on the conformation of a transcriptionally inactive allele and was not inducible by IL-3 or anisomycin. It is possible that selection for growth factor independence not only required retention of high level virus expression but also called for expression of other, as yet unidentified cellular functions. This is not the case in fibroblast cells, where no selection was involved in the generation of v-*myc* expressor cells. All such cells were negative

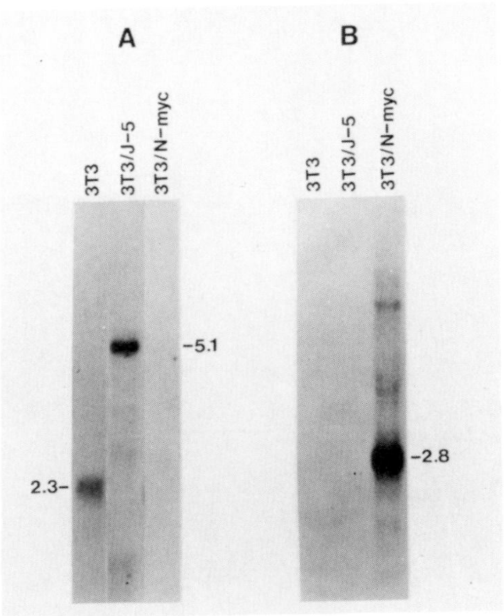

Fig. 10. Downregulation of c-*myc* expression in NIH 3T3 cells transfected with an N-*myc* expression vector. Total poly(A) RNA was prepared, blotted (5 μg), and hybridized with c-*myc* (A)- and N-*myc* (B)-specific probes. Blots were exposed for 20 h. Sizes of RNAs are given in kb.

for c-*myc* expression, independent of the virus used for infection, and examination of single cell clones of infected cultures showed it to be a uniform property of infected cells. Downregulation of c-*myc* in fibroblasts was reversible by anisomycin treatment, suggesting an active role for v-*myc* not only in induction, but also in maintenance of c-*myc* suppression. Moreover, a deletion analysis of v-*myc* has narrowed down the structural requirements for suppression by v-*myc* and established that removal of 109 amino acids of coding exon-derived sequences was sufficient to abolish suppression.

Comparison of expression patterns of the different members of the *myc* oncogene family, such as c-*myc*, N-*myc* and L-*myc* reveals that, generally, expression is mutually exclusive in various lineages (Zimmerman *et al.* 1985, DePinho *et al.* 1985). Such a pattern arouses the suspicion that each member of the family may be able not only to autoregulate its own expression, consistent with our data on v-*myc* regulation of c-*myc* and earlier speculations (Leder *et al.* 1983, Taub *et al.* 1984, Siebenlist *et al.* 1984), but that there might also be a cross-regulation of the entire family by each member. This type of mechanism might be involved then, in either an active or a passive fashion, in the programming of lineage-specific differentiation. To test the hypothesis for cross-regulation, we have obtained an N-*myc* expression vector (Schwab *et al.* 1985) and found that its expression is associated with a downregulation of c-*myc* expression in transfected NIH 3T3 cells (Figure 10).

ACKNOWLEDGMENT

Research sponsored by the National Cancer Institute, DHHS, under contract NO. N01-C0-23909 with Littion Bionetics, Inc. The contents of this publication do not necessarily reflect the views or policies of the Department of Health and Human Services, nor does mention of trade names, commercial products, or organizations imply endorsement by the U.S. Government.

REFERENCES

Cleveland, J. L., Jansen, H. W., Bister, K., Fredrickson, T. N., Morse III, H. C., Ihle, J. N. & Rapp, U. R. (1986) Interaction between *raf* and *myc* oncogenes in transformation *in vivo* and *in vitro*. *J. Cell. Biochem. 30,* 195–218.

DePinho, R. A., Legouy, E., Feldman, L. B., Kohl, N. E., Yancopoulos, G. D. & Alt, F. W. (1986) Structure and expression of the murine N-*myc* gene. *Proc. Natl. Acad. Sci. USA 83,* 1827–31.

Leder, P., Battey, J., Lenoir, G., Moulding, C., Murphy, W., Potter, H., Stewart, T. & Taub, R. (1983) Translocations among antibody genes in human cancer. *Science 222,* 765–71.

Morse III, H. C., Hartley, J. W., Fredrickson, T. N., Yetter, R. A., Majumdar, C., Cleveland, J. L. & Rapp, U. R. (1986) Recombinant murine retroviruses containing avian v-*myc* induce a wide spectrum of neoplasms in newborn mice. *Proc. Natl. Acad. Sci. USA 83*, 6868–72.

Rapp, U. R., Bonner, T. I., Moelling, K., Jansen, H. W., Bister, K. & Ihle, J. (1985c) Genes and gene products involved in growth regulation of tumor cells. In: *Recent Results in Cancer Research,* pp. 221–236.

Rapp, U. R., Cleveland, J. L., Brightman, K., Scott, A. & Ihle, J. N. (1985a) Abrogation of IL-3 and IL-2 dependence by recombinant murine retroviruses expressing v-*myc* oncogenes. *Nature 317,* 434–38.

Rapp, U. R., Cleveland, J. L., Fredrickson, T. N., Holmes, K. L., Morse III, H. C., Jansen, H. W., Patschinsky, T. & Bister, K. (1985b) Rapid induction of hemopoietic neoplasms in newborn mice by a *raf(mil)/myc* recombinant murine retrovirus. *J. Virol. 55,* 23–33.

Schwab, M., Varmus, H. E. & Bishop, J. M. (1985) Human N-*myc* gene contributes to neoplastic transformation of mammalian cells in culture. *Nature 316,* 160–62.

Siebenlist, U., Henninghausen, L., Battey, J. & Leder, P. (1984) Chromatin structure and protein binding in the putative regulatory region of the c-*myc* gene in Burkitt lymphoma. *Cell 37,* 381–91.

Stanton, L. W., Watt, R. & Marcu, K. B. (1983) Translocation, breakage and truncated transcripts of c-*myc* oncogene in murine plasmacytomas. *Nature 303,* 401–06.

Taub, R., Moulding, C., Battey, J., Murphy, W., Vasicek, T., Lenoir, G. M. & Leder, P. (1984) Activation and somatic mutation of the translocated c-*myc* gene in Burkitt lymphoma cells. *Cell 36,* 339–48.

Zimmerman, K. A., Yancopoulos, G. D., Collum, R. G., Smith, R. K., Kohl, N. E., Denis, K. A., Nau, M. M., Witte, O. N., Toran-Allerand, D., Gee, C. E., Minna, J. D. & Alt, F. W. (1986) Differential expression of *myc* family genes during murine development. *Nature 319,* 780–83.

DISCUSSION

LOWY: Do you think this is a direct effect on the *myc* gene or an indirect effect of the protein?

RAPP: We do not have experiments at this point that would distinguish between a direct and indirect mode of action of the *myc* protein in this down-regulation.

SCHLESSINGER: What portion of the *myc* protein is required for the down-regulation.

RAPP: When we took out 70 amino acids from the 5′ end of the second coding exon, we lost down-regulation.

SCHLESSINGER: So, how many exons are required for down-regulation?

RAPP: We do not know yet. We have only made a few deletions at this point. This does not allow us to come up with a minimal size of the protein that would be required for down-regulation. At this point we can only say how soon we loose down-regulation.

SCHLESSINGER: But you cannot say how much of deletion ...

RAPP: I cannot tell you how much of what we take out would be sufficient for down-regulation. We did not do this experiment.

CORY: I forget whether there is a protein being made in those cells when you introduce that deletion.

RAPP: There is a stable deleted protein made in those cells, at a level completely parallel to the levels of RNA that I have shown you.

CORY: Like you, I have been looking to see whether or not the endogenous c-*myc* gene is turned off by a virally controlled c-*myc* gene. In my case I have worked mainly with the c-*myc* gene and not so much with the v-*myc* gene, and my distinct impression is that the silencing of the endogenous gene is dependent on the level of expression achieved by the virally controlled *myc* gene. It seems possible that lower levels of v-*myc* are needed than c-*myc*, owing to the sequence changes. Different cell types may also require different levels of deregulated *myc* expression before the endogenous alleles are silenced.

RAPP: I would like to confirm what you said. We have also been looking at a large number of c-*myc* virus-infected clones. We did observe correlation between the levels of virally expressed c-*myc* and the degree to which c-*myc* was downregulated in NIH 3T3 cells. Also, in terms of cell type dependency of this phenomenon, I might add here that we see reversibility of c-*myc* down-regulation in 3T3 cells with anisomycin. You could do the same experiment with serum removal and readdition. In other words, treatment of these down-regulated cells with growth factor or serum brings c-*myc* expression back on, and can transiently override the down-regulation. If we look at the chicken cell question, with this in mind, it may be relevant that these cells tend to be grown in a growth medium that is much richer than the one in which NIH 3T3 cells are grown. Moreover, with chicken cells you are looking at primary cells and it might be that there is heterogeneity within that cell population in terms of their responsiveness to c-*myc* induction by this enriched growth medium.

WONG-STAAL: Can you microinject either the v-*myc* or c-*myc* protein and get the same down-regulation? There is no direct evidence that it is the protein and not the message that regulates.

RAPP: Actually, I have not shown it, but we have since made a v-*myc* mutant virus, where v-*myc* is taken out of reading frame. However, you do get the original level of transcripts made, and there you do not see the down-regulation. But these protein injection experiments could also be done.

LOWY: How would you see the effect on c-*myc*?

RAPP: What you could do is to take v-*myc* protein and look for the c-*myc* protein. You can distinguish the two with specific antibodies in fluorescence assays.

SÜMEGI: You do not know whether *myc* protein made from the truncated *myc* gene is active or not. You remove protein coding sequences from the 3′ end of the foreign *myc* gene. If protein is made in the cells from such a construct it may be functional or it may not be functional. Only those cells are able to proliferate which have an active c-*myc* gene.

RAPP: It is not active in down-regulating c-*myc* expression. That is all I can say about its activity.

SÜMEGI: The cells require an active c-*myc* gene for their proliferation and in your experiment you select for cells with an active, functional *myc* gene.

RAPP: I see what you are saying now. If you look at growing cells, there might be a constraint in the functional analysis because the cells have to maintain some level of a functional *myc* protein. But we would have to be sure about avoiding a selective situation? For example we would have to re-do that experiment in acute infections and look after fluorescence with v-*myc*/c-*myc* differentiating antibodies in those cells.

WEINBERG: You mentioned a situation with the FD cells where you could not turn on *fos* or *myc* by adding IL3. This poses the question whether the IL3 receptor is defective, or whether there is another element of defectiveness. I do not think you answered the puzzles that were created by that data. In other words, why is *fos* not inducible by IL3, and is the IL3 signalling pathway intact in those cells?

RAPP: We do not know. I shall tell you what we are doing along those lines. First of all we want to test if *fos* is inducible. We have done anisomycin experiments on the FD cells, and *fos* RNA comes storming back. So, it is not that there is something principally wrong with *fos* inducibility. However, it is not easy to conceive of a way for checking out the signal transduction pathway from the IL3 receptor. The other thing we can do is to check on the inducibility of the c-*myc* gene is those same FD/J2 cells by coming with another growth factor, GMCSF which those cells happen to respond to. So we can check out the inducibility question of the *myc* gene is another way. I do not quite see how to check out at this point the ability of the receptor to signal to the nucleus.

SCHLESSINGER: Did you try to see what happens to the IL3 receptor in these cells, is there a change in number or down-regulation?

RAPP: They do not change at all in number as judged by IL3 binding. Nothing is changed in binding characteristics. Of course, we do not know the IL3 receptor. We do not know whether it is a tyrosine kinase type of receptor or whatever it is which makes it more difficult to check out this point.

LEPPARD: Did you show whether the half-life of the viral *myc* protein was different in the FD cells from the 3T3. You seemed to be suggesting that there was some difference in the state of inactivation of the c-*myc* gene in FD compared with 3T3, because you could reactivate it. I wonder whether the difference was that protein decayed less fast.

RAPP: Yes, this is a possibility. We do not yet have the v-*myc* half-life data on the FD cells.

VII. Signal Pathways

Functional Expression of Human EGF-Receptor and its Mutants in Mouse 3T3 Cells

J. Schlessinger, E. Livneh, A. Ullrich+ & R. Prywes[1]*

Epidermal growth factor (EGF) is a potent mitogenic polypeptide that binds to a specific receptor which is expressed on the surface of various cultured cells (for reviews see Carpenter & Cohen 1979, Schlessinger *et al.* 1983). Binding experiments with ^{125}I-EGF indicate that the receptor is present on the cell surface in two forms, of high and of low affinity for EGF (Shechter *et al.* 1978, King & Cuatrecasas 1982, Rees *et al.* 1984). The relationship and importance of these two forms is not known. After binding, the receptor internalizes rapidly into cells via clathrin-coated pits (reviewed in Schlessinger *et al.* 1983). Despite elucidation of these early responses it is not clear how they are related to the ability to induce DNA synthesis, inasmuch as EGF must be present in the extracellular medium for more than 6–10 h in order to have a mitogenic effect (reviewed in Carpenter & Cohen 1979, Schlessinger *et al.* 1983).

The purification (Yarden *et al.* 1985 a,b) and subsequent cDNA cloning of the EFG receptor resulted in the deduction of the complete primary structure of the protein (Downward *et al.* 1984, Ullrich *et al.* 1984). It is comprised of an extracellular cysteine-rich EFG-binding domain, a single hydrophobic transmembrane domain and an intracellular tyrosine kinase domain among other possible structures. The sequence also revealed that the EGF receptor is highly related to the v-erbB oncogene of Avian Erythroblastosis Virus and is therefore probably equivalent to the proto-oncogene c-erbB.

The v-erbB product is a truncated form of the EGF receptor; however, amplification of the EGF-receptor gene has also been associated with a neoplastic

---

[1]Dept. of Chemical Immunology, The Weizmann Institute of Science, Rehovot, Israel and +Genentech, Inc., South San Francisco, CA 94080, U.S.A. *Current Address: Biotechnology Research Center, Meloy Laboratories, 4 Research Court, Rockville, Maryland 20850, U.S.A.

VIRAL CARCINOGENESIS, Alfred Benzon Symposium 24,
Editors: N. O. Kjeldgaard, J. Forchhammer, Munksgaard, Copenhagen 1987.

phenotype (Ullrich *et al.* 1984, Merlino *et al.* 1984, Lin *et al.* 1984, Libermann *et al.* 1985).

We have utilized the cDNA clone of the receptor to analyze which sequences of the receptor are responsible for its biological properties. The cDNA has been cloned into a retrovirus expression vector and transfected into NIH/3T3 fibroblast cells. Mutations have been made in the transfected plasmid to study the importance of specific sequences. The ability of cytoplasmic sequences to affect high versus low affinity binding of the extracellular domain, EGF-induced internalization, tyrosine kinase activity and mitogenicity is reported here.

RESULTS AND DISCUSSION

We utilized a murine retrovirus vector, pZIPNeoSV(X) (Cepko & Mulligan 1984), to express the EGF receptor (EGF-R) cDNA. The cDNA is expressed directly from the retroviral LTR promoter while the Geneticin G418 (neomycin) resistance gene (NeoR) is expressed via a sub-genomic message using the natural Moloney murine leukemia virus (M-MuLV) 5' and 3' splice sites as described by Prywes *et al.* (1986).

The full coding region of the EGF receptor was constructed using clones λ HER64 and λ HER21 from Ullrich *et al.* (1984). The inserted EGF-R cDNA stretched from an Sstl site 20 bases before the initiation codon (AUG) to an Xmnl site 150 nucleotides past the termination codon for translation. A minimal amount of the 5' untranslated region was included so as not to interfere with translational initiation within the framework of the retrovirus message (Prywes *et al.* 1986).

The general strategy was to transfect the expression vector into ψ-2 cells and select for cells resistant to neomycin. ψ-2 cells are NIH/3T3 cells transfected with a defective clone of M-MuLV (Mann *et al.* 1983). These cells allow for the rescue of virus containing the transfected genome, free of the M-MuLV helper virus. We were able to harvest a recombinant virus from ψ-2 cells transfected with pER which passaged both the NeoR gene as well as the EGF-receptor gene. This virus, however, was present at very low titers (10^2-10^3 neomycin-resistant colony forming units per ml) most probably due to inefficient splicing to the NeoR gene when the EGF-R insert is present. Due to these difficulties we have not pursued viral studies using this construct. Nevertheless, the vector was able to efficiently express EGF-R following transfection as demonstrated by the expression of an active EGF-receptor.

The following mutants of EGF receptor were constructed *in vitro* in the expression vector pER.

pER ΔPst: In order to understand the role of the cytoplasmic part of the EGF-R a deletion was made so as to leave only 16 amino acids following the membrane spanning domain. The coding region then proceeds for 65 amino acids out of frame as predicted from the nucleotide sequence (Ullrich *et al.* 1984).

pERB2: Insertion mutants were made in the kinase domain of EGF-R in order to determine the role of the kinase domain on other biological properties of the receptor. Synthetic SalI linker DNA (5'GGTCGACC3') was inserted at a BamHI site near the beginning of the kinase domain. The method of insertion results in an in-frame insertion of 4 amino acids (GRPI) after amino acid number 708.

pERB3: Similar to pERB2, a synthetic SalI linker was inserted at another BamHI site near the 3' end of the kinase domain resulting in an insertion of 4 amino acids (VDRS) after amino acid 888, as shown in:

pRAB: This is a fusion gene of EGF-R and the kinase domain of v-*abl.* The v-abl oncogene of Abelson Murine leukemia virus (A-MuLV) encodes a product with protein tyrosine kinase activity similar to that of EGF-R (Witte *et al.* 1980). This is also reflected in the amino acid homology between each of their kinase domains (Bishop 1983, Ullrich *et al.* 1984). We were interested in determining the effect of deletion of the EGF-R sequences on its activities, as well as in studying whether the external EGF-binding domain of EGF-receptor could modulate the activity of a heterologous kinase domain.

EGF Binding

The EGF receptor is present on the surface of cells in two forms with different affinites for EGF - on average a high affinity of $KD = 2 \times 10^{-10}M$ and a low affinity of $KD = 5 \times 10^{-9}M$. The difference between these two forms is unknown. The reconstitution of both forms on the cell surface after transfection with EGF-R cDNA demonstrates that 1 gene codes for both forms.

We made mutations in the cytoplasmic portion of the receptor to determine whether internal interactions and activities of the receptor might be responsible for these two forms. Indeed, deletion of all but 16 amino acids of the cytoplasmic sequences, in pERΔPst, resulted in a receptor exhibiting only low affinity binding sites. In addition, a chimeric gene of the EGF receptor from the N-terminus up to 44 amino acids on the cytoplasmic side of the membrane, fused to the v-abl kinase domain also displayed only low affinity sites. A cytoplasmic function or structure is thus required to have high affinity binding sites.

This may not be the kinase activity *per se* since a mutant, pERB2, with a 4 amino acid insertion in the kinase domain (at amino acid 708) and with undetectable kinase activity *in vitro* exhibits both high and low affinity receptors. Another insertion mutant, pERB3, also within the kinase domain (at amino acid 888) and lacking kinase activity *in vitro,* exhibits only low affinity sites. This insertion may disrupt a structure or other activity required for obtaining high affinity sites. The mutated protein, though entirely intact except for the 4 amino acid insertion, exhibits a slightly faster mobility on SDS-polyacrylamide gel electrophoresis. This may be due to differences in post-translational modification. If the latter were the case, it may be these modifications which are required to have high affinity sites, rather than the specific sequence or structure interrupted by the insertion mutation. The reason for the difference in mobility is not known.

A loss of high affinity receptor sites has been observed previously after treatment of cells with the tumor promoter 12-0-tetradecanoylphorbol 13-acetate (TPA) (Brown *et al.* 1979, Shoyab *et al.* 1979, Magun *et al.* 1980, King & Cuatrecasas 1982). TPA is able to activate the calcium- and phospholipid-dependent protein kinase C. This enzyme in turn can phosphorylate the EGF receptor, predominantly on a threonine residue (amino acid number 654) which resides 10 amino acids on the cytoplasmic side of the membrane (Hunter *et al.* 1984, Davis & Czech 1985). This suggests that phosphorylation of threonine-654 results in a loss of high affinity sites. However, it is clear that TPA causes phosphorylation of additional sites on the EGF-receptor which could also play a role in the regulation of the affinity of the receptor for EGF (Davis & Czech 1985). Indeed, phosphorylation of threonine-654 does not appear to be sufficient to explain the loss of high affinity sites in our mutants. We have analyzed a point mutant at threonine-654 which suggests that factors, in addition to phosphorylation of Thr654, may be responsible for the effects of TPA on the affinity of EGF-receptor (Livneh *et al.* 1986). Thus, the mechanisms of control of high and low affinity forms appear to be more complex.

Yarden & Schlessinger (1985a, b) have recently described an allosteric aggregation model for the activation of the EGF-receptor kinase. Moreover, evidence was provided that the receptor exists in an equilibrium between monomeric and dimeric forms. EGF induces dimers to form and it is proposed that these dimers would have higher affinity for EGF, with EGF binding stabilizing that form. In addition, EGF binding and dimerization result in activation of the kinase activity. Using this as a working model it is possible to explain the phenotype of the mutants in terms of their ability to dimerize. Cytoplasmic sequences deleted in pER ΔPst and pRAB, or interrupted in pERB3 would therefore be required for the efficient formation of dimers. In addition, protein kinase-C phosphorylation might act by shifting the equilibrium toward monomers.

Internalization

The EGF-receptor internalizes via clathrin-coated pits upon EGF binding (reviewed in Carpenter & Cohen 1979, Schlessinger *et al.* 1983). This route of internalization is specific and excludes many surface proteins. The receptors for polymeric IgA/IgM, asialoglycoproteins, transferrin, insulin and low density lipoproteins (LDL) among others also all enter the cell via clathrin-coated pits (reviewed in Brown *et al.* 1983). The mechanism of localization within coated pits is not known; however, Lehrman *et al.* (1985) have demonstrated that mutations within the cytoplasmic domain of LDL receptor strongly reduce the ability of receptors to internalize. We can now confirm these results in the EGF receptor system where deletion of cytoplasmic sequences also lowers the ability of the EGF receptor to internalize.

There are several important differences between the LDL and EGF receptors. The LDL receptor has only 50 amino acids in its cytoplasmic domain and no known enzymatic activity, while the EGF receptor has 542 amino acids encoding a tyrosine kinase activity and specific autophosphorylation sites. In addition, the EGF receptor requires EGF to internalize and is not recycled, while the LDL receptor is constitutively internalized with or without LDL and can be recycled to the cell surface (Brown *et al.* 1983). Therefore, in the EGF receptor system there are at least two questions of interest with regard to internalization:

Which domains are essential for EGF binding to transmit the signal for the receptor to internalize and which domains are required for internalization specifically into clathrin-coated pits? At present we cannot distinguish between these sequences, but can conclude that cytoplasmic sequences are required for at least one of these functions since a deletion of all but 16 amino acids of the cytoplasmic domain, as well as other mutations introduced into this region, lead to endocytosis-defective receptor mutants (Livneh *et al.* 1986 a,b).

The protein-tyrosine kinase activity, however, may not be required for internalization since a linker insertion mutant at amino acid 708 (pERB2), with defective kinase activity *in vitro,* internalizes like "wild type" receptor. Another linker insertion mutant at amino acid 888 (pERB3) also lacks tyrosine kinase activity *in vitro,* but does not internalize efficiently. This mutant, which also lacks high affinity binding sites, may provide a clue as to which specific structures are required for efficient internalization. More mutants must be made to precisely define these sequences.

The internalization mutants we have described do internalize partially. The initial kinetics are faster than EGF receptor turnover without EGF ($t1/2 = 10$ h in human fibroblasts; Stoscheck & Carpenter 1983). It will be of interest to determine whether this internalization is via coated pits or utilizes another mechanism.

Mitogenicity

The transfected EGF receptor gene was able to reconstitute the ability of NIH/3T3 cells to respond mitogenically to EGF. Increased DNA synthesis was measured by the incorporation of 3H-thymidine into trichloroacetic acid precipitable material after incubation of cells with or without EGF. Surprisingly, the cell line ψB2-2, containing a kinase-defective protein, showed an EGF-stimulated incorporation of 3H-thymidine, i.e., responded mitogenically to EGF. ψB2-2 is "wild type" for EGF binding and internalization. The deletion mutant, in ψ4-P, and insertion mutant, in ψB3-2, are deficient in these activities as well as in tyrosine kinase activity and do not respond mitogenically to EGF. Finally, neither of the control cell lines ψ-2 nor NIH/3T3 cells respond mitogenically to EGF.

On the basis of the positive result with ψB2-2 it is possible to suggest that kinase activity is not required for the mitogenic signal and that some other unknown function is necessary: However, there is an extensive literature supporting the importance of the tyrosine-specific kinase. There is conservation of this domain and activity through a large number of growth factor receptors and oncogene products (reviewed in Bishop 1983, Hunter & Cooper 1985). In addition, in several oncogenic retroviruses containing these related oncogenes, temperature-sensitive or transformation-defective mutants have been found that are temperature-sensitive or defective in protein-tyrosine kinase activity (reviewed in Bishop 1983, Parsons *et al.* 1984, Stone *et al.* 1984, Prywes *et al.* 1985).

Another possible explanation is that, while defective in our *in vitro* assays for autophosphorylation and exogenous substrate phosphorylation, the mutated protein is still active in the intact cells. Namely, the insertional mutation at the kinase domain renders the immunoprecipitated enzyme unstable, while in intact cells this receptor mutant possesses kinase activity capable of phosphorylating the physiologically important substrates. It would be of interest to assay the *in vivo* kinase activity of the receptor; however, this is complicated by the low phosphotyrosine level in the cells and the unknown nature of physiologically important substrates. While an EGF-induced increase in phosphotyrosine levels can be observed in A431 cells which overexpress the EGF receptors (Hunter & Cooper 1981) it is difficult to observe a change in cells with lower EGF-R expression such as human foreskin fibroblasts (1×10^5 receptors/cell; Yarden 1985), N22 cells and the cells expressing the various EGF-receptor mutants. Further experiments are currently being performed in order to resolve this question.

Receptor-v-abl chimera

The chimera between the EGF binding and transmembrane domains of EGF receptor and the v-abl kinase domain of Abelson murine leukemia virus (A-MuLV) was able to morphologically transform Rat-1 fibroblast cells. The chimeric protein is precipitable with both EGF-R and v-abl-specific antisera, has autophosphorylation activity, and binds EGF. EGF, however, cannot activate the heterologous kinase activity nor can it induce the endocytosis of the chimeric molecule.

The v-abl kinase domain alone, expressed in a retrovirus construct, is not sufficient to transform NIH/3T3 fibroblast (Prywes 1984) nor Rat-1 cells (Peter Jackson, personal communication). This appears to be due to the requirement for the viral structural (gag) sequences normally fused to the N-terminus of v-abl in A-MuLV. These sequences provide a site for covalent attachment of the fatty acid myristate to the amino terminal amino acid glycine. Myristylation of an amino terminal glycine is present in pp60src, the transforming protein of Rous Sarcoma Virus (Buss & Sefton 1985). Mutants that do not myristylate, while retaining tyrosine kinase activity, do not transform fibroblasts. In addition, the mutant proteins are no longer membrane-associated as is the "wild type" protein (Cross *et al.* 1984, Pellman *et al.* 1985). Thus, by analogy, the EGF receptor sequence appears to activate the transforming potential of the v-abl kinase by localizing the kinase to the inner membrane surface. It will be intersting to learn whether the bcr gene fused to c-abl in chronic myelogenous leukemia cells (Groffen *et al.* 1984) is also activating c-abl by localizing it to the membrane surface.

It is noteworthy that EGF is unable to enhance the tyrosine kinase activity and internalization of the chimeric molecule. Hence, the tailoring of a kinase domain to ligand binding region does not necessarily maintain the regulatory properties of a "wild type" receptor. Still, it is possible that with other chimeric constructs, the ligand will be able to stimulate the cytoplasmic abl-kinase function.

ACKNOWLEDGMENTS

We gratefully acknowledge the help of Stephen Felder in developing the program for analysis of the binding data and Nachum Reiss for help with the exogenous substrate assay. This work was supported by grant CA-25820 from the National Institutes of Health and a grant from the U.S.-Israel Science Foundation.

REFERENCES

Bishop, J. M. (1983) Cellular oncogenes and retroviruses. *Ann. Rev. Biochem. 52,* 301–54.

Brown, K. D., Dicker, P. & Rozengurt, E. (1979) Inhibition of epidermal growth factor binding to surface receptors by tumor promoters. *Biochem. Biophys. Res. Commun. 86,* 1037–43.

Brown, M. S., Anderson, R. G. W. & Goldstein, J. L. (1983) Recycling receptors: the round-trip itinerary of migrant membrane proteins. *Cell 32:* 663–7.

Buss, J. E. & Sefton, B. M. (1985) Myristic acid, a rare fatty acid, is the lipid attached to the transforming protein of Rous sarcoma virus and its cellular homolog. *J. Virol 53,* 7–12.

Carpenter, G. & Cohen, S. (1976) 1251-labelled human epidermal growth factor (hEGF): binding, internalization and degradation in human fibroblasts. *J. Cell Biol. 71,* 159–71.

Cross, F. R., Garber, E. A., Pellman, D. & Hanafusa, H. (1984) A short sequence in p60src N-terminus is required for p60src myristsylation and membrane association, and for cell transformation. *Mol. Cell Biol. 4,* 1834–42.

Davis, R. J. & Czech, M. P. (1985) Tumor-promoting phorbol diesters cause the phosphorylation of epidermal growth factor receptors in normal human fibroblasts at threonine-654. *Proc. Natl. Acad. Sci. USA 82,* 1974–8.

Downward, J., Yarden, Y., Mayes, E., Scrace, G., Totty, N., Stockwell, Ullrich, A., Schlessinger, J. & Waterfield, M. D. (1984) *Nature 307,* 521–7.

Groffen, J., Stephenson, J. R., Heisterkamp, N., deKlein, A., Bartram, C. R. & Grosveld, G. (1984) Philadelphia chromosome breakpoints are clustered within a limited region, bcr, on chromosome 22. *Cell 36*, 93–9.

Hunter, T. & Cooper, J. A. (1981) Epidermal growth factor induces rapid tyrosine phosphorylation of proteins in A431 human tumor cells. *Cell 24,* 741–52.

Hunter; T. & Cooper, J. A. (1985) Protein-tyrosine kinases. *Ann. Rev. Biochem. 54,* 897–930.

Hunter, T., Ling, N. & Cooper, N. A. (1984) Protein Kinase-C phosphorylation of the EGF-receptor at a threonine residue close to the cytoplasmic face of the plasma membrane. *Nature 314,* 480–3.

King, A. C. & Cuatecasas, P. (1982) Resolution of high and low affinity epidermal growth factor receptors: inhibition of high affinity component by low temperature, cydoherimide and phorbol esters. *J. Biol. Chem. 257,* 3053–60.

Lehrman, M. A., Goldstein, J. C., Brown, M. S., Russell, D. W. & Schneider, W. J. (1985) Internalization-defective LDL receptors produced by genes with nonsense and frameshift mutations that truncate the cytoplasmic domain. *Cell 41,* 735–43.

Libermann, T. A., Nussbaum, H. R., Razon, N., Kris, R., Lax, I., Soreq, M., Whittle, N., Waterfield, M. D., Ullrich, A. & Schlessinger, J. (1985) Amplification, enhanced expression and possible rearrangement of the EGF-receptor gene in primary human brain tumors of glial origin. *Nature 313,* 144–7.

Lin, C. R., Chen, W. S., Kruiger, W., Stolarsky, L. S., Weber, W., Evans, R. M., Verma, I. M., Gill, G. N. & Rosenfeld, M. G. (1984) Expression cloning of human EGF receptor complementary DNA: Gene amplification and three related messenger RNA products in A-431 cells. *Science 224,* 417–9.

Livneh, E., Prywes, R., Dull, T., Ullrich, A. & Schlessinger, J. (1986b) Release of phorbol ester induced mitogenic block by mutation at Thr654 of EGF receptor. (Submitted).

Livneh, E., Prywes, R., Kashles, O., Reiss, N., Sasson, I., Mory, Y., Ullrich A. & Schlessinger, J. (1986a) Reconstitution of human EGF receptors and its deletions mutants in cultured hamster cells. *J. Biol. Chem.* (in press).

Magun, B. E., Matrisian, L. M. & Bowden, G. T. (1980) Epidermal growth factor: ability of tumor promoter to alter its degradation, receptor affinity and receptor number. *J. Biol. Chem. 255,* 6373–81.

Mann, R., Mulligan, R. C. & Baltimore, D. B. (1983) construction of a retrovirus packaging mutant and its use to produce helper-free defective retrovirus. *Cell 33,* 153–9.

Merlino, G. T., Xu, Y., Ishii, S., Clark, A. J. L., Semba, K., Toyoshima, K., Yamamoto, T. & Pastan, I. (1984) Amplification and enhanced expression of the epidermal growth factor gene in A-431 human carcinoma cells. *Science 224,* 843–8.

Parson, J. T., Bryant, D., Wilkerson, V., Gilmartin, G. & Parsons, S. J. (1984) Site-directed mutagenesis of Rous sarcoma virus pp60src: identification of functional domains required for transformation. p.37–42. In: *Cancer cells. Oncogenes and viral genes.* eds. Vande Woude, G. F., Levine, A. J., Topp, W. & Watson, J. D.. Cold Spring Harbor Laboratory, Cold Spring Harbor, N.Y.

Pellman, D., Garber, E. A., Cross, F. R. & Hanafusa H. (1985) Fine structural mapping of a critical N-terminal region of p60src. *Proc. Natl. Acad. Sci. USA 82,* 1623–7.

Prywes, R. (1984) In vitro mutagenesis of Abelson murine leukemia virus. M.I.T. Ph.D. thesis.

Prywes, R., Foulkes, J. G., Rosenberg, N. & Baltimore, D. (1983) Sequences of the A-MuLV protein needed for fibroblast and lymphoid cell transformation. *Cell 34,* 569–79.

Prywes, R., Livneh, E., Ullrich, A. & Schlessinger, J. (1986) Mutations in the cytoplasmic domain of EGF-receptor affect EGF-binding and receptor internalization. *EMBO J.* (in press).

Rees, A. R., Gregoriou, M., Johnson, P. & Garland, P. B. (1984) High affinity epidermal growth factor receptors on the surface of A-431 cells have restricted lateral diffusion. *EMBO J. 3,* 1843–7.

Schlessinger, J., Schreiber, A. B., Levi, A., Lax, I., Libermann, T. & Yarden, Y. (1983) Regulation of cell proliferation by epidermal growth factor. *CRC Crit. Rev. Biochem. 14,* 93–111.

Shechter, L., Hernandez, L. & Cuatrecasas, P. (1978) Epidermal growth factor: biological activity requires persistent occupation of high affinity cell surface receptors. *Proc. Natl. Acad. Sci. USA 75,* 5788–91.

Shoyab, M., DeLarco, J. E. & Todaro, G. J. (1979) Biologically active phorbol esters specifically alter affinity of epidermal growth factor receptors. *Nature 279,* 387–91.

Stone, J. C., Atkinson, T., Smith, M. & Pawson, T. (1984) Identification of functional regions in the transforming protein of Fujinami sarcoma virus by in-phase insertion mutagenesis. *Cell 37,* 549–58.

Stoscheck, L. M. & Carpenter, G. (1984) Down regulation of epidermal growth factor receptors: Direct demonstration of receptor degradation in Human fibroblasts. *J. Cell Biol. 98,* 1048–53.

Ullrich, A., Coussens, L., Hayflick, J. S., Dull, T. J., Gray, J. A., Tam, A. W., Lee, J., Yarden, Y., Libermann, T. A., Schlessinger, J. Downward, J., Mayes, E. L. V., Whittle, N., Waterfield, M. D. & Seeburg, P. H. (1984) Human epidermal growth factor receptor cDNA sequence and aberrant expression of the amplified gene in A-431 epidermoid carcinoma cells. *Nature 30,* 418–25.

Witte, O. N., Dasgupta, A. & Baltimore, D. (1980) *Nature 283,* 826–31.

Yarden, Y. (1985) Ph. D. thesis. Weizmann Institute of Science.

Yarden, Y., Harari, I. & Schlessinger, J. (1985) *J. Biol. Chem. 260,* 315–9.

Yarden, Y. & Schlessinger, J. (1985) Self-phosphorylation of EGF-receptor: Evidence for a model of intermolecular allosteric activation. (Submitted).

Yarden, Y. & Schlessinger, J. (1985) EGF induces rapid, reversible aggregation of the purified EGF-receptor. (Submitted).

DISCUSSION

CORY: I was confused about the curve you showed us with the threonine 654 experiment because they both seem to have the lower affinity, but are lacking the high affinity.

SCHLESSINGER: Yes, I should say it is an intermediate affinity. The high affinity corresponds to a certain value, and the low affinity corresponds to a lower value for most cells. In this mutant we have detected an intermediate value.

CORY: So you have modified the affinity?

SCHLESSINGER: It does alter the affinity, but you can further reduce the affinity by TPA.

WEINBERG: You have argued on many occasions that receptor clustering is essential for receptor kinase activation. What are the basic experiments or basic data that could argue for and against that hypothesis.

SCHLESSINGER: The basic experiments in favor of receptor clustering as a mechanism for the activation of the kinase are as follows: cross-linking antibodies activate the kinase. Monovalent Fab fragment would not stimulate the kinase. The kinase activity has a parabolic concentration profile. A parabolic concentration dependence is consistent with an intermolecular activation step. More direct evidence involves the utilization of cross-linking agent and demonstration that EGF induces the dimerization of the purified receptor. We hope that direct proof will come with our studies with the EGF receptor mutants.

SHERR: If you express the human analogue of v-*erb*B in a cell that expresses the human EGF receptor, do you see transmodulation? Do you see down-regulation of the normal receptor, and do you see phosphorylation?

SCHLESSINGER: The cells which express the double truncated receptor do express also the normal receptor, and they exist side by side. We did not look as yet into the communication between the double truncated receptor and the normal receptor.

LOWY: Could you go into a little bit more of an explanation of the difference between what you say is positive for DNA synthesis and positive for transformation?

SCHLESSINGER: Many cell types overexpress the EGF receptor, such as in A431 cells and many glioblastomas. So, we thought that overexpression of EGF receptor may lead to transformation. It does not. We overexpressed these EGF receptors in Rat 1 cells and we did not detect transformation.

LOWY: In your biological characterization of these cells, when you say there is increased DNA synthesis, is the assay performed on cell confluent monolayer?

SCHLESSINGER: Yes, this is in a confluent monolayer.

BRUGGE: Have you looked at whether there are different sites of phosphorylation in the TPA treated mutant?

SCHLESSINGER: We are analyzing these sites now.

BRUGGE: Dr. Stanley Cohen showed recently that the phosphorylation of the 35 K substrate of the EGF receptor *in vivo* does not peak until around 4 hours after treatment with EGF even though within minutes after treatment with EGF you have down-regulated the receptor. How do you interpret those results in intact cells?

SCHLESSINGER: It is probably phosphorylated inside the cell somewhere in the endocytic pathway. This reinforces what I was saying earlier, that it is very important to compare the localization of v-*erb* B kinase activity and the EGF receptor in trying to understand their action.

BRUGGE: What is the time before complete loss of the receptor?

SCHLESSINGER: There is never complete loss. You lose not more than 80% of the receptor. Down-regulation is not more than 80% of the surface receptors.

BRUGGE: Do any of your antibodies precipitate a truncated version of the receptor that would be a candidate intermediate in the endocytic pathway?

SCHLESSINGER: That is one of the things we wanted to see. We thought that maybe what happens in normal responses is a transient appearance of a v-*erb* B-like receptor, because the extracellular domain would be degraded, say in lysosomes. So, what we did was to use two antibodies, one against the C-terminal peptide and one against N-terminal peptide, to see whether we could detect a transient appearance of a v-*erb* B-like protein in normal cells. We could not detect such protein probably because the receptor is degraded very quickly.

GAMMELTOFT (from the audience): It has been suggested that autophosphorylation of tyrosine residues regulates the kinase activity of EGF receptors. In your construct you remove the putative phosphorylation site. What happens to the kinase activity?

SCHLESSINGER: It was suggested for the insulin receptor, and it seems to be the case for the insulin receptor. For the EGF receptor autophosphorylation does not seem to regulate the kinase activity.

Conversion of a Putative Growth Factor Receptor into an Oncogene Protein

Cornelia I. Bargmann[1,2], *David F. Stern*[1], *Mien-Chie Hung*[1] & *Robert A. Weinberg*[1,2]

A variety of experimental procedures have been used over the past decade to detect genetic sequences in tumor genomes that are responsible for one or another aspect of the malignant phenotype. Among these strategies, the technique of gene transfer or transfection has proved to be particularly useful. DNAs prepared from a wide variety of tumor cells have been found to be able to induce malignant transformation of recipient cells into which they have been introduced by transfection (Varmus 1984). In our own laboratory, we used this technique to detect a number of oncogenes in chemically-induced animal tumor DNAs as well as in the DNAs of spontaneously arising tumors of human origin.

In 1980, we began experiments designed to detect the proteins encoded by these transfected oncogenes. The transfected, transformed cells, deriving from the NIH3T3 mouse fibroblast line, were grown up into mass cultures and inoculated in semi-syngeneic NFS mice in which they grew out into fibrosarcomas. We anticipated that the introduced oncogene would induce the synthesis of novel antigens that would prove immunogenic in the tumor-bearing mice. Consequently, we tested the serum from these animals for reactivity with proteins that were present in these transfectants but absent from control untransfected NIH3T3 cells.

This exercise failed in the great majority of cases. In retrospect, we realize that many of these transfected cells had acquired *ras* oncogenes. These oncogenes, as originally shown by Scolnick and colleagues (Ellis *et al.* 1982), specify proteins of 21 000 daltons that are highly conserved between species and are poorly immunogenic. The exception to this was presented by transfectants bearing

---

[1]Whitehead Institute for Biomedical Research, Nine Cambridge Center, Cambridge, Massachusetts, 02142 and [2]Department of Biology Massachusetts Institute of Technology, Cambridge, Massachusetts, U.S.A.

VIRAL CARCINOGENESIS, Alfred Benzon Symposium 24,
Editors: N. O. Kjeldgaard, J. Forchhammer, Munksgaard, Copenhagen 1987.

oncogenes that originated in the DNAs of a series of ethyl nitrosourea-induced rat neuroblastomas (Shih *et al.* 1981, Padhy *et al.* 1982).

These transfected cells displayed an antigen of 185 kilodaltons which was strongly immunogenic in the mice. Indeed, the display of this antigen was absolutely correlated with the presence of an acquired neuroblastoma oncogene. Cells bearing other oncogenes showed no trace of this protein.

We termed the oncogene *neu*, to reflect its detection in the DNAs of neuroblastomas, and assumed that the associated p185 antigen was encoded by this gene. However, in the absence of molecular clones of the oncogenes, there was little prospect of proving this directly. The gene seemed unrelated to the *ras* oncogenes and we had developed no clear strategy for isolating it by molecular cloning.

Clues concerning the nature of the *neu* oncogene were provided by the properties of the associated p185 antigen. It was a cell surface glycoprotein, the phosphorylation of which could be modulated by a number of pharmacologically active agents including phorbol esters and cyclic AMP analogues.

Most of these difficulties were dispelled by the findings of Downward, Waterfield, Schlessinger and their colleagues who discovered that a protein similar to p185, the epidermal growth factor receptor, is encoded by a gene closely related to the *erb*B oncogene transduced in the genome of Avian erythroblastosis virus (Downward *et al.* 1984). Their finding held great general interest, in that it showed that a normal growth factor receptor gene could become converted into an oncogene protein. In our own case, this relationship between oncogene protein and receptor gene seemed highly analogous to that observed with the p185 and the *neu* gene.

Use of nucleic acid hybridization soon confirmed that the *neu* oncogene shared extensive homology with the *erb*B oncogene (Schechter *et al.* 1984). We concluded tentatively that the *neu* oncogene was a mutated, oncogenic version of the normal rat cellular gene encoding the EGF receptor. This suspicion was strengthened by experiments that showed that the antisera prepared against the EGF receptor possessed reactivity against the p185 antigen as well.

The similarities between the antigens became less compelling upon closer examination. Most obvious was the size difference between the 2 proteins: the EGFr migrated in gel electrophoresis with the mobility of a protein of 170 kilodaltons, this being apparently 15 kilodaltons less than that of the *neu*-encoded protein (Schechter *et al.* 1984). Chromosomal mapping of the *neu* gene showed it to be localized to a different human chromosome than the one known to carry *erb*B (Schechter *et al.* 1985).

Of additional interest was the analysis of the proteins synthesized by Rat-1 cells, which are immortalized but carry none of the known oncogenes in activated form. These cells display both the EGFr and p185 on their surface. This normal rat p185 seemed to be similar in structure to that expressed on the surface of oncogene-bearing cells (Stern *et al.* 1986).

This accumulation of evidence strongly suggested that the p185 of the normal cell functions as a growth factor receptor in normal rat cells. As described below, the p185 and EGFr proteins show extensive structural homology, providing further support for this supposition. We have consequently undertaken a search for the ligand that is recognized by p185 and the EGFr, using a variety of tests to check initially whether p185 functions as an alternative receptor for EGF. Such tests included assays of the ability of EGF to induce down-modulation and autophosphorylation of p185.

The results of these assays suggested that EGF has little, if any, affinity for p185. The most compelling result came from a study of NIH3T3 cells that had been transfected with multiple copies of a cosmid clone of the normal rat *neu* gene. These cells expressed very high levels of normal p185 molecules, comparable in number to the EGFr molecules displayed on the surface of the frequently studied A431 human epidermoid carcinoma cells. Yet these p185 overexpressors adsorbed no more radiolabelled EGF than did untransfected NIH3T3 cells (Stern *et al.* 1986).

This led us to analyze other commonly studied growth factors. These also showed no special affinity for p185 (Stern *et al.* 1986). Such results, when taken together, suggest that p185 is a ligand for an as yet unidentified mitogenic growth factor. Attractive candidates are the EGF-like peptides that may derive from post-translational cleavage of the high molecular weight polyprotein that also serves as precusor to EGF.

While the extracellular domain of the normal p185 seems to function in growth factor binding, this portion of the oncogenic version of the protein can be viewed in a totally different context – it serves as a tumor-specific antigen in transfected NIH3T3 cells. This particular antigen can be viewed differently most other antigen markers found on the surface of tumor cells. The display of these other antigens would appear unnecessary for the maintenance of the transformed phenotype of the tumor cell. Because of this, removal of these other antigens should have little effect on the growth properties of the tumor cell, while removal of p185 from transfected cells should have a profound effect on cell physiology since its display presumably serves to maintain the malignant phenotype.

These predictions were borne out by use of monoclonal antibodies against p185 that were made by Jeffrey Drebin and Mark Greene (Drebin *et al.* 1984). When placed in culture medium, they cause a reversion of the anchorage-independent growth phenotype of *neu*-transfected NIH3T3 cells (Drebin *et al.* 1985). This reversion appears to be due to internalization and degradation of the p185 that is secondary to cell surface clustering induced by the bound antibody molecules.

A more recent line of work has followed the molecular mechanism that was responsible initially for the creation of the *neu*-oncogene. Circumstantial evidence suggested that this oncogene arose during the outgrowth of the neuroblastoma cells, perhaps in the single cells that served as ancestors of the tumor cell clone. Because the carcinogen that provoked these tumors, ethylnitrosourea, is a point mutagen, we suspected that the differences between the *neu* oncogene and the corresponding protooncogene would be minor. This speculation was confirmed by analysis of the cosmid clones of the normal and oncogenic alleles of *neu*. Their restriction enzyme cleavage site maps were indistinguishable, excluding the involvement of gross structural changes in the activation of the gene (Hung *et al.* 1986).

These apparently subtle changes in gene structure could have had one of two effects on the gene: they might have caused deregulation of expression or change in the structure of the encoded protein. Several results appeared to rule out deregulation as a mechanism of activation. The neuroblastomas whose transformation was due to the *neu* oncogene expressed low levels of p185 yet appeared highly transformed. This was also true for a number of oncogene-transformed cell lines. Most important, the NIH3T3 cells expressing very high amounts of the normal p185 appeared to be fully normal in their growth properties (Hung *et al.* 1986).

Such results supported the notion that changes in the structure of the p185 protein were essential for oncogenic activation. Because of this we sought mutations in the structure-encoding portion of the *neu* oncogene, undertaking analysis of cDNA versions of this gene. These cDNA clones were biologically active: both normal and transforming versions induced synthesis of p185 after fusions to a strong transcriptional promoter and transfection into NIH3T3 cells. Sequence analysis of the oncogenic cDNA showed that it encoded a protein with many structural homologies with the EGF receptor (Bargmann *et al.* 1986a), a result also obtained by others who analyzed the human homologue of *neu* (Coussens *et al.* 1985, Yamamoto *et al.* 1986). Thus the p185 protein was seen to have a cysteine-rich extracellular domain, a hydrophobic transmembrane sequence and an intracellular tyrosine kinase domain.

Although we could have subjected the entire normal cDNA clone to sequence analysis, we chose instead to localize the activating lesion by functional analysis of recombinants between the 2 alleles. This was managed by replacing restriction fragments of the normal cDNA with homologous fragments from the oncogenic clone. Some reciprocal recombinants were constructed as well. All these recombinants were subjected to functional testing by transfection onto NIH3T3 monolayers, followed by screening for foci of transformants (Bargmann *et al.* 1986b).

These various tests rapidly converged on a single region of the oncogene that appeared to carry a mutation that was both necessary and sufficient for transforming activity. This segment specified the region surrounding and including the transmembrane domain of p185. It remained to subject the segments of the oncogene and protooncogene in this region to comparative sequence analysis.

The outcome was the discovery of a single nucleotide difference between the 2 allelic forms of the gene in this region. This difference indicated that a transversion had occurred during the creation of the oncogene, causing a T to be replaced by an A (Bargmann *et al.* 1986b).

This in turn leads to the replacement of a valine residue that is present in the hydrophobic transmembrane domain of the normal p185 protein by a glutamic acid residue. This was a quite unexpected result, in that the transmembrane domain of surface glycoproteins has generally been thought to play a passive role in anchoring the protein in the lipid bilayer. This observed change suggests instead that this domain may have a role in signal transduction by the normal protein. One might speculate that the observed amino acid change places the oncogenic protein in a state of constitutive activation in which it sends mitogenic signals into the cell, even in the absence of bound ligand. By contrast, one postulates that such signalling is only undertaken by the normal protein upon ligand binding.

This analysis was only performed on 1 of the 4 different *neu* oncogenes that we had detected in the DNAs of 4 independently arising rat neuroblastomas. We undertook to see whether the other 3 oncogenes carried similar lesions. To do this, we synthesized short oligonucleotide probes that would anneal preferentially to either the normal allele or the initially characterized oncogenic allele. The sequences of these oligonucleotides spanned the region around the point mutation.

Use of these probes soon showed that all 4 *neu* oncogenes carry the identical point mutation (Bargmann *et al.* 1986b). The agent which we believe to have induced these mutations, ethylnitrosourea, is a point mutagen that appears to act promiscuously by alkylating millions of different bases in the genomes of target cells. Yet the identical mutations appear repeatedly in the subsequently arising tumors. This echoes the results of other studying methyl nitrosourea-induced rat mammary carcinomas, in which a point mutation affecting the 12th codon of the Ha-*ras* gene is repeatedly found (Sukumar *et al.* 1983). Such results suggest strong chemical and biochemical forces which select a small number of mutations from among the many that must initially be created by these alkylating agents.

These findings concerning the *neu* oncogene provoke more questions than they answer in the areas of receptor function and chemical caracinogenesis. Many of these questions can be addressed with the reagents that have become available in recent years.

REFERENCES

Bargmann, C. I., Hung, M.-C. & Weinberg, R. A. (1986a) The *neu* oncogene encodes an epidermal growth factor receptor-related protein. *Nature 319*, 226–30.

Bargmann, C. I., Hung, M.-C. & Weinberg, R. A. (1986b) Multiple independent activations of the *neu* oncogene by a point mutation altering the tramsmebrane domain of p185. *Cell 45*, 649–57.

Coussens, L., Yang-Feng, T. L., Liao, Y.-C., Chen, E., Gray, A., McGrath, J., Seeburg, P. H. Libermann, T. A., Schlessinger, J., Francke, U., Levinson, A. & Ullrich, A. (1985) Tyrosine kinase receptor with extensive homology to EGF receptor shares chromosomal location with *neu* oncogene. *Science 230*, 1132–9.

Downward, J., Yarden, Y., Mayes, E., Scrace, G., Totty, N., Stockwell, P., Ullrich, A., Schlessinger, J. & Waterfield, M. D. (1984) Close similarity of epidermal growth factor receptor and v-*erb*B oncogene protein sequences. *Nature 307*, 521–7.

Drebin, J. A., Stern, D. F., Link, V. L., Weinberg, R. A. & Greene, M. I. (1984) Monoclonal antibodies identify a cell-surface antigen associated with an activated cellular oncogene. *Nature 312*, 545–8.

Drebin, J. A., Link, V. C., Stern, D. F., Weinberg, R. A. & Greene, M. I. (1985) Down-modulation of an oncogene protein product and reversion of the transformed phenotype by monoclonal antibodies. *Cell 41*, 695–706.

Ellis, R. W., Lowy, D. R. & Scolnick, E. M. (1982) The viral and cellular p21(*ras*) gene family. In: *Advances in Viral Oncology*, ed. Klein, G., pp. 107–26. Raven Press, N.Y.

Hung, M.-C., Schechter, A. L., Chevray, P.-Y. M., Stern, D. F. & Weinberg, R. A. (1986) Molecular cloning of the *neu* gene: absence of gross structural alteration in oncogenic alleles. *Proc. Natl. Acad. Sci. USA 83*, 261–4.

Padhy, L. C., Shih, C., Cowing, D., Finkelstein, R. & Weinberg, R. A. (1982) Identification of a phosphoprotein specifically induced by the transforming DNA of rat neuroblastomas. *Cell 28*, 865–71.

Schechter, A. L., Hung, M.-C., Vaidyanathan, L., Weinberg, R. A., Yang-Feng, T., Francke, U., Ullrich, A. & Coussens, L. (1985) The *neu* gene: an *erb*B homologous gene distinct from and unlinked to the gene encoding the EGF receptor. *Science 229*, 976–8.

Schechter, A. L., Stern, D. F., Vaidyanathan, L., Decker, S. J., Drebin, J. A., Greene, M. I. & Weinberg,

R. A. (1984) The *neu* oncogene: an *erb*B-related gene encoding a 185,000 M_r tumor antigen. *Nature 312*, 513–6.

Shih, C., Padhy, L. C., Murray, M. & Weinberg R. A. (1981) Transforming genes of carcinomas and neuroblastomas introduced into mouse fibroblasts. *Nature 290*, 261–4.

Stern, D. F., Heffernan, P. A. & Weinberg, R. A. (1986) p185, product of the *neu* proto-oncogene, is a receptor-like protein associated with tyrosine kinase activity. *Mol. Cell. Biol. 6*, 1729–40.

Sukumar, S., Notario, V., Martin-Zanca, D. & Barbacid, M. (1983) Induction of mammary carcinomas in rats by nitrosomethylurea involves malignant activation of H-*ras*-1 locus by single point mutations. *Nature 306*, 658–61.

Varmus, H. E. (1984) The molecular genetics of cellular oncogenes. *Ann. Rev. Genet. 18*, 553–612.

Yamamoto, T., Ikawa, S., Akiyama, T., Semba, K., Nomura, N., Miyajima, N., Saito, T. & Toyoshima, K. (1986) Similarity of protein encoded by the human c-*erb*B-2 gene to epidermal growth factor receptor. *Nature 319*, 230–4.

DISCUSSION

GALLO: Has anybody examined other receptor genes to see if there are other mutations than what is done by your assay systems? In other words: What is their argument for or against selectivity?

WEINBERG: There certainly are facts which suggest, at least for procaryotes, that with chemical carcinogens there are preferential sites of mutation. But it is not an absolute preference, and if we recognize that this gene of 33 000 bases represents only 1 in 10^5 of the total cellular genome, we can recognize that even in spite of there being slight preferences as far as the chemical reactivity of the various sequences in the *neu* gene, this hardly explains the fact that out of 3 billion bases in the genome, one always has the same single base altered in these different tumors. There is something very interesting going on: There must be some very strong selectivity for this particular outcome. Of course, we are biasing things in a certain way and are only looking for transfectable oncogenes here. Therefore, we have to ask the question, how many of the protooncogenes, if treated by point mutagens, could become oncogenes? That is another interesting question. We now know of 30, maybe 40 protooncogenes from a variety of sources. How many of them, if they were altered in a single nucleotide, would become an oncogene? I suspect it is a rather small percentage of the known protooncogenes. Certainly *ras* and *neu*, but maybe not very many others. So, there are strong selective pressures here favoring a certain kind of genes, a very small number of genes. Why it is always in the same site in this particular gene is something I cannot tell you.

CORY: It would be interesting to take other tyrosine kinases and deliberately introduce such mutations to see if their activity was changed.

WEINBERG: Cori Bargmann has been trying in fact to introduce the analogous mutation into normal EGF receptor cDNA. To date nothing interesting has come out, I regret to say. It would be nice to generalize this. We are thinking that perhaps it is just not sufficient to put any kind of hydrophilic amino acid in the transmembrane domain. I guess I did not say explicitly that the point mutation causes valine to be replaced by glutamic acid. You could say that any kind of hydrophilic amino acid residue in the transmembrane domain of the EGF receptor might make an oncogene. To date our results have been that this is not the case. So maybe there is something very subtle going on here.

HANAHAN: Could you elaborate a little on the tissue distribution of *neu* and, in particular, is it coexpressed with EGF receptor. Are they always in the same cells?

WEINBERG: The tissue distribution of normal *neu* is something which has been studied by Gail Martin. We gave her probes perhaps a year ago, but I do not really know what she has found except that *neu* is expressed rather widely, and it is not always coexpressed with EGF receptor. There are certain tissues where one is up and the other is down, but there is not any narrow window expression as has been seen with *fms*.

HANAHAN: Is there any indication that the two receptors could coassociate together?

WEINBERG: You mean, does the EGF receptor associate with *neu*? No, but there is one interesting kind of crosstalk. Cells that we have created by introduction of the normal *neu* gene into NIH 3T3 cells have a high copy numbers of p185, which cause an increased level in the basal phosphorylation of p185 protein and of EGF receptor protein. This suggests some kind of crosstalk. But there is no reason to think that there is any other kind of special relationship between the two receptors aside from an evolutionary relatedness that probably dates back 3–4–500 million years. Probably, each is doing its own thing now and has been doing that for a rather long time.

YANIV: When you immunize the mouse with the vaccine, you do not get any autoimmune disease?

WEINBERG: We do not, because of something that I should have pointed out explicitly. We are always immunizing with a rat antigen, and to the extent that it is immunogenic in mice, this must be a consequence of determinants that distinguish the rat from the homologous mouse proteins. We have done the same thing now on rats, immunizing them with vaccinia virus and there we have not got any good immune response. We were hoping firstly for an autoimmune disease, and secondly for an ability to reject the rat neuroblastomas. Neither of those has happened yet. It may be that the rat protein is totally non-immunogenic in rats. It is certainly poorly so, if it is at all.

RAPP: With the neuroblastoma you would get your antibody to the site of the tumor?

SHERR: Neuroblastomas are not beyond the blood brain barrier. They are peripheral.

WEINBERG: I think they are retroperitoneal.

SHERR: I think the activating mutation here is very interesting. Several investigators have done experiments where they systematically deleted portions of transmembrane spanning regions. Even though the transmembrane spanning fragment may be some 20 amino acids, deletions to as few as 12 amino acids can still leave a functional transmembrane spanning region. So, I wonder whether this glutamic acid insertion, by shortening the transmembrane spanning segment, actually displaces the kinase domain in the cytoplasm.

WEINBERG: It may. It certainly does not affect the gross localization of the protein, i.e. the protein is still displayed on the cell surface, but whether its positioning in the plasma membrane is slightly shifted is something that we cannot say.

SHERR: Because, in a way that may be strikingly analogous to the kind of model that Joseph Schlessinger presented, where binding of the ligand to its receptor necessitates a push-pull mechanism for activation.

SCHLESSINGER: Not necessarily. Once you add a charged residue to the transmembrane region, you may have many effects. One of them is that the extra hydrophobic amino acid would remain in the extracellular environment. It is of course interesting to put the same charged amino acids in different positions. If you try systematically to alter the position of the glutamic acid in the transmembrane domain 20 times, and then see whether it forms transforming mutants. This way, you may get a clue into what is required. Whether it is receptor aggregation or another mechanism.

WEINBERG: The protein that is responsible for transformation is clearly on the cell surface, since we can revert much of the transformed phenotype with monoclonal antibodies applied to intact cells. In fact, we can cause virtually all of the protein to disappear from the oncogene-transformed cell by adding monoclonal antibody to intact cells, suggesting that there is not a large compartment of protein elsewhere. I suspect that the mutation has something to do with either push-pull or that it affects clustering, but we really do not have any good answers right now.

WONG-STAAL: I assume the monoclonal antibody you have crossreacts with the normal p185. Have you tried putting that into normal cells expressing p185 to see what happens?

WEINBERG: We have not found that it is particularly mitogenic, but I suspect that we did not do that carefully enough.

WONG-STAAL: What is the function of the normal p185?

WEINBERG: I do not know. We suspect that the ligand for the normal p185 is not present in tissue culture medium, and as such, we are now looking for the normal ligand via some of the techniques I mentioned here. I assume that the function of normal p185 is very similar to the EGF receptor. It just recognizes a different growth stimulating hormone, but it has very similar effector functions, I imagine.

WONG-STAAL: If you can revert transformed phenotype with an antibody, then presumably you can also crosslink the normal p185 with your antibody, and the question is, what happens to the cell if you do that.

WEINBERG: One concentration of monoclonal antibody clearly induces crosslinking and internalization of protein. It could be that lower concentrations may induce clustering of the normal protein and effect some kind of mitogenic response, but we do not have the latter result today.

BALTIMORE: I presume the antibodies do not cross react with the human protein. So, you cannot tell whether the human tumors that have an amplified *neu* are actually something using that as an oncogenic ...

WEINBERG: Correct, although we assume that genetic changes in proto-oncogenes are selectively advantageous for tumors, we have not proven this.

BALTIMORE: Two of the cell lines where you had those changes were T24 and

HL60. The number of oncogenes in HL60 at this time must be uncountable?

WEINBERG: The counts are 4 or 5.

BALTIMORE: I cannot count that high.

WEINBERG: Still, one has the faith that changes in certain specific genes that are observed in tumor cells, especially when those genes are proto-oncogene, are changes which confer some selective advantage on the tumor cells.

BALTIMORE: That is why I ask: is there any way to determine the role that a given oncogenic change plays in a given tumor cell? This is one of the few systems where you can begin to ask that question, because the proteins are on the outside of the cell. But you need antibodies that can see the proteins.

WEINBERG: Yes, and they do not yet exist.

GALLO: What about primary tumors, why do you not examine some primary tumors?

WEINBERG: We did. One tumor was called Wednesday and one tumor was called Thursday. That is because the record keepers in our lab got them on Wednesday and Thursday and did not note the initials of the patients.

GALLO: And what happened?

WEINBERG: Those had gross rearrangements.

GALLO: What kind of tumors were they?

WEINBERG: Transitional cell carcinomas of the bladder. It could be that many human tumors also have point mutations of this gene. We would not have picked that up, because we are only surveying the human tumor DNA by Southern blotting.

BALTIMORE: The other question has to do with this mutation in the transmembrane region. That might have an interesting consequence that I have not thought of before. If you put hydrophilic, charged amino acids into a transmembrane domain, the charge has to be neutralized somehow, and it could either be neutralized as a negative charge by binding to phospholipids or to positively charged residue of some other protein, either one of which might give you a change in the mobility of the protein. The question is, can you cap the protein with any greater or lesser ease in the mutant than the normal p185?

WEINBERG: To quote three or four previous speakers, "that is being done this week". The journals are going to be flooded with results 4–5 months from now! The fact is that it could be that the glutamic acid paradoxically causes clustering or dimerization, simply in order to dissipate the negative charge in the hydrophobic domains.

BRUGGE: Was there a difference in the kinase activity either in an immune complex or in the cytoplasmic membrane between the mutant and normal forms of p185 protein?

WEINBERG: Not a clear difference yet.

LEPPARD: If the mutation alters the length of the hydrophobic domain in membranes that would alter the position of the phosphorylatable threonine residue relative to the membrane. Is that residue still phosphorylatable in the mutant protein?

WEINBERG: We know that TPA-inducible phosphorylation of this protein is possible. We do not know whether that particular threonine which happens to be present in p185 is still accessible.

SHERR: The way in which you assay the kinase activity is important. Did you measure it in immune complexes or did you measure it in membranes?

WEINBERG: We measured in immune complexes which I do not trust. Therefore, I do not know if our lack of any observed difference in the kinase activity is meaningful, which is why I sort of waffled when asked.

SHERR: Yes, but in our case we see a lot of kinase activity with the c-*fms* glycoprotein in immune complexes, but we do not see the activity in membranes in the absence of ligand. Have you tested c-*neu* and activated *neu* kinase in a membrane phosphorylation reaction? It might be different.

SCHLESSINGER: Yes, if you occupy *neu* with its ligand.

SHERR: No, in the absence of a ligand you might see a big difference. The transforming gene codes for an activated kinase whereas the normal gene might not.

WEINBERG: The hypothesis that everybody is implying here for kinases is that these are constitutively activated as a consequence of the point mutation, which is possible but hardly proven.

CORY: Do you have any information yet on the changes that have occurred in those tumors where you got rearrangements of the *neu* gene? Is the *neu* gene transmembrane domain well conserved between rats and humans? Can you make a survey of a wider range of human tumors to see if some of them are point mutations?

WEINBERG: First of all, we have not cloned any of the human rearrangements yet, in part because they came from biopsies and DNA was not available in sufficient amounts. Secondly, the domain is not that well conserved between rats and humans, and it is even conceivable, given the nucleotide sequence of the human gene, that methyl nitrosourea exposure in a human being would not create that point mutation and would not create an oncogene. It is possible that you can only create this oncogene in rats. We will see. I only suggest it as a possibility, which would be rather unusual, but may be just a reflection of the constellation of nucleotides which are used to encode the transmembrane domain.

SCHLESSINGER: The tissue specificity seems to be broader than EGF receptors. *Neu* seems to be expressed in more different cell types, some express the EGF receptor and some do not.

WEINBERG: I had forgotten that Joseph Schlessinger had done the survey perhaps better than anyone else has.

SCHLESSINGER: And also found that it is amplified in gastric tumors.

A Mutational Analysis of *ras* Function

Berthe M. Willumsen[1], Hsiang-fu Kung[2], Morten Johnsen[1] & Douglas R. Lowy[3]

Ras genes are widely conserved in eukaryotes, from yeast to humans (Shih & Weeks 1984, Gibbs *et al.* 1985). Mammalian cells contain at least 3 *ras* genes: c-*ras*^{H}, c-*ras*^{K}, and c-*ras*^{N}; each of these genes encodes similar 189 amino acid protein products which have been called p21 because they migrate in gels as molecules that are approximately 21 kd. The *ras* genes were first identified as the viral oncogenes of Harvey murine sarcoma virus (v-*ras*^{H}) and Kirsten (Ki) MuSV (v-*ras*^{K}), which are highly oncogenic versions of their normal cellular counterparts.

Members of this multigene family appear to serve essential, growth-related, physiologic functions in eukaryotes (De-Feo-Jones *et al.* 1985, Kataoka *et al.* 1985, Mulcahy *et al.* 1985). They have also been implicated in the pathogenesis of a variety of human and animal tumors (Marshall 1985). *Ras* genes can induce tumors *in vivo* (Tabin & Weinberg 1985) and tumorigenic transformation of tissue culture cells *in vitro*. The mammalian *ras* proteins can induce morphologic transformation of NIH 3T3 cells by overproduction of the normal *ras* protein product, by amino acid deletion (Chipperfield *et al.* 1985), or by single amino acid substitution. In tumors, missense mutations appear to be the commonest mechanism by which the genes become activated; the v-*ras*^{H} and v-*ras*^{K} genes of Ha-MuSV and Ki-MuSV, respectively, contain two missense mutations, either of which can independently activate the gene.

It has not yet been determined how the *ras* proteins carry out their normal physiological functions, nor has the pathway by which they induce cellular transformation been elucidated. In the yeast *S. cerevisiae, ras* apparently functions primarily by stimulating adenylate cyclase (Broek *et al.* 1985), but this does not seem to be the case for mammalian cells (Beckner *et al.* 1985, Levitzki *et al.* 1986) or for a different yeast (*S. pompe;* Fukui *et al.* 1986). Several presumably relevant

[1]University Microbiology Institute, Øster Farimagsgade 2A, DK-1353 Copenhagen, Denmark, [2]Hoffman-La Roche, Inc., Nutley, New Jersey 07110, and [3]Laboratory of Cellular Oncology, National Cancer Institute, Bethesda, Maryland 20892, U.S.A.

VIRAL CARCINOGENESIS, Alfred Benzon Symposium 24,
Editors: N. O. Kjeldgaard, J. Forchhammer, Munksgaard, Copenhagen 1987.

features of the *ras* proteins have been identified. They non-covalently bind guanosine nucleotides (GDP and GTP) and possess a GTPase activity that is analogous to that of the regulatory G proteins, with which they share some sequence homology (Gibbs *et al.* 1985). Activated versions of many *ras* proteins are associated with a significantly reduced GTPase activity, and genes activated by mutation of amino acid 59 from alanine to threonine possess an autophosphorylation activity, reflecting the transfer of the gamma phosphate of the bound GTP to the hydroxyl group of the threonine at position 59.

It has also been determined that the primary *ras* translation product (pro-p21) is synthesized in the cytosol and undergoes post-translational processing (to mature p21), leading to its translocation from the cytosol to the plasma membrane (Shih & Weeks 1984). The mature p21 protein has a slightly faster mobility and contains palmitic acid linked near the C-terminus of the protein (Chen *et al.* 1985, Buss & Sefton 1986). Genetic studies of v-ras^{H} have shown that both the processing and membrane association depend on a conserved cysteine residue (at amino acid 186) that, in the primary translation product, is located 4 amino acids from the C-terminus (Willumsen *et al.* 1984a, b). Mutants that encode a protein lacking cysteine-186 cannot transform NIH 3T3 cells, their proteins remain in the cytosol, do not change their migration rate, and fail to bind lipid. The biochemical nature of the processing event has not been elucidated; however, there is evidence that, in yeast and in mutant *ras* proteins in mammalian cells (Fujiyama & Tamanoi 1986, Lowy *et al.* 1986), lipid attachment and the faster migration rate are separable events.

We are carrying out structure-function studies of *ras* via a mutational analysis of the Ha-MuSV v-ras^{H} oncogene (Willumsen *et al.* 1984a, b, 1985, 1986). The protein encoded by v-ras^{H} is identical to that encoded by the normal human or rodent c-ras^{H} gene except for the two independently activating mutations, which are located at amino acid residues 12 and 59. These studies have enabled us to map segments in v-ras^{H} that are dispensable for morphological transformation, others that are essential for this function and for guanine nucleotide binding (see Willumsen *et al.* 1986 for a more detailed report of these results).

METHODS

*v-*ras^{H} *mutants and expression*

We have previously described the technique used to generate in frame v-ras^{H} mutants by deletion and linker insertion mutagenesis (Willumsen *et al.* 1984a). In summary, the mutants are constructed by the combination of two sequenced

TABLE I
Transforming activity of deletion mutants

Mutant number	Amino acid structure†	focus formation*
pBW601	wild type	+
pBW1303	63SDQ73	+
pBW1418	68ADQ77	+
pBW1304	63SDQ77	–
pBW1404	85TDQ87	+
pBW1267	92LIR96	+
pBW1271	100LIR104	+
pBW1248	101PDQ109	+
pBW1244	106ADQ112	low
pBW1197	110LIR112	–
pBW1220	119LIR126	low
pBW1237	123LIR130	+
pBW1238	123LIR132	+
pBW1239	130LIR139	+

† The numbers indicate the v-*ras*[H] amino acids encoded by front- and tail-ends, respectively, the letters indicate the 3 amino acids specified by the oligonucleotide linker (indicated with the one letter amino acid code) which join the front- and tail-ends.

* A transforming activity equivalent with that of the wild-type gene in the same vector is indicated by +: approximately 2000 focus forming units (ffu) per microgram of DNA. Low is 50 to 200 ffu/μg DNA. No foci obtained with 0.2 μg DNA is indicated by –.

parts (N-terminal front ends and C-terminal tail ends) of v-*ras*[H] through a BclI oligonucleotide linker. This linker results in the addition of 3 novel amino acids within the protein at the site of the deletion, as shown in Table I, for each mutant. Two vectors, one for expression of the mutant genes in NIH 3T3 cells and one for expression in *E. coli* have been used. The eukaryotic vector used for most of the mutants was developed by C. Jhappan, G. Vande Woude, and T. Robins (see Willumsen *et al.* 1986). In this vector, the v-*ras*[H] gene is located upstream from the SV40 sequences and neo[R] gene of pSV*neo*; these 2 genes are flanked by Moloney MuLV LTR. The use of the prokaryotic vector (pJCL-30), which places the *ras* mutants under control of the lambda pL promoter and initiates *ras* protein synthesis from its authentic initiation AUG, has been described previously (Willumsen *et al.* 1986). The *E. coli* cells that harbor the expression vector contain a temperature-sensitive allele of the lambda repressor gene, cI857.

The NIH 3T3 cells and DNA transfection procedure have been previously described (Willumsen *et al.* 1984a). For immunoprecipitation, cultures transfected with Ha-MuSV mutants selected either for focus formation or for G418 resistance were metabolically labeled with ^{35}S-methionine (250 μCi/ml) in methionine-free medium. Extracts of whole cells were prepared and precipitated as previously described with a *ras* monoclonal antibody – either Y13-238 or Y13-259 (Willumsen *et al.* 1984a). Mutants were expressed in coli and purified as described (Willumsen *et al.* 1986).

RESULTS AND DISCUSSION

Construction of mutants

The *ras* proteins can be divided into at least three functional domains. The extreme C-terminus, which we call the membrane anchoring domain, is required for posttranslational processing and membrane localization. A 20 amino acid segment that is called the major heterogenous region lies just upstream from the C-terminus (amino acid residues 165-185); it is highly divergent among different *ras* genes. The N-terminal 160 amino acids, which are highly conserved among mammalian *ras* proteins, represents the catalytic domain of the protein.

Because of the oligonucleotide linker used in constructing the mutants, each mutated v-*ras*^{H} gene encoded 3 novel amino acids at the site of the deletion. The designation of the proteins encoded by the various mutants, as XNNNY, specifies that the protein has the normal sequence from amino acid 1 to and including amino acid X, encodes 3 amino acids specified by the oligonucleotide linker which is indicated with the one letter amino acid code and continues with the normal protein sequence from and including amino acid Y to 189 (the terminal amino acid). The eukaryotic vector into which the mutants were placed contained a linked selectable marker (a neomycin-resistance gene). The selectable marker enabled us to study in NIH 3T3 cells the *ras* proteins encoded by all mutants, irrespective of their transforming capacity. Representative mutants were also placed in a prokaryotic expression vector to analyze the GDP binding activities of purified mutant *ras* proteins produced in bacteria.

Mapping domains that are required and dispensable for cell transformation and nucleotide binding

The capacity of the mutant v-*ras*^{H} genes (promoted by a viral Long Terminal Repeat Sequence LTR) to induce focal transformation of NIH 3T3 cells is shown in Figure 1. Some mutants were transformation-competent and induced foci with an efficiency similar to that of the wild-type v-*ras*^{H} gene, while the transforming capacity of other competent mutants was significantly reduced, and some mutants

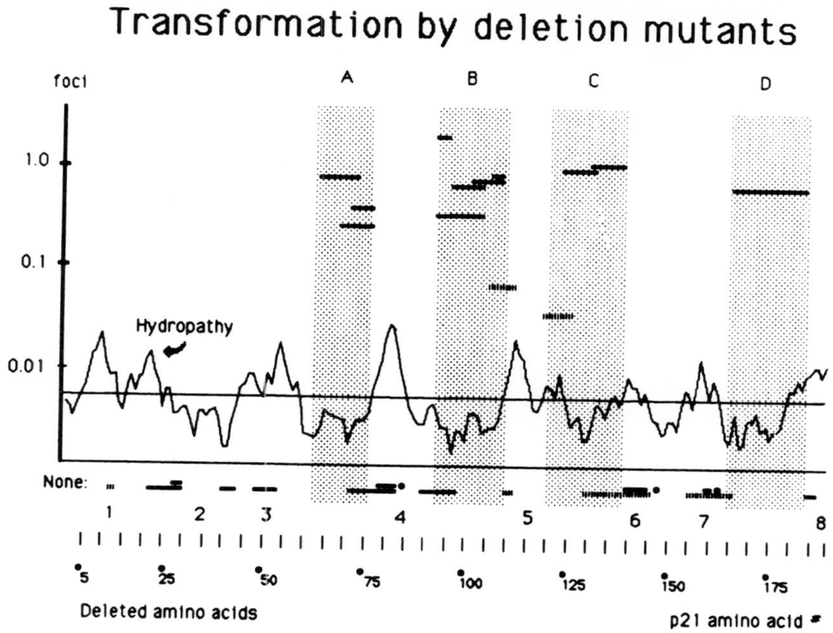

Fig. 1. Transforming activity of representative *ras* mutants and hydropathic index of the *ras* protein (Willumsen *et al.* 1986). The main horizontal axis represents the amino acid (1–189) in the *ras* protein; the vertical axis represents the relative NIH 3T3 focus-forming activity of the mutants. Each mutant is represented by a horizontal line (solid for those mutant proteins that bind GDP, interrupted for those mutants whose proteins purified from *E. coli* do not bind detectable levels of GDP); this line corresponds to the location and extent of the deletion. Segments designated A–D represent sequences that are not essential for transformation. Regions designated 1–8 contain sequences that are essential for transformation. Data for mutants in A–C and 1–6 are taken from Willumsen *et al.* (1986); data from mutants in D, 7, and 8, from Willumsen *et al.* (1984a). The hydropathic index has been plotted according to Kyte & Doolittle (1982); hydrophobic regions are above the axis of the midpoint line, hydrophilic below.

were transformation-defective (any mutant whose transforming acitivity was more than 3 orders of magnitude lower than that of the wild-type gene will score as defective in this assay). The foci induced by mutants possessing a transforming efficiency less than 20% that of the wild-type gene were generally detected later and remained smaller than those mutants that induced foci with an efficiency similar to that of the wild type gene.

Within the catalytic domain, we noted three different segments where deletions did not abolish the transforming activity of the gene (labeled A–C in Figure 1). Each non-essential segment was quite large; A, B, and C were at least 13 (residues

64–76), 16 (93–108), and 19 (120–138) amino acids long, respectively. Segment D in Figure 1 corresponds to the majority of the heterogeneous region; the dispensable nature of this region for transformation has been documented previously (Willumsen *et al.* 1985).

Given the evolutionary conservation of most v-*ras*^{H} amino acids, it might be expected that most deletions outside the heterogeneous region would abolish the transforming activity of the gene. Indeed, six different regions within the catalytic domain were apparently essential for transformation (labeled 1–6 in Figure 1) since lesions in each region rendered the genes defective. The precise boundaries of these essential regions are not yet defined. This is because the defective phenotype represents loss of a function, and we did not determine which of the variant amino acids in each mutant were responsible for the phenotype.

When the *in vitro* GDP-binding activities of the mutants were determined (from mutant protein synthesized in bacteria; Willumsen *et al.* 1986), proteins from three of the six essential regions (1, 5 and 6) were negative (any mutant that bound less than 1% as much GDP as the wild-type protein registered as negative in this assay). Two GDP-binding-negative mutants from essential region 5 were exceptional in that they had some transforming activity, but it was markedly impaired (Figure 1). When protein was isolated from cells from the few foci induced by the 2 mutant genes there was a small amount of phosphorylated p21 present, indicating that these mutant proteins retained some GTP-binding activity *in vivo*. The few mutants from essential regions 3 and 4 had low (2% of wild-type) binding activity. Mutants from essential region 2 retained high GDP-binding activity. These results are consistent with the hypothesis that GDP binding is necessary, but not sufficient, for efficient *ras*-mediated transformation.

When an attempt was made to characterize the proteins expressed in NIH 3T3 cells by the transformation-defective mutants, it was found that only cells expressing mutants in essential region 2 contained detectable levels of p21 protein. Since the epitope(s) recognized by the antibody used for detection of p21 is not deleted from mutants in the other essential regions (measured on the proteins as expressed in *E. coli*), we assume that the failure to detect such p21 proteins is due to an increased instability of the mutated proteins in mammalian cells. By contrast, those cells carrying mutants from essential region 2 contained readily detectable mutant protein as had been found previously for td mutants with lesions at the C-terminus. Sub-cellular localization studies indicated further that these mutant proteins migrated normally to the membrane (Willumsen *et al.* 1986).

ras *protein structure*

Figure 1 also plots the hydropathic index of the wild-type protein, according to the method of Kyte & Doolittle (1982). For soluble globular proteins, interior portions generally map to the hydrophobic side of the midpoint line, while exterior portions are usually found on the hydrophilic side. The regions required for *ras*-transforming function tend to fall on the hydrophobic side, with the notable exception of region 2, which is hydrophilic.

When the transforming and biochemical activities of the mutants are considered in conjunction with the p21 structure model derived from bacterial elongation factor EF-Tu (Jurnack 1985, McCormick *et al.* 1985), the hydropathic index suggests several topological and functional features of the p21 protein. Most of the required regions appear to be hydrophobic, implying that disruption of interior portions alter the protein sufficiently to inactivate it biologically. This could occur either by rendering the protein unable to interact appropriately with other cellular components, or by partially denaturing the protein, thereby reducing its half-life in the cells. Each of the non-essential segments falls within hydrophilic regions, suggesting that these segments are located on the exterior of the protein, which in the EF-Tu model are composed of alpha-helices. It therefore appears that these putative alpha helices can be substituted by adjacent amino acid and/or linker sequences, showing that there is no sequence-specificy requirements in these regions. There is, however, a requirement for a certain number of amino acids, since enlarging the deletions to span the entire essential region usually results in a non-transforming protein (for example 63SDQ73 and 68ADQ77 are both transforming, but 63SDQ77 is transforming-defective, Table I). Evolutionarily, region A is practically identical in all *ras* proteins, while regions B and C are less tightly conserved. For all or some of these segments it remains possible that the specific sequences are necessary for interactions pertaining to the normal function of p21, but are not essential in the activated, transforming protein.

P21 is known to interact with guanosine nucleotides, and the EF-Tu based structural model of p21 is derived from the considerable homology between p21 and Tu of the amino acid sequences thought to be involved in GDP binding. In order for p21 to mediate the growth signals that lead to transformation, it is assumed that the protein interacts with other cellular components. Catalytically active sites in proteins are often thought to be situated in the parts of the protein where the structural components (the alpha-helices and beta-strands) join each other. Therefore, it may be of interest to analyze our mutants in the context of the joining regions as defined by the EF-Tu-derived structure model. Such an examination suggests that the regions in the part of p21 modelled after Tu that might be involved in sequence-specific interactions with other cellular components are extremely limited. In the model, the amino acid sequences believed to take part in the four beta-strands are 5–11, 76–83, 111–118 and 138–142 (M. Kjeldgaard, personal communication), and the alpha helices are made from amino acids 16–34, 63–69, 94–100 and 129–133 (McCormick *et al.* 1985). For this analysis, the joining regions are considered to be the amino acids 11–16, 35–62, 70–75, 84–93, 101–110, 119–128 and 134–137.

Amino acids 11–16 are presumed to be part of the GDP-binding pocket (McCormick *et al.* 1985). In joining region 34–63, none of our mutants has been found to retain transformation acitivity (these mutants will be discussed in greater detail in the next section). Between positions 69 and 76, all amino acids have been deleted without a loss of transformation activity, and in region 83–94 some distortion is tolerated since mutants 85SDQ87 and 92LIR96 are transformation-competent (Table I). We do not as yet have data on the possible necessity for amino acids 83–85 or 87–92. In region 100–111, 100LIR104 and 101ADQ109 give transformation-competent proteins; however, deleting amino acids 110–112 results in impairment or loss of transformation (Willumsen *et al.* 1986, Table I). Mutants informative for joining region 118–129 are 119PDQ126 and 123LIR132, both of which are transforming; however, 119PDQ126 transforms 10 to 30 times less than wild-type. Amino acids 116–119 have been presumed to bind to the purine ring in GDP (Jurnack 1985, McCormick *et al.* 1985); therefore, the poor transformation activity of mutant protein 119PDQ126, which has no detectable guanosine nucleotide binding *in vitro*, is most likely to be explained by this impairment. Mutant 130LIR139 is fully transforming; therefore, neither position 123–129 nor 133–138 require a particular amino acid sequence.

Thus, with the exception of amino acids 34–63 and possibly amino acids 110–112, there is little evidence that these joining regions, as defined by the Tu model, are involved in functions that are essential for transformation. However, results obtained recently indicate that amino acids 142 to 165, which lie outside those assigned a special structure in the Tu-derived model, are essential for transformation. The mutants in this region have not yet been fully analyzed.

An essential region that may interact with the ras *target*

Essential region 2, which is highly conserved among *ras* proteins, apparently represents a required exterior portion of the protein. Although there is no sequence homology between *ras* and EF-Tu in this region, acylated tRNAs are believed to bind to the analogous region of EF-Tu-GTP (Kaziro 1978). The *ras* mutants with lesions in essential region 2 are transformation-defective, despite possessing all the known important biochemical features of *ras* proteins in that they synthesize a protein that is stable in NIH 3T3 cells, localize to the membrane, and have GDP-binding, GTPase, and autophosphorylating activities that are similar to that of the wild-type v-ras^{H} protein. These mutants argue for the existence of an essential *ras* function that has not yet been defined, and we speculate that the region 2 mutant proteins are defective because they fail to interact with the putative target of the normal *ras* protein. Using information obtained with these mutants, we hope it will be possible to further elucidate the steps involved in the transmission of the growth signal given off by p21.

SUMMARY

We have used linker insertion-deletion mutagenesis to study the Harvey murine sarcoma virus v-ras^{H} transforming protein. The mutants were characterized with respect to their ability to induce morphologic transformation of NIH 3T3 cells and the capacity of their proteins to bind guanosine nucleotides, undergo post-translational processing, and localize to the plasma membrane. We have identified four non-overlapping segments that are dispensable for morphologic transformation of NIH 3T3 cells, several segments that are required for transformation and stability in mammalian cells as well as guanosine nucleotide binding by protein synthesized in *E. coli*. One essential segment that does not affect guanine nucleotide binding or stability, which appears to lie on the exterior of the protein and therefore may interact with the putative *ras* protein target, has been identified.

ACKNOWLEDGMENTS

Parts of this work have been supported by the Danish Cancer Society (84-068), the Danish Medical Research Council (12-5345 and 12-4663), the Danish Natural Science Research Council (11-5095) and NATO (84/165).

REFERENCES

Beckner, S. K., Hattori, S. & Shih, T. Y. (1985) The *ras* oncogene product p21 is not a regulatory component of adenylate cyclase. *Nature 317*, 71–73.

Broek, D., Samily, N., Fasano, O., Fujiyama, A., Tamanoi, F., Northrup, J. & Wigler, M. (1985) Differential activation of yeast adenylate cyclase by wild-type and mutant *ras* proteins. *Cell 41*, 763–769.

Buss, J. E. & Sefton, B. M. (1986) Direct identification of palmitic acid as the lipid attached to p21[ras]. *Mol. Cell Biol. 6*, 116.

Chen, Z.-O., Ulsh, L. S., DuBois, G. & Shih, T. Y. (1985) Posttranslational processing of p21 *ras* proteins involves plamitylation of the C-terminal tetrapeptide containing cysteine-186. *J. Virol. 56*, 607–612.

Chipperfield, R. G., Jones, S. S., Lo, K.-M. & Weinberg, R. (1985) Activation of Ha-ras p21 by substitution, deletion and insertion mutations. *Mol. Cell. Biol. 5*, 1809–1813.

DeFeo-Jones, D., Tatchell, K., Robinson, L. C., Sigal, I. S., Vass, W. C., Lowy, D. R. & Scolnick, E. M. (1985) Mammalian and yeast ras gene products: Biological function in their heterologous systems. *Science 228*, 179–184.

Fujiyama, A. & Tamanoi, F. (1986) Processing and fatty acid acylation of RAS1 and RAS2 proteins *Saccharomyces cerevisiae*. *Proc. Nat. Acad. Sci. USA 83*, 1266–1270.

Fukui, Y., Kozasa, T., Kaziro, Y., Takeda, T. & Yamamoto, M. (1986) Role of a *ras* homolog in the life cycle of Schizosaccharomyces pompe. *Cell 44*, 329–336.

Gibbs, J. B., Sigal, I. S. & Scolnick, E. M. (1985) Biochemical properties of normal and oncogenic *ras* p21. *Trends Biochem. Sci. 10*, 350–353.

Jurnack, F. (1985) Structure of the GDP domain of EF-Tu and lication of the amino acids homologous to *ras* oncogene proteins. *Science 230*, 32–36.

Kataoka, T., Powers, S., Camaron, S., Fasano, O., Goldfarb, M., Borach, J. & Wigler, M. (1985) Functional homology of mammalian and yeast RAS genes. *Cell 40*, 19–26.

Kaziro, Y. (1978) The role of guanosine 5′-triphosphate in polypeptide chain elongation. *Biochim. Biophys. Acta 505*, 95–127.

Kyte, J. & Doolittle, R. F. (1982) A simple method for displaying the hydropathic character of a protein. *J. Mol. Biol. 157*, 105–132.

Levitzki, A., Rudick, J., Pastan, I., Vass, W. C. & Lowy, D. R. (1986) Adenylate cyclase activity of NIH 3T3 cells morphologically transformed by *ras* genes. *FEBS Lett. 197*, 134–138.

Lowy, D. R., Papageorge, A. G. Vass, W. C. & Willumsen, B. M. (1986) in UCLA symposium on Molecular and Cellular Biology: *Cellular and Molecular Biology of Tumors and Potential Clinical Applications*. J. Minna and M. Kueni (eds.). (In press).

Marshall, C. (1985) Human Oncogenes. In [Weiss, R., et al. (eds.)] *RNA Tumor Viruses. Molecular Biology of Tumor Viruses*. New York: Supplement to 2nd edition, Cold Spring Harbor Laboratory, pp. 487–558.

McCormick, F., Clark, B. F. C., La Cour, T. F. M., Kjeldgaard, M., Norskov-Lauritsen, L. & Nyborg, J. (1985) A model for the tertiary structure of p21, the product of the *ras* oncogene. *Science 230*, 78–82.

Mulcahy, L. S., Smith, M. R. & Stacey, D. W. (1985) Requirements for *ras* proto-oncogene function during serum stimulated growth of NIH 3T3 cells. *Nature 313*, 241–243.

Shih, T. Y. & Weeks, M. O. (1984) Oncogenes and cancer: The p21 *ras* genes. *Cancer Invest. 2*, 109–123.

Tabin, C. J. & Weinberg, R. A. (1985) Analysis of viral and somatic activations of the cHa-*ras* gene. *J. Virol 53*, 260–265.

Willumsen, B. M., Christensen, A., Hubbert, N. L., Papageorge, A. G. & Lowy, D. R. (1984a) The p21 *ras* terminus is required for transformation and membrane association. *Nature 310*, 583–586.

Willumsen, B. M., Norris, K., Papageorge, A. G., Hubbert, N. L. & Lowy, D. R. (1984b) Harvey murine sarcoma virus p21 *ras* protein: biological and biochemical significance of the cysteine nearest the carboxy terminus. *EMBO J. 3*, 2581–2585.

Willumsen, B. M., Papageorge, A. G., Hubbert, N. L., Bekesi, E., Kung, H.-F. & Lowy, D. R. (1985) Transforming p21 *ras* protein: flexibility in the major variable region linking the catalytic and membrane-anchoring domains. *EMBO J. 4*, 2893–2896.

Willumsen, B. M., Papageorge, A. G., Kung, H.-F., Bekesi, E., Robins, T., Johnsen, M., Vass, W. C. & Lowy, D. R. (1986) Mutational analysis of a *ras* catalytic domain. *Mol. Cell. Biol. 6*, 2646–2654.

DISCUSSION

YANIV: The mutations that are present in your gene at position 12 or 61, do they change anything in the properties of hydrolysis of GTP or in the affinity constants?

WILLUMSEN: The dissociation constants have been measured for the binding of GTP and no changes between the normal version and the activated versions have been reported. Generally, transforming, activating mutation leads to low GTPase activity. However, there are low GTPase mutants that do not become transforming. Here, we could argue that besides the GTPase activity being destroyed, something else is also destroyed, so the active conformation is not obtained.

DYNAN: Would you comment on the prospects for obtaining actual, rather than hypothetical structural information?

WILLUMSEN: I have no direct information concerning attempts to crystallize the p21.

WEINBERG: You said that the effector domain or the signal output domain is in regions amino acids 20–45, but there could be other domains of the protein which are important for putting out the signal. You alluded very briefly to there being more C-terminal regions, maybe equally important.

WILLUMSEN: We have data suggesting that sequences between 142 and 163 are important for transformation. Again, those regions might be important for the overall structure and integrity of the protein, and since our analysis is not complete, we cannot distinguish between these possibilities.

WEINBERG: But you assign more of a secondary role for structural integrity to those domains as opposed to the 20–45 region to which you seem to attach a greater functional role.

WILLUMSEN: The basis for that interpretation is our inability to find the protein in mammalian cells once we have interfered with components in the beta sheet. When a mutant gene does not give accumulation of the protein product in the cells, it is not possible to assign a particular function important for transformation to the mutated region.

It is very possible that some of the regions that we have found to be dispensable are only dispensable in an activated version of the protein. It could be that these regions are important in, say, taking the GTP off, or are important in interacting with the incoming growth signal. Either those putative alpha helices lying on the outside of the protein or the heterogeneous region of the C-terminal part of p21 could be involved. We do not have an assay for that, since we only look for the disappearance of transforming function.

SCHLESSINGER: With this rather small protein, I think you should be rather talking about conformational changes and propagation of conformational change introduced by point mutations rather than communication between domains, because domains are usually used for larger proteins.

WILLUMSEN: Perhaps you are right. Anyway, with regard to the Tu model, what is depicted is considered a single domain of the EF-Tu. We are thinking about the word in terms of functional domains, and we may have to find a different word.

BALTIMORE: What is the role of that cysteine that is so important for membrane binding? Is it really physically attached to the membrane?

WILLUMSEN: It has been shown that palmitic acid is covalently bound to the protein. There can also be isolated a peptide corresponding to the C-terminal peptide which is present when the palmitic acid has been hydrolyzed off, but is not present when the peptide is still bound to palmitic acid. That is the C-terminal end. The reason why I am so careful in stating it is because we have the shift in electrophoretic migration rate that occurs when the protein matures, and that shift is dissociable from the lipid binding. It is not the attachment of the lipid which makes the protein migrate differently, because we have mutants that dissociate the two events. We can isolate mutants that undergo processing, but which do not have lipid bound and are not located in the membrane. Also, in yeast it is normal that you can isolate both processed, non-lipid-bound as well as processed, lipid-bound protein. Our feeling is that the processing involves a proteolytic cleavage of the three C-terminal amino acids which exposes the cysteine, this enables it to bind lipid, whereafter the lipid draws the p21 into the plasma membrane. The cDNA for the transducing gene has been sequenced, and it has such a cysteine located in an identical position. The report containing the amino acid sequence as determined by direct amino acid sequencing did not give cysteine and three downstream amino acids. This is suggesting that in some instances there may be cleavage of some amino acids and modification of a cysteine.

BALTIMORE: But the lipid may not be attached through the sulphur linkage but rather through internal carboxyl groups?

WILLUMSEN: The chemical sensitivity of the linkage is consistent with its being a thioester.

RAPP: Your most transforming version of the *ras* protein still has GTPase activity?

WILLUMSEN: It has some residual low GTPase activity.

RAPP: So, would you not expect further up-transformation, up-mutations, to be possible?

WILLUMSEN: Perhaps. However, it may be that such dead negative mutations result in other characteristics, leading them to be non-transforming. No GTPase-less mutant has been described, when the protein is isolated it purifies with GDP, indicating that the GTP has been hydrolyzed.

CUZIN: How does your model account for the observation that overexpression of the normal protein is capable of transforming cells?

WILLUMSEN: Overexpressing the normal p21 would automatically lead to an increased concentration of the GTP-bound species, and if it is the concentration of that species which leads to the increased proliferation of the cells, then that would explain it.

CUZIN: Is it not possible that some mutations, by changing the stability of the protein, could increase or decrease its steady state level?

WILLUMSEN: The stability of oncogenic versions versus the normal version in the overexpressed situation has been investigated, and there is no difference. We have reason to believe that some of our mutants have decreased stability. These mutants do not transform cells. The mutants that have been hit in membrane-binding or in the region of aa 30–40 are as stable as the wild type protein.

BALTIMORE: It occurs to me that I have seen the model that you presented before. The GTP/GDP relationships and binding implies that in the ground state the protein should bind GDP and in the activated state it should bind GTP, and yet, if I remember correctly, the binding studies on the transforming version of the protein show more or less equal binding of GTP and GDP.

WILLUMSEN: That is correct.

BALTIMORE: If you look at the non-transforming version, does that change?

WILLUMSEN: No, it does not. The dissociation constants that have been reported all have the same difference between the oncogenic and the non-oncogenic versions, and it is of the same order of magnitude. However, the intracellular concentration of GTP is probably 3 to 10 times higher than the intracellular concentration of GDP. A naked p21 molecule, given the choice between this mixture of GTP and GDP would pick up, I assume, GTP, and it would become activated that way. That is implying that the initial stimulus is the removal of the GDP. That has not been demonstrated.

BALTIMORE: So, there is no intrinsic difference in the affinity for the two ligands, it is rather a replacement of one rather than the other?

WILLUMSEN: Yes, and it is the mechanism of that replacement that we need to understand.

BALTIMORE: But the binding of GTP is supposed to activate, and if most of what binds is GTP, the protein should be in general an active configuration without having received the signal for hydrolysis.

WILLUMSEN: One hypothesis is that it is the exchange of GTP with GDP that is regulated and which requires an interaction with, for example, an activated growth factor-bound receptor.

WEINBERG: All this is a bit moot because you do not know whether the specificity of nucleotide binding is totally different *in situ* and in the membrane when the protein may be in complex with many other proteins.

WILLUMSEN: There is an approximation of an *in vitro* assay for the mammalian p21, utilizing its ability to stimulate adenylate cyclase in yeast membrane preparations, and there are some experiments which utilize an unhydrolyzable analogue of GTP. That analogue gives the maximal response, and under conditions where GDP cannot be kinased to GTP, GDP results in an essentially inactive protein. Only GTP-bound protein will stimulate the adenylate cyclase in that system.

RIGBY: Do you think this stimulation of yeast adenylate cyclase is valid as an assay? My understanding of Wigler's work is that in vertebrates you do not see that effect, and I gather that in *Schizosaccharomyces* the *ras* gene has nothing to do with the adenylate cyclase system at all. Is this assay not a questionable thing to base conclusions on?

WILLUMSEN: We know that mutants that affect the amino acid 30–35 region (mutants that we conclude are defective in recognition of the target) are also defective in this *in vitro* property.

RIGBY: But that could just be because you are dissecting out the central nucleation site of the protein, and so you abolish the main activity.

WILLUMSEN: I think that it is probable that all these proteins have part of the same structure. Nature evolved a way of regulating things, a universal on and off switch, and of course the parts of the molecules have evolved just to stimulate different kinds of proteins, so that the incoming signals as well as the outgoing signals can be different. The interactions are basically the same though there will naturally be subtle differences. It is clear that that mammalian *ras* proteins work in *Saccharomyces cerevisiae*, and it is clear that yeast protein, if slightly modified, can transform mammalian cells, so even though it does not interact with cyclase in mammalian cells, whatever it does, they can replace one another in their different systems. The structure is sufficiently similar so they can mimic each other.

LOWY: Some comments to Peter Rigby's remarks. The mutants that Dr. Willumsen is talking about that map to the putative effector region encode a protein that is stable in NIH-3T3 cells, while those that do not bind GTP are unstable in the cells. I think that the *in vitro* adenylate cyclase stimulation assay using *Saccharomyces cerevisiae* membranes does represent a valid enzymatic assay for mammalian p21. That is not to say, however, that that is the major function of p21 protein in species other than *Saccharomyces*.

Regulation of the Expression of the Cellular *src* Proto-Oncogene Product

Joan S. Brugge

Investigations of the mechanisms of oncogenic transformation by retroviruses has led to the identification of a large number of cellular genes which are capable of eliciting cellular transformation when their expression is altered by association with retroviral genomes (Bishop 1985). One of the best-characterized viral oncogenes is the v-*src* gene that is encoded by Rous sarcoma virus (RSV). The protein product of the v-*src* gene, denoted $pp60^{v\text{-}src}$, was the first oncogene identified in retrovirus-transformed cells, and the first transforming protein shown to possess an enzymatic activity. The $pp60^{v\text{-}src}$ protein possesses an intrinsic protein kinase activity that specifically phosphorylates tyrosine residues. It is now recognized that the v-*src* gene product is one member of a large family of viral transforming proteins that are tyrosine-specific protein kinases. Multiple cellular tyrosyl-kinases have been identified, several of which are the receptors for growth factors; i.e. epidermal growth factor, platelet-derived-growth-factor, colony-stimulating factor I, and insulin.

Normal cells contain a highly conserved cellular gene that is homologous to the viral *src* gene (Hanafusa 1986). The function of this gene product in normal cells is unknown. The v-*src* gene in RSV-transformed fibroblasts causes many alterations in the phenotype of the host cell, including changes in cellular morphology, metabolism, transport of small metabolites, secretion of extracellular proteases, dependence on exogenous growth factors, and tumorigenicity. It is considered unlikely that the c-*src* gene is responsible for the regulation of these events in normal cells since expression of the c-*src* protein in fibroblasts at levels comparable to those of the v-*src* protein in RSV-transformed cells does not lead to cellular transformation, or to alterations in cell morphology and growth behavior similar to those found in RSV-transformed cells (Hanafusa 1986). These

---

Department of Microbiology, State University of New York, Stony Brook, New York, 11794, U.S.A.

VIRAL CARCINOGENESIS, Alfred Benzon Symposium 24,
Editors: N. O. Kjeldgaard, J. Forchhammer, Munksgaard, Copenhagen 1987.

results indicate that the quantitative differences (15- to 20-fold) in expression of the cellular and viral *src* gene products in normal and RSV-transformed fibroblasts are not responsible for the differences in the behavior of the cellular and viral forms of the *src* gene in fibroblasts.

There are multiple amino acid differences between $pp60^{c\text{-}src}$ and $pp60^{v\text{-}src}$ that were generated by mutations in the viral *src* gene. These mutations are responsible for qualitative differences in the expression of the v-*src* protein compared to the c-*src* protein. The tyrosyl-kinase activity of the $pp60^{v\text{-}src}$ is 20- to 100-fold greater than that of $pp60^{c\text{-}src}$. This results in quantitative differences in the phosphorylation of cellular proteins on tyrosine in normal and RSV-transformed fibroblasts. Many different cellular proteins are phosphorylated on tyrosine in RSV-transformed cells. Although the phosphorylation of many of these proteins may be of no consequence to fibroblasts, it is likely that some of these phosphorylation events are responsible for eliciting the multiple phenotypic changes in the host cell which accompany expression of the v-*src* gene product. It is possible that many of the substrates that are phosphorylated in RSV-transformed cells may represent the substrates of normal cellular tyrosyl-kinases. The phosphorylation of these substrates in specific cellular environments may be responsible for triggering the events which are mediated by the v-*src* gene product in fibroblasts. For instance, the phosphorylation of substrates of the epidermal-growth-factor receptor in fibroblasts could cause stimulation of cell proliferation and ruffling of the plasma membrane. While these substrates are not phosphorylated in normal fibroblasts by the c-*src* protein, it is possible that the activated, unregulated v-*src* protein could constitutively phosphorylate these substrates at high levels, causing a constitutive stimulation of cell proliferation. A diagram illustrating such a model is shown in Figure 1. This model is supported by the evidence that many of the substrates of the v-*src* protein overlap with those of the epidermal-growth-factor- and platelet-derived-growth-factor-receptors (Hunter & Cooper 1985).

Implicit in the rationale for this model is the assumption that cellular tyrosine kinases are specifically involved in events associated with the transduction of extracellular signals. The cellular *src* protein can be distinguished from the growth factor receptor tyrosine kinases in the nature of its association with the plasma membrane. Unlike the growth factor receptors, which are transmembrane proteins with ligand binding sites on their extracellular domain, $pp60^{c\text{-}src}$ and several other tyrosine kinases are exclusively localized on the cytoplasmic face of the plasma membrane. If these kinases are involved in extracellular signal transduc-

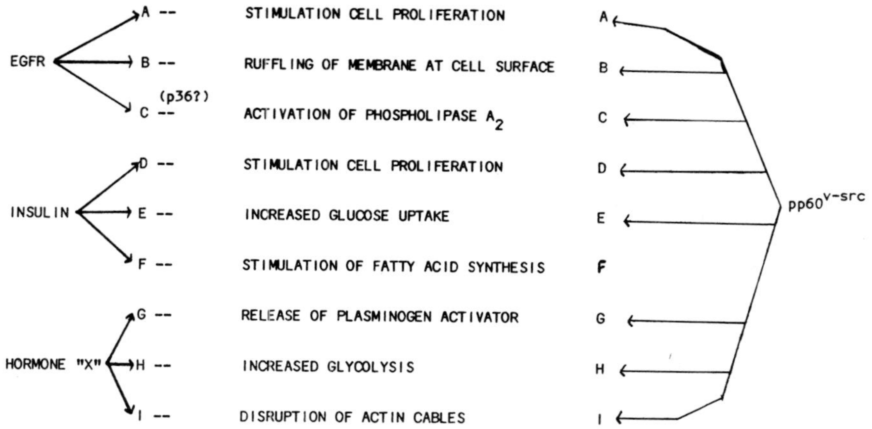

Fig. 1. Model describing a proposed mechanism for $pp60^{v\text{-}src}$-mediated alterations in host cell physiology. The letters A–I represent the substrates of normal cellular tyrosine kinases which mediate the phenotypic changes described in the middle of the diagram.

tion, one must assume that these kinases communicate with transmembrane receptors in a manner similar to receptors which interact with adenyl cyclase (see Figure 2). Since such a receptor which might communicate with the c-*src* kinase has not been identified, there are no obvious clues to the role of the cellular *src* protein in any cell type.

In an attempt to identify the type of cells in which the c-*src* protein might provide a specific function, we have examined the expression of c-*src* protein in a variety of embryonic and adult tissues, in peripheral blood cells, and in cells transformed by other tumor viruses. These studies have revealed that the expression of the c-*src* protein is regulated qualitatively and quantitatively in different cellular environments. The lowest levels of c-*src* protein were found in fibroblasts, which have previously served as the standard source of c-*src* protein for comparison between $pp60^{c\text{-}src}$ and $pp60^{v\text{-}src}$. The highest levels of c-*src*-specific kinase activity were detected in 3 types of cells: neurons, platelets, and polyma virus-transformed cells. This report will focus on studies of the c-*src* protein in neural tissues, with a brief summary on the c-*src* protein from platelets and polyoma-transformed cells.

EXPRESSION OF THE $pp60^{src}$ PROTEIN IN NEURAL TISSUES

All neural tissues that have been examined from avian and mammalian embryos, including brain, retina, and spinal ganglia, express 8- to 10-fold higher levels of

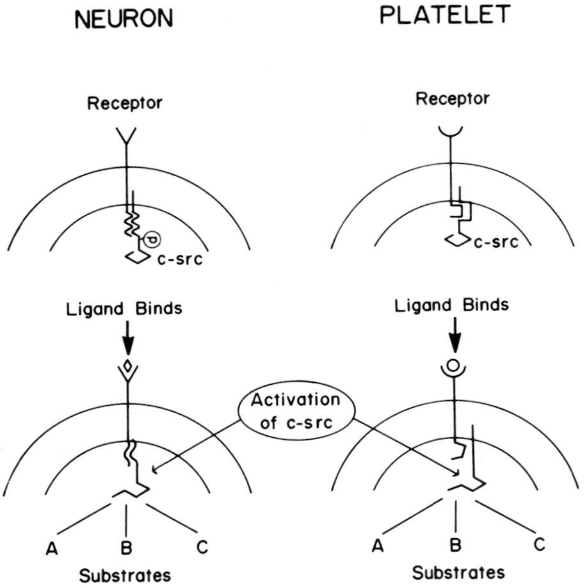

Fig. 2. Model showing the proposed alignment of pp60$^{c\text{-}src}$ in the plasma membrane in close association with a cellular receptor. The amino-terminal region of pp60$^{c\text{-}src}$ is associated with the membrane. A structural difference is shown in this region of the c-*src* protein from neurons and platelets to reflect the differences detected by peptide analysis. P represents the unique site of phosphorylation in the amino-terminal region of the neuronal pp60$^{c\text{-}src}$.

the c-*src* protein than other tissues (Cotton & Brugge 1983, Schartl & Barnekow 1982, Levy *et al.* 1984, Sorge *et al.* 1984). The levels of pp60$^{c\text{-}src}$ are generally lower in adult, compared to embryonic neural tissues, yet they are still elevated compared to most other tissues. The neural specificity of pp60$^{c\text{-}src}$ expression has been conserved in lower eukaryotic species. *In situ* hybridization of adult *Drosophila* RNA using *src*-specific probes revealed elevated levels of c-*src*-related transcription in the brain and eye region of the fly (Simon *et al.* 1985).

In screening the expression of pp60$^{c\text{-}src}$ in embryos, we noted that the c-*src* protein expressed in avian and mammalian neural tissues displayed a retarded electrophoretic mobility on SDS-polyacrylamide gels. Crucial to an understanding of the function of this unique form of the c-*src* protein in neural tissues was the identification of the specific types of neural cells which express high levels of this altered form of the *src* protein. Several lines of evidence suggest that neurons specifically synthesize high levels of this protein. 1) The period in embryonic development at which elevated levels of the modified c-*src* protein are first observed corresponds to the stage when neurons begin to differentiate. 2) Sorge and coworkers (1984) found the strongest pp60$^{c\text{-}src}$ immunoreactivity in the neuronal cells from chicken embryo neural retinal cells. In the cerebellum, pp60$^{c\text{-}src}$ immunoreactivity was also strongest in the process-rich molecular layer (Fults *et al.* 1985). 3) The neural form of pp60$^{c\text{-}src}$, denoted pp60*, was exclusively found in cultures of pure neuronal cells from rat embryos at levels 15- to 20-fold higher than the c-*src* protein in fibroblasts (Brugge *et al.* 1985). This form of the c-*src* protein was not detectable in astrocytic cultures.

The pp60$^{c\text{-}src}$ expressed in neuronal cells also displays a higher specific activity than the protein expressed in astrocytes or fibroblasts (see Figure 3). Although similar levels of the c-*src* protein were expressed in the neuronal and astrocytic cultures, the pp60$^{c\text{-}src}$ molecules from neurons showed 6- to 10-fold higher levels of enolase and auto-phosphorylation than the protein from astrocytic cultures (Brugge *et al.* 1985). Thus, the structurally distinct form of pp60$^{c\text{-}src}$ expressed in neuronal cultures possesses an activated tyrosine kinase activity.

The partial proteolytic cleavage profile shown in Figure 4 demonstrates that the amino-terminal region of the neuronal form of pp60* contains the alteration responsible for the retarded mobility of this protein. The V4 and V3 peptides, which represent the amino-terminal 16- and 18-kDa of pp60$^{c\text{-}src}$, migrate more slowly than the corresponding region of the astrocytic form of pp60$^{c\text{-}src}$. Cleavage with a reagent that hydrolyses tryptophan residues indicated that the first 118 amino acids of the neuronal pp60* is structurally distinct from the corresponding region of the c-*src* protein from other cell types (Cotton and Brugge, unpublished results).

The nature of the modification responsible for the altered electrophoretic mobility of pp60* is not known. The peptide maps shown in Figure 4 indicate that pp60* contains an additional site of phosphorylation which is not detected in the astrocytic form of pp60$^{c\text{-}src}$ and which maps within the V3 and V4 fragments of pp60$^{c\text{-}src}$ derived by partial proteolysis with Staphyloccus V8 protease (Brugge, Lipsich and Lustig, unpublished results). However, this modification is not believed to be responsible for the shifted electrophoretic mobility of pp60*. Removal of 95% of the phosphate from neuronal pp60* using potato acid phosphatase does not cause a shift in its electrophoretic mobility (Lustig and Brugge, unpublished results), and the protein translated *in vitro* using rabbit reticulocyte lysates programmed with RNA from embryonic chicken brains displays the same electrophoretic mobility as the protein expressed *in vivo* (Coussens and Brugge, unpub-

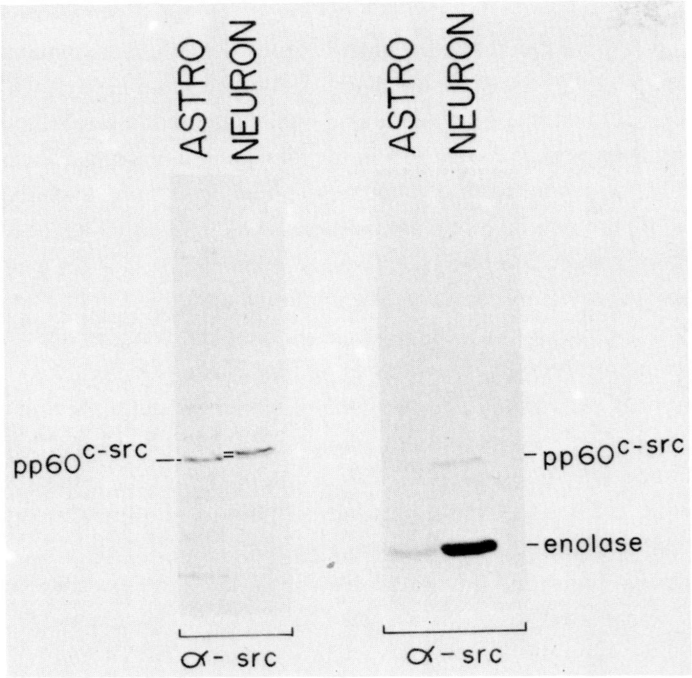

Fig. 3. Specific activity of pp60$^{c\text{-}src}$ expressed in neurons and astrocytes. Pure cultures of neurons and astrocytes were prepared by D. Nonner, J. Barrett, and R. Keane as described (Brugge *et al.* 1985). The cells were labeled with ^{35}S-methionine and lysates were immunoprecipitated with monoclonal antibody to pp60$^{c\text{-}src}$. The immunoprecipitates were either eluted with SDS-mercaptoethanol (Panel A) or incubated with enolase and [γ-^{32}P]ATP as described (Brugge *et al.* 1985) before addition of SDS and mercaptoethanol (Panel B).

lished results). These results suggest the possibility that the c-*src* protein expressed in neuronal cells may be translated from a unique messenger RNA species, possibly resulting from a modified pattern of post-transcriptional splicing. While we cannot exclude the possibility that the protein detected in neuronal cultures represents a gene product that is highly related, yet distinct from the c-*src* gene product, several lines of evidence strongly suggest that the pp60 protein is encoded by the c-*src* gene: 1) The monoclonal antibody employed for immunoprecipitation of this protein does not recognize the related v-*yes*, v-*ros*, v-*fps*, or v-*fgr* proteins (Bishop 1985). 2) The peptide maps of the ^{32}P-labeled protein, precipitated with the monoclonal antibody from neuronal and astrocytic cultures by cleavage with trypsin, BNPS-skatole, cyanogen bromide, and Staph. V8 protease, are similar

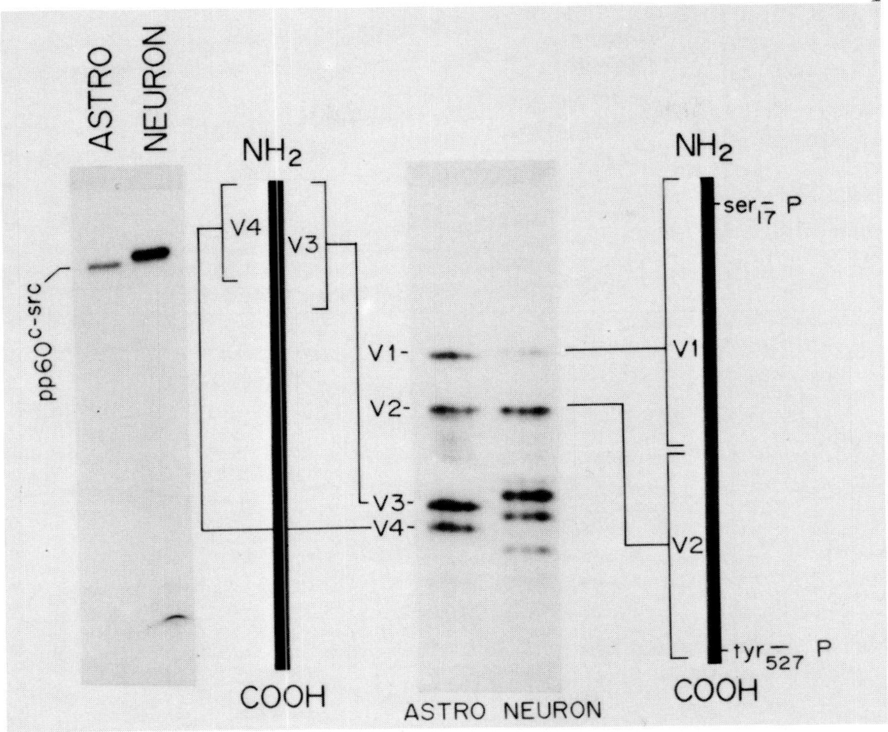

Fig. 4. Cleavage profile of the pp60$^{c\text{-}src}$ molecules from astrocytes and neurons. pp60$^{c\text{-}src}$ was immunoprecipitated from ^{32}P-labeled cultures of neurons and astrocytes and subjected to electrophoresis on SDS polyacrylamide gels (left panel). The pp60$^{c\text{-}src}$ bands were excised and subjected to electrophoresis in the presence of 200 nanograms of Staphylococcus V8 protease.

except for the single tryptic phosphopeptide which is specifically associated with the neuronal form of the c-*src* protein (Cotton, Coussens, and Brugge, unpublished results).

In summary, these results indicate that a structurally unique, activated form of the c-*src* protein is expressed at elevated levels in neuronal cells. This form appears to be translated from a unique mRNA species which is processed in a cell-type-specific pattern. It is not known whether this structural alteration is solely responsible for the activation of the c-*src* protein since the neuronal form of pp60$^{c\text{-}src}$ is also phosphorylated on a unique site within the amino-terminal 118 amino acids which could influence the tyrosyl-kinase activity of this protein.

Figure 5 shows a diagram summarizing the results from studies of the cellular *src* protein in normal neurons, platelets and fibroblasts, and in RSV and polyoma

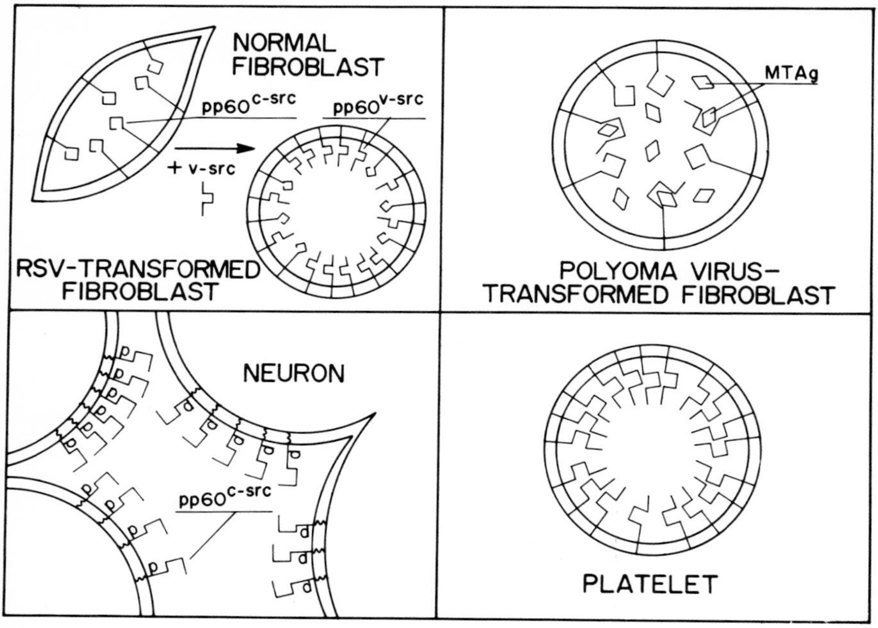

Fig. 5. Diagram illustrating qualitative and quantitative differences in the expression of the c-*src* protein in different cell types.

(Py) virus-transformed fibroblasts. Our current understanding of the regulation of the c-*src* protein in Py virus-transformed cells and in platelets is described below.

ACTIVATION OF THE *src* GENE PRODUCT IN POLYMA VIRUS-TRANSFORMED CELLS

The $pp60^{c\text{-}src}$ protein expressed in cells transformed by Py virus, a DNA tumor virus, is subject to a novel mechanism of activation. The middle tumor antigen transforming protein (MTAg) encoded by Py virus associates in a complex with $pp60^{c\text{-}src}$ in Py-infected and -transformed cells (Courtneidge *et al.* 1983, Lipsich *et al.* 1983). The association of the MTAg with $pp60^{c\text{-}src}$ activates the tyrosine kinase activity of $pp60^{c\text{-}src}$ 30- to 50-fold. Only a small percentage of intracellular $pp60^{c\text{-}src}$ molecules are bound to MTAg in Py-transformed cells, and MTAg binding is necessary for the activation of $pp60^{c\text{-}src}$ protein kinase activity. The basis for the activation of $pp60^{c\text{-}src}$ by MTAg is not known; however it is of interest that the *in vivo* and *in vitro* sites of phosphorylation of the $pp60^{c\text{-}src}$ molecules bound to middle T can be distinguished from those of c-*src* proteins from unifected cells. *In vivo*, pp60$^{c\text{-}src}$ molecules that are bound to MTAg are phosphorylated primarily on tyrosine residue 416, whereas the pp60$^{c\text{-}src}$ molecules that are not associated with MTAg are phosphorylated on tyrosine 527 (Cartwright *et al.* 1986). In immune complex protein kinase assays the MTAg-bound molecules of pp60$^{c\text{-}src}$ are phosphorylated on a novel tyrosine residue within the aminoterminal 16 000 kd of the protein, while the unbound c-*src* protein molecules are phosphorylated exlusively on tyrosine 416 (Yonemoto *et al.* 1985). The role of the phosphorylation at either tyrosine 416 or the aminoterminal tyrosine site is under investigation. It is difficult to distinguish events that are *responsible* for the activation, from those which are a *consequence* of activation.

These results suggest that transformation by Py virus might involve events which are mediated by activation of the tyrosyl kinase activity of the cellular *src* protein. Such a strategy represents a novel mechanism of activation of a cellular proto-oncogene product. The importance of this activation awaits the identification of the phosphorylated substrates that are responsible for the phenotypic changes which accompany Py transformation. We have recently found that many of the substrates that are phosphorylated in RSV-transformed cells can be detected in Py-transformed cells following treatment with sodium orthovanadate, an inhibitor of cellular phosphotyrosine phosphatases (Yonemoto *et al.*, unpublished results). It is likely that tyrosine phosphorylation of these protein is undetectable in untreated Py-transformed cells because the dosage of activated *src* in these cells is approximately 100-fold lower than that in RSV-transformed cells, since only about 10–15% of the fibroblastic *src* protein is activated in Py-transformed cells (and RSV-transformed cells contain 15- to 20-fold higher levels of pp60$^{v\text{-}src}$ than pp60$^{c\text{-}src}$).

EXPRESSION OF THE CELLULAR *src* GENE PRODUCT IN PLATELETS

Peripheral blood platelets contain approximately 10-fold higher levels of the c-*src* protein than those found in RSV-transformed cells or 150-fold more than fibroblasts (Golden *et al.* 1986). The specific activity of the platelet *src* protein appears to be similar to that of the viral *src* protein, and at least 20-fold higher than fibroblastic pp60$^{c\text{-}src}$. The platelet form of the *src* protein does not display the retarded electrophoretic mobility of the neuronal form of pp60$^{c\text{-}src}$, and the basis for activation is not understood. Interestingly, most stable lines of megakaryocytes (the precursors of platelets), also posses high levels of the c-*src* protein; however, the specific activity of the megakaryotic form of this protein is low, comparable to fibroblastic pp60$^{c\text{-}src}$ (Golden, Morgan and Brugge, unpublished results). These preliminary results suggest that c-*src* kinase activity might by activated during platelet maturation. Analysis of the phosphorylation profile of isolated platelet plasma membranes following *in vitro* kinase reactions revealed that tyrosine phosphorylation represented 80–90% of the total protein phosphorylation. At least major phosphotyrosine-containing proteins were detected. These results suggest that the $pp60^{c\text{-}src}$ might provide a function in events that take place in platelet membranes, possibly platelet activation.

CONCLUSIONS

The cellular *src* protein has been shown to be regulated at several different levels. Figure 6 shows the relative levels of *src* protein/mg of cell protein and the specific activity of the *src* protein in different cell types relative to fibroblasts which contain the lowest levels of $pp60^{c\text{-}src}$. The steady-state levels of $pp60^{c\text{-}src}$ were the highest in peripheral blood platelets (75–150 ×). Neurons, astrocytes, and megakayocyte cells also express high levels of the c-*src* protein (15–20 ×). In platelets, the specific activity of the c-*src* protein is similar to that of the viral *src* protein, yet the mechanism of activation is not known. Neurons contain 15–20 × levels of a structurally distinct form of the *src* protein that might be synthesized from a uniquely processed c-*src* mRNA molecule that encodes novel aminoterminal amino acid sequences. The specific activity of this protein is approximately 6- to 10-fold higher than fibroblastic $pp60^{c\text{-}src}$. Astrocytes and megakaryocyte cells also contain 15–20 × levels of $pp60^{c\text{-}src}$; however, the specific activity of the c-*src* protein is similar to that of fibroblastic $pp60^{c\text{-}src}$. Py-transformed cells contain 1 × levels of $pp60^{c\text{-}src}$, 10–15% of which is bound in a complex with the MTAg. The specific activity of the MTAg-bound form of $pp60^{c\text{-}src}$ is similar to that of $pp60^{v\text{-}src}$.

The identification of high levels of the c-*src* protein in post-mitotic, terminally differentiated cells, like neurons and platelets, was unexpected. Expression of the viral *src* protein in immature stem cells has been shown to stimulate cell proliferation and to interfere with the differentiation of these cells in culture. This behavior of the viral *src* protein led to the speculation that the cellular *src* gene product might be involved in the regulation of normal cell growth and proliferation. The evidence that neurons and platelets express high levels of functionally activated forms of the c-*src* protein forces one to consider other possible functions for these proteins. It is possible that the c-*src* protein is involved in triggering differentiation along specific pathways, and the high levels of $pp60^{c\text{-}src}$ in terminally differentiated cells represents irrelevant remnants. However, it is

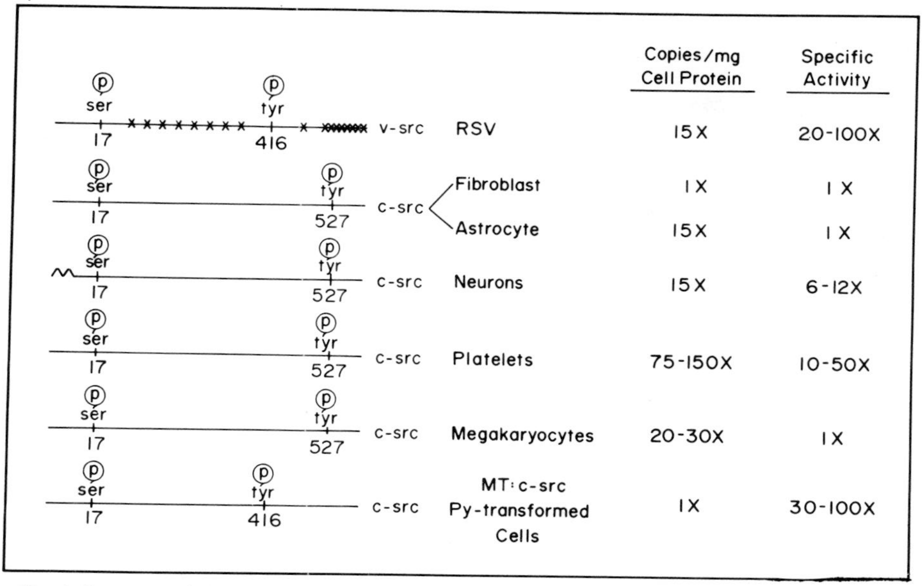

Fig. 6. Summary of the differences in the expression of the c-*src* protein phosphorylation sites, copy number, and specific activity. All values are given relative to those of fibroblastic pp60$^{c\text{-}src}$.

also possible that the c-*src* protein plays a specific role in mature, post-mitotic cells. Tyrosine kinases may be involved in a variety of different cellular events that take place within the plasma membrane. Elucidation of the cellular functions that are regulated by pp60$^{c\text{-}src}$ awaits demonstration of the cellular proteins which interact with pp60$^{c\text{-}src}$.

REFERENCES

Bishop, J. M. (1985) Viral Oncogenes. *Cell 42*, 23–8.

Brugge, J. S., Cotton, P. C., Queral, A. E., Barrett, J., Nonner, D. & Keane, R. (1985) Neurones express high levels of a structurally modified, activated form of pp60$^{c\text{-}src}$. *Nature 316*, 524–6.

Cartwright, C. A., Kaplan, P. L., Cooper, J. A., Hunter, T. & Eckhart, W. (1986) Altered sites of tyrosine phosphorylation in pp60$^{c\text{-}src}$ associated with polyomavirus middle tumor antigen. *Mol. Cell. Biol. 6*, 1562–70.

Cooper, J. A. & Hunter, T. (1985) Protein-tyrosine kinases. *Ann. Rev. Biochem. 54*, 897–930.

Courtneidge, S. & Smith, A. E. (1983) Polyomavirus transforming protein associates with the product of the c-*src* cellular gene. *Nature 303*, 435–9.

Fults, D. W., Towle, A. C., Lauder, J. M. & Maness, P. F. (1985) pp60$^{c\text{-}src}$ in the developing cerebellum. *Mol. Cell. Biol. 5*, 27–32.

Golden, A., Nemeth, S. P. & Brugge, J. S. (1986) Blood platelets express high levels of the pp60$^{c\text{-}src}$-specific tyrosine kinase activity. *Proc. Natl. Acad. Sci. USA 83*, 852–6.

Hanafusa, H. (1986) Activation of the c-*src* gene. *EMBO J.* (in press).

Levy, B. T., Sorge, L. K., Meymandi, A. & Maness, P. (1984) pp60$^{c\text{-}src}$ kinase is in chick and human embryonic tissues. *Developmental Biology 104*, 9–17.

Lipsich, L. A., Yonemoto, W., Bolen, J. B., Israel, M. A. & Brugge, J. S. (1984) Structural and functional studies of Rous sarcoma virus transforming protein, pp60$^{c\text{-}src}$. In: *Cancer Cell II: Oncogenes and Viral Genes*, eds., VandeWoude, G. F., Levine, A. J., Topp, W. C. & Watson, J. D., pp. 43–52. Cold Springer Harbor Laboratory Press, New York.

Schartl, M. & Barnekow, A. (1982) The expression in eukaryotes of a tyrosine kinase which is reactive with pp60$^{v\text{-}src}$ antibodies. *Differentiation 23*, 109–14.

Simon, M., Drees, B., Kornberg, T. & Biship, J. M. (1985) The nucleotide sequence and the tissue-specific expression of Drosophila c-*src*. *Cell 42*, 831–40.

Sorge, L. K., Levy, B. T. & Maness, P. F. (1984) pp60$^{c\text{-}src}$ is developmentally regulated in the neural retina cell. *Cell 36*, 249–56.

Yonemoto, W., Jarvis-Morar, M., Brugge, J. S., Bolen, J. & Israel, M. (1985) Novel tyrosine phosphorylation within the amino-terminal domain of pp60$^{c\text{-}src}$ molecules associated with polyoma virus middle-sized tumor antigen. *Proc. Natl. Acad. Sci. USA 82*, 4568–72.

DISCUSSION

YANIV: Did you try to do immunofluorescence on brain thin sections to see if there are certain neurons that have high levels and others not?

BRUGGE: We have not done that. Patricia Maness, University of North Carolina, has done immunohistochemical staining in the neuroretina from developing chicken embryos, and she found that pp60 c-*src* immunoreactivity was detected in the neuroretina at the point at which neuronal cells developed and that it was specifically found associated with the neuronal layer in the neuroretina. She has looked in the cerebellum and found specific immunoreactivity in the fibre region or molecular layer of the cerebellum, but there are also astrocytic fibres, in that region as well. Bob Keane, our collaborator from the University of Miami, has looked at cultured neurons to see if there is a specific localization of pp60 c-*src*, i.e., at the termini of the exons, but he could not find any specific localization, merely generalized staining throughout the membrane.

YANIV: In your case, you can take brain and show that there is a high concentration of *src* and that it has high kinase activity?

BRUGGE: Yes.

YANIV: So it is in non-dividing cells?

BRUGGE: Yes.

BALTIMORE: In the *in vivo* translation, it looked like you had two bands from the brain?

BRUGGE: In the *in vivo* labelling there were two bands. In the *in vitro* translation there is one. The brain consisting of mixed population of different types of cells, we usually see a mixture of both forms of the *src* protein in the brain; however, when we use cultured neurons, 90% of pp60 c-*src* migrates with a retarted mobility. In the *in vitro* translation of brain RNA we also observe only the slower moving form. It is not clear what percentage of the cells in the brain actually express high levels of *src* protein, and what the mass ratio of those cells compared to the other cells is, so it is difficult to know what the quantitation means.

FORCHHAMMER: You speculated on the function of c-*src* in the platelets. Since they are related to wound healing and we know there also is PGDF, have you tried to add *src* protein for instance to quiescent fibroblast to see whether this has any growth factor activity?

BRUGGE: Do you mean extracellularly?

FORCHHAMMER: Yes. It could easily be released, at least when the platelets break down.

BRUGGE: We do not see any secretion of pp60 c-*src* when platelets break down. The *src* protein seems to be firmly associated with the membrane in platelets, so I do not think it would be secreted when the platelets break down. I have never checked.

RIGBY: If we could go back to polyoma, did you say that you see the same spectrum of proteins phosphorylated in polyoma-transformed cells as in a *ras*-transformed cell.

BRUGE: It is similar. There are some differences. I can show you that.

RIGBY: But if my memory is right, there are some old experiments by Hunter, which show that total phosphotyrosine levels in polyoma-transformed cells are not elevated.

BRUGGE: The only way you can detect the increased level of tyrosine phosphorylation is to treat the cells with vanadate. In untreated cells we detected no increase in phosphotyrosine in py-transformed cells. In vanadate-treated py-transformed cells we see 16-fold higher levels of total cell phosphotyrosine, with only a two-fold increase in the normal control.

SCHLESSINGER: The model concerning tyrosine kinases, which are cytoplasmic and part of the transmembrane signal processes mediated by receptor, is an appealing one and David Baltimore also talked about a similar mechanism for the Ableson oncogene. They probably thought about using crosslinking agents and try to see whether you can fish out something which will interact with *src*.

BRUGGE: We wanted to avoid using crosslinking agents until it was clear that we could identify interactive proteins by solubilization with detergent.

SCHLESSINGER: That would probably not work and I will tell you why. We are used to very strong interactions when we think about chemistry in solutions. Very weak interactions in the plane of the membrane can be very relevant, because of the high local concentration in two dimensions in the membrane. So, you probably need to do the crosslinking experiments *in situ* and then solubilize, otherwise you would lose these low affinity interactions which are important when the protein is in two dimensions.

RAPP: Does treatment of cells with a mitogen that reacts with a tyrosine kinase-negative receptor actually modulate *src* kinase in any cell? Is *src* in T cells?

BRUGGE: The level of *src* in T cells seems to be very low, although there are high levels of another tyrosine kinase.

RAPP: Did anybody look what happens after IL2 treatment?

SCHLESSINGER: Yes, we thought about that hypothesis, and we tried to see whether a tyrosine kinase activity was stimulated by IL2 without success.

BALTIMORE: Did you do a vanadate treatment?

SCHLESSINGER: Yes, we did add vanadate.

BRUGGE: In the experiments with vanadate you have to be very selective in the type of cells that you choose for analysis, because there are a variety of normal cell lines that show elevated levels of phosphotyrosine upon treatment with vanadate. It is likely that these cells have elevated levels of an unidentified tyrosine kinase.

Structure and Function of Normal Cellular and Mutant Viral *erbA* Oncogenes

Klaus Damm, Hartmut Beug, Jan Sap, Thomas Graf & Björn Vennström

Tumorigenic conversion of normal cells is thought to involve several genetic and possibly also epigenetic changes induced by mutagens and tumor promoters. Recently, some of these processes have been reproduced *in vitro* by introducing two activated oncogenes into normal primary cells.

The avian erythroblastosis virus AEV-ES4 provides another example of oncogene cooperativity: both of its oncogenes, v-*erbA* and v-*erbB,* contribute to the transformed phenotype of infected erythroid cells. The v-*erbA* oncogene has no transforming activity on its own, while v-*erbB* is sufficient for transformation of both erythroblasts and fibroblasts *in vivo* and *in vitro* (Frykberg *et al.* 1983, Sealy *et al.* 1983). However, erythroblasts transformed by v-*erbB* alone differentiate spontaneously and require specific culture conditions for propagation *in vitro,* whereas erythroblasts expressing v-*erbA* in addition to v-*erbB* are completely arrested at an early stage of differentiation and grow in standard tissue culture media (Frykberg *et al.* 1983, Kahn *et al.* 1986).

The v-*erbA* part of the *gag-erbA* 75kd hybrid protein was recently shown to bear homology to receptors for steroid hormones (Weinberger *et al.* 1985, Green *et al.* 1986). However, no ligand has yet been assigned to this protein; in fact, it is not clear whether or not the *gag-erbA* protein needs a ligand for activity.

V-*erbB* encodes a trans-membrane glycoprotein with deregulated tyrosine kinase activity and the protein represents part of the receptor for epidermal growth factor (EGF): both the amino terminal ligand binding domain as well as some of the carboxyterminal autophosphorylation sites have been truncated (Ullrich *et al.* 1984).

We have previously described a transformation-defective mutant of AEV, de-

---

Differentiation Program European Molecular Biology Laboratory, Meyerhofstrasse 1, 6900 Heidelberg, Fed. Rep. of Germany.

VIRAL CARCINOGENESIS, Alfred Benzon Symposium 24,
Editors: N. O. Kjeldgaard, J. Forchhammer, Munksgaard, Copenhagen 1987.

TABLE I
Properties of td*359 and* r*12 AEVs*

Virus	Transformation[a)] of erythroblasts	fibroblasts	v-erbA[b)] protein (kd)	v-erbB[b)] protein (kd)
wt	+	+	75	gp66–68
*td*359	–	+	74	gp54–56
*r*12	+	+	75.5	gp54–56

a) determined both by *in vitro* transformation and after injection of virus into young chicks.
b) virus-transformed cells were labeled with ^{35}S-methionine, protein extracts were immunoprecipitated with *a-erbA* or *a-erbB* antisera, and analyzed by SDS-polyacrylamide gel electrophoresis.

noted *td*359, which has lost the capacity to cause leukemia and to transform erythroblasts, although it still transforms fibroblasts and induces sarcomas (Royer-Pokora *et al.* 1979, Beug *et al.* 1980). However, in rare instances *td*359-infected chicks develop erythroleukemia, and from one such animal we isolated a revertant virus (designated *r*12AEV) that had regained erythroblast transforming capacity. To localize and characterize the lesions in *td*359 and in *r*12, we cloned and sequenced both viral genomes and constructed v-*erbA* and v-*erbB* recombinants from *td*359, *r*12 and *wt* in all possible combinations. Tests of the erythroid transforming capacity revealed that both *erb* oncogenes are defective in *td*359-AEV, and that *r*12AEV recovered is erythroid transforming potential due to further changes in v-*erbA*. Furthermore, the v-*erbA*^{r12} gene is more efficient than the *wt erbA* gene in blocking spontaneous differentiation and in altering the specific growth requirements of v-*erbB*-transformed erythroblasts. This shows that the v-*erbA* gene can rescue the erythroid transforming capacity of a defective v-*erbB* gene.

Comparison of the nucleotide sequence of the v-*erbA* gene with that of c-*erbA* shows that v-*erbA* has suffered a 12 amino acid truncation in the amino terminus, 12 point mutations that have led to amino acid changes, and a short (11 amino acids) deletion close to the carboxy-terminus. The deduced amino acid sequence of the v-*erbA* polypeptide gives no clues to the function of this steroid hormone receptor-like protein.

RESULTS

Characterization of td*359 AEV and* r*12 AEV*

We first determined the transforming properties of the mutant viruses, and compared their v-*erb* proteins to determine if a correlation between oncogenicity

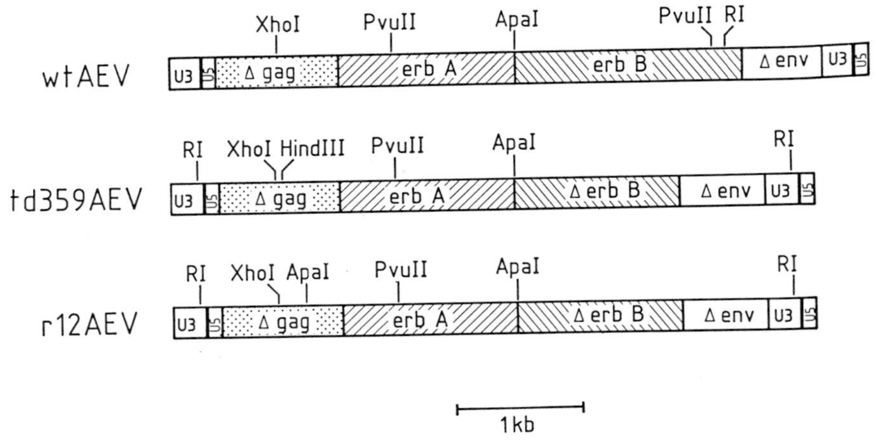

Fig. 1. Structure of the *wt, td*359 and *r*12AEV genomes. The *td*359 genome has a novel HindIII restriction site in its *gag* sequence, whereas *r*12 has an ApaI site in this region. In addition, the mutant viruses lack 2 cleavage sites in the 3' region of their v-*erbB* genes. The mutant viral proviruses were excised from chromosomal DNA of transformed cells and cloned in lambda phage vectors. Their structures were validated by restriction mapping, hybridization with helper-virus-derived probes and partial nucleotide sequence analysis.

and protein expression could be found. Table I shows that td359 AEV encoded a 74kd v-*erbA* and a gp56kd v-*erbB* protein, suggesting that *td*359 had undergone mutations in both of its oncogenes. Also *r*12AEV encoded a gp56kd v-*erbB* protein; its v-*erbA* protein, however, had a molecular weight of 75.5 kd, indicating that the recovery of full transforming capacity was due to back mutations in v-*erbA*.

Molecular cloning and nucleotide sequence analysis of td*359AEV and* r*12AEV*

To allow a determination of the nature of the mutations in *td*359AEV and the backmutation(s) in *r*12AEV, both viral genomes were molecularly cloned and characterized. Figure 1 demonstrates that the genome size and restriction map of both *td*359AEV and *r*12AEV differ from *wt*AEV (clone AEV-11, Vennström *et al.* 1980).

Nucleotide sequencing the v-*erbB* genes of the cloned proviral genomes showed that both *td*359 and *r*12AEV have a deletion of 306 nucleotides as compared to the same region in AEV-ES4 (Figure 2). A direct repeat of 11 nucleotides is present in the *wt erbB* sequences at the junctions of the deletion, suggesting that

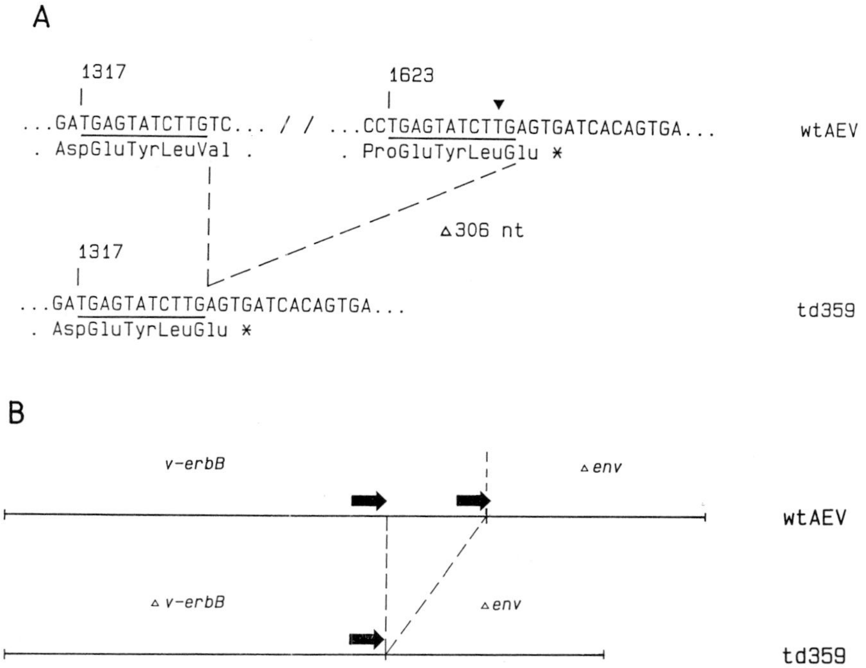

Fig. 2. Site of deletion in the v-*erbB* gene of *td*359 AEV. The nucleotide sequence of the 3' end of the *td*359 v-*erbB* gene was determined by the chain termination method. The 306nt deletion is flanked in *wt* v-*erbB* by an 11nt direct repeat, suggesting that the deletion occurred during reverse transcription of viral RNA. A: nucleotide sequence of the endpoint of the deletion. Nucleotide number 1 refers to the first nucleotide after the splice acceptor site for v-*erbB*. B: Schematic illustration of A.

the deletion may have been generated during minus-strand DNA synthesis by a reverse transcriptase "jumping" or "slipping" over the sequence flanked by the direct repeats, resulting in the deletion of one of these direct repeats and the v-*erbB* region in between.

Nucleotide sequence analysis of the *gag-erbA* region of the *td*359, *r*12 and *wt* AEVs revealed several differences between the mutant and *wt* sequences. Figure 3 shows that the *td*359 *gag-erbA* gene had 3 amino acid changes in the *gag* domain, and 2 in the v-*erbA* domain. The *gag* domain of *r*12AEV had reverted to a *wt* sequence in 2 of the 3 positions mutated in *td*359 (Figure 4), but had also undergone a further amino acid change (Gly 94) as well as an insertion of 31 amino acids of unknown origin, accompanied by a 4 amino acid deletion. Most notably, however, *r*12 AEV had reverted to a *wt* amino acid at position 144 in v-*erbA*.

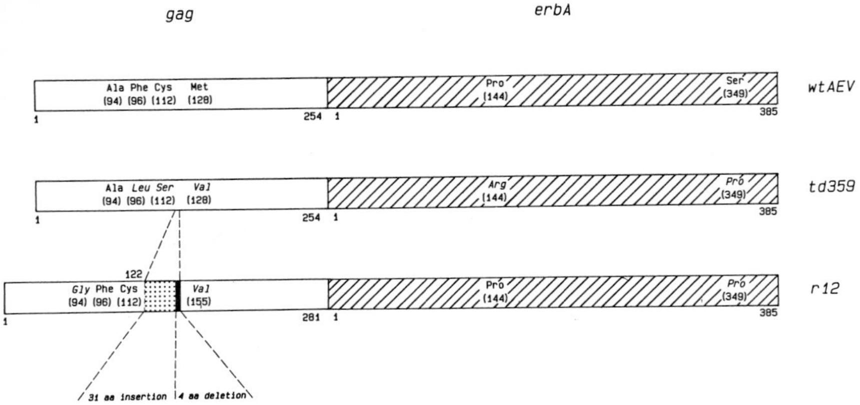

Fig. 3. Schematic illustration of mutations found in the *gag-erbA* proteins of *td*359 and *r*12. The nucleotide sequences between the XhoI and ApaI sites (see Figure 1) were determined by the chain termination method. In addition, several regions of the *wt gag-erbA* gene were resequenced, as the previously published sequence contained 2 sequencing errors close to the carboxy terminus (Debuire *et al.* 1983). Number 1 in *gag* refers to the first amino acid in RSV *gag* sequence of Schwartz *et al.* 1983.

Comparison of the v-erbA *with* c-erbA

To provide information on the relationship between v-*erbA* and c-*erbA*, and to understand the significance of the amino acid changes in the mutant v-*erbA* genes, we isolated cDNA clones for c-*erbA* and determined their nucleotide sequences. Figure 4 shows that the deduced c-*erbA* protein is 12 amino acids longer than v-*erbA* in the amino terminus, and that v-*erbA* has suffered an 11 amino acid deletion, accompanied by a 2 amino acid insertion, close to the carboxy terminus. In addition, 12 other nucleotide differences leading to amino acid changes were found. Interestingly, the proline at position 144 in *wt* v-*erbA* was conserved in c-*erbA,* whereas at position 349 c-*erbA* contained a proline, as in the mutant v-*erbA* genes. This suggests that the critical mutation in v-*erbA* of *td*359 is the Pro→Arg change at position 144, although at this point we cannot rule out that the mutations found in the *gag* domain of *td*359 also influence the activity of the *td*359 protein.

A comparison of the c-*erbA* nucleotide sequence to that of *gag* in Rous sarcoma virus (RSV) showed a 20/22 nucleotide homology at the point where *gag* is fused to v-*erbA* in AEV (Figure 4). This suggests that c-*erbA* was captured by an avian retrovirus by homologous recombination, either at the DNA level or during reverse transcription.

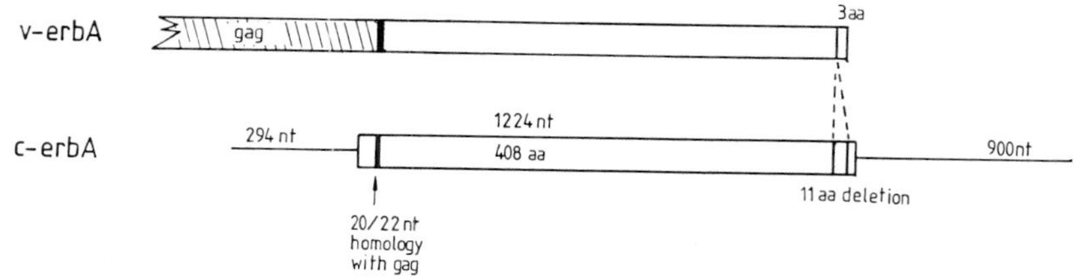

Fig. 4. Comparison of v-*erbA* to chicken c-*erbA*. c-*erbA* cDNA clones were isolated from a lambda phage library, and their nucleotide sequences determined. Thin line: noncoding sequence; boxed areas: coding sequence. The exact 5' and 3' ends of the c-*erbA* mRNA have not yet been defined.

Transforming capacities of the v-erbA *and* v-erbB *oncogenes of* td*359AEV* *and* r*12AEV*

To study the biological activity of the v-*erbA* and v-*erbB* genes of *td*359 and *r*12AEV, and to determine how the respective mutations in these oncogenes contribute to the observed mutant and revertant phenotypes, recombinant viruses were constructed containing the 9 possible combinations of the *wt, td*359 and *r*12 v-*erbA* and v-*erbB* oncogenes. A schematic illustration of the resulting recombinant genomes is shown in Figure 5. To test the biological activity of the constructed recombinant viral genomes, chicken embryo fibroblasts (CEF) were transfected with the construct DNAs together with cloned RAV-1 helper virus DNA, passaged two to three times, and then assayed for focus formation. All virus constructs transformed fibroblasts and gave rise to foci, which were isolated and expanded in liquid culture.

In a first set of experiments the erythroid transforming capacities of the virus constructs were determined by cocultivating the virus-producing, transformed fibroblast cultures with chick bone marrow cells in CFU-E medium containing serum from anemic chicks (as a source of erythropoietin). This medium promotes growth of erythroblasts transformed by v-*erbB* in the absence of v-*erbA* (Frykberg *et al.* 1983, Beug *et al.* 1985). Seven of the nine constructs were able to generate proliferating mass-cultures of transformed erythroblasts (Figure 5). The two constructs that failed to transform erythroblasts both contained the v-*erbA* gene from *td*359, suggesting that this gene, in combination with either of the two mutant v-*erbB* genes, is incapable of supporting transformation of erythroblasts.

Differentiation phenotypes of erythroblasts transformed by the recombinant viruses

To characterize the biological properties of the erythroblast – transforming viral constructs in more detail, transformed erythroblasts were analyzed for their

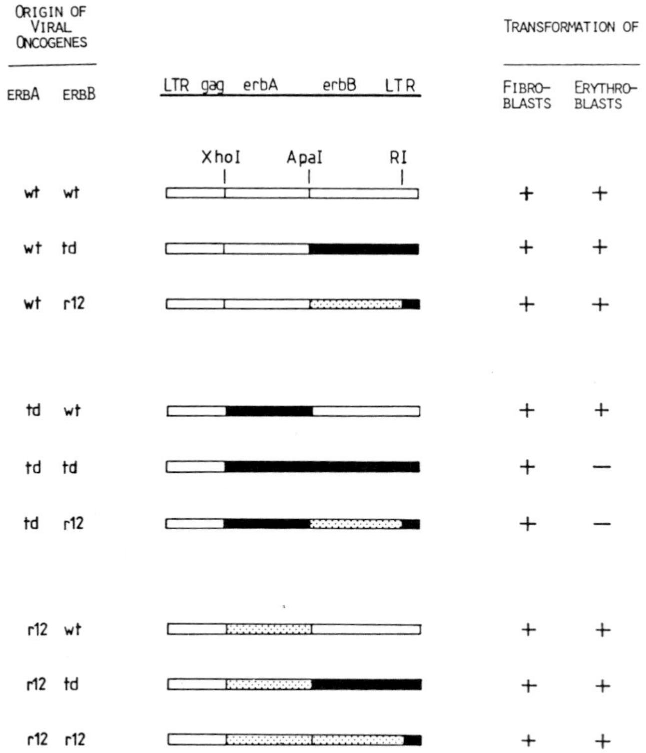

Fig. 5. Transforming capacities of recombinant viruses carrying mutant and/or *wt erb* oncogenes. For explanations, see text.

capacity to undergo spontaneous maturation to erythrocytes. For each virus, a number of individual transformed erythroblast clones were isolated from methocel cultures and tested separately for a set of erythroid differentiation markers including benzidine staining at acid and neutral pH and staining of the cells with antibodies for erythrocyte antigen.

The data (Figure 6) allow several conclusions to be drawn. Firstly, erythroblasts transformed by the $erbA^{td}erbB^{wt}$ virus resemble cells transformed by v-*erbB* alone, in that they exhibit a high proportion of spontaneously differentiating cells: in 11 of 24 clones analyzed, more than 20% of the cells were positive for hemoglobin. The similarity of $erbA^{td}erbB^{wt}$ erythroblasts with cells transformed by v-*erbB* alone was confirmed in a control experiment using several clones transformed by the

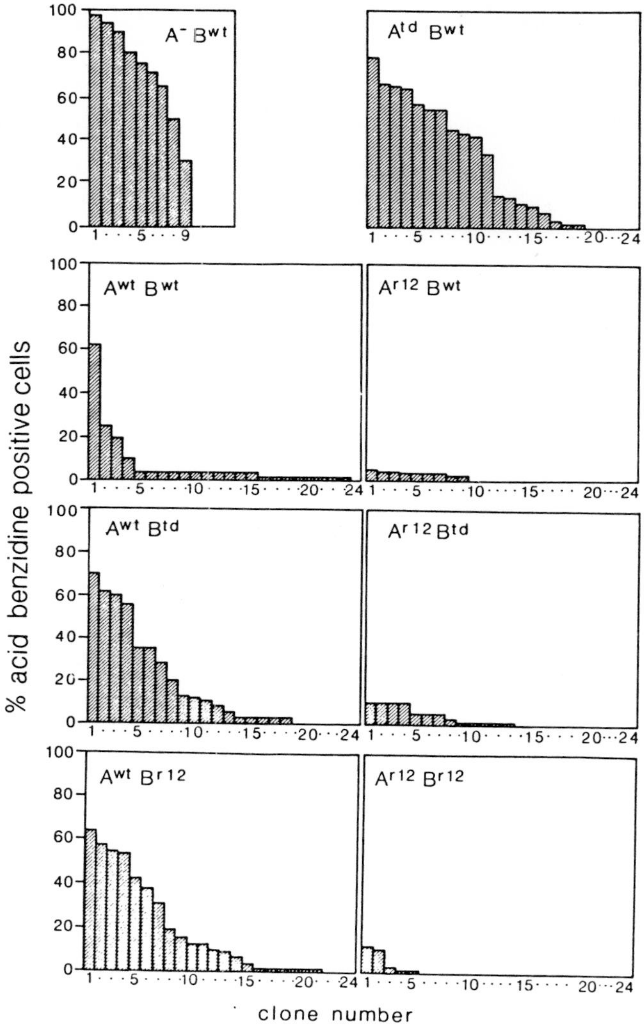

Fig. 6. Analysis of hemoglobin content in erythroblast clones transformed by recombinant viruses. Bone marrow cells were infected with virus, seeded into methocel, and 24 individual cell clones for each were stained with benzidine at acid pH to determine percentage of hemoglobinized cells in each cell clone.

v-*erbB* containing virus AEV-H. The results indicate that the v-*erbA*^{td} protein is biologically inactive or grossly impaired in its function.

Secondly, a similar, but lower level of spontaneous differentiation was found in clones transformed by viruses containing the v-*erbA*^{wt} gene in combinations with the *td*359 or *r*12 v-*erbB* genes. Seven of the 24 clones tested for each virus exhibited a high proportion of hemoglobin-positive cells. This indicates that v-*erbA*^{wt} partially restores the erythroblast-transforming ability of the defective v-*erbB* genes of *td*359 and *r*12AEV, suggesting that the *td*359 or *r*12 *erbB* genes are impaired but not inactive in their erythroblast-transforming ability.

Thirdly, cells transformed by the *wt* virus derived from the molecular clone AEV-11 were identical to those obtained with the standard *wt* AEV stock used previously: only a small proportion of the cells expressed markers of mature erythroid cells.

Finally, all erythroblast clones transformed by virus constructs containing the v-*erbA*^{r12} gene in combination with any of the v-*erbB* genes showed virtually no spontaneous differentiation, as only a few clones contained hemoglobin-positive cells. This suggests that v-*erbA*^{r12} has an even stronger capacity than v-*erbA*^{wt} in blocking spontaneous differentiation, and thus complements the transformation-defective v-*erbB* genes of *td*359 and *r*12AEV, resulting in a *wt* transformation phenotype.

The conclusions drawn from the experiments with acid benzidine staining were fully supported by analysis of the other differentiation markers mentioned above, as well as by the finding that v-*erbA* from *r*12 but not *td*359 supports the growth of v-*erbB*-transformed erythroblasts in standard tissue culture media (data not shown).

DISCUSSION

How does an active v-erbA *complement a defective* v-erbB *in transformation of erythroblasts?*

The molecular and biological analysis of *td*359AEV and its revertant *r*12AEV presented here demonstrates that both viruses have identical mutations in their v-*erbB* genes. An independent mutational event in v-*erbA* renders *td*359 AEV defective in erythroblast transformation, whereas a recovery of a functional v-*erbA* compensates the defective v-*erbB*^{r12} gene, thus restoring full leukemogenic potential. These results demonstrate that the cooperative action of 2 oncogenes, a functional v-*erbA* and a defective v-*erbB*, can lead to transformation of a cell type which is not transformed by either oncogene alone.

It has previously been shown that v-*erbB* alone transforms immature erythroblasts by inducing self-renewal, and that it is unable to induce this effect in more mature, "committed" cells (Beug *et al.* 1982, Beug & Hayman 1984). Assuming that mutant v-*erbB* genes, like v-*erbB*^{td}, are impaired but not entirely inactive in inducing self-renewal in erythroid precursors, a population of v-*erbB*^{td}-infected cells would transiently both proliferate and differentiate. If the rate of self-renewal in such a population is lower than the rate of commitment to differentiation, the culture would eventually consist entirely of mature cells. Such transiently proliferating cultures were indeed observed with v-*erbB*^{td}-infected cells (not shown). Furthermore, v-*erbA*^{wt} blocks the spontaneous differentiation exhibited by *wt* and *td* v-*erbB*-transformed erythroblasts. It is therefore conceivable that the low ability of v-*erbB*^{td} to stimulate self-renewal suffices to build up a population of transformed cells when expressed in concert with a differentiation-inhibiting v-*erbA* gene. By assuming that v-*erbA*^{r12} is more effective than v-*erbA*^{wt} in arresting differentiation, the differences in phenotypes observed between the *erbA*^{wt}/*erbB*^{td} and the *erbA*^{r12}/*erbB*^{td} transformed erythroblasts are readily explained.

What is the nature of mutations in td*359 and* r*12?*

Sequence determination of the genomes of *td*359 and *r*12 has identified several mutations in both viruses. The v-*erbB* genes of the mutant viruses have the same 306nt deletion in their 3' termini, explaining the identical biological properties of the two genes. The mutant *gag-erbA* genes differed from each other and from the *wt* gene. However, the mutation at position 349 close to the carboxy terminus in v-*erbA*^{td} has no effect on v-*erbA* function, as determined by exchanging restriction enzyme fragments between *wt* and *td*359 AEV (not shown). In contrast, several items of circumstantial evidence suggest that the Pro-Arg change at position 144 is the mutation that inactivates *gag-erbA* of *td*359; i) the 2 amino acids differ markedly in their properties; ii) c-*erbA* and v-*erbA*^{r12} have a proline in this position; and iii) glucocorticoid and estrogen receptors mutated in this region, located between the DNA-binding and ligand-binding regions, are defective (G. Ringold and P. Chambon, personal communication). However, we cannot exclude that the mutations in *gag* also have some influence on the activity of the *td*359 v-*erbA* protein.

In contrast, the high activity of the *r*12 *gag-erbA* protein appears to be due to the insertion of unknown sequences found in the *gag* domain. Construction of additional recombinants between *wt* and *td*359 *erbA* is in progress to provide unambiguous identification of those mutational events that caused the loss of *erbA* activity in *td*359 and its recovery and enhancement in *r*12AEV.

How does v-erbA *exert its function?*

Homologies in the amino acid sequence between the v-*erbA* protein and receptors for steroid hormones were recently described (Weinberger *et al.* 1985, Green *et al.* 1986). The homology was most pronounced in a region of the receptors assumed to bind DNA, suggesting that *erbA* could encode a protein with similar functions, and it is therefore possible that *erbA* represents an as yet unidentified hormone receptor or DNA-binding protein. As such, it could regulate the transcription of genes that are involved in the cellular differentiation process.

V-*erbA* is a truncated version of c-*erbA* and may represent an activated hormone-receptor complex or a constitutively active regulatory protein. The expression of v-*erbA* in erythroid cells may then activate the transcription of c-*erbA*-regulated genes that arrest erythroblast maturation. On the other hand, v-*erbA* may have lost its transcription-activating function while still binding to regulatory regions on the DNA, thereby inhibiting the expression of genes required for differentiation. If the 3 different *erbA* genes described in this paper have different binding affinities to regulatory regions, their distinct biological effects are readily explained.

REFERENCES

Beug, H. & Hayman, M. J. (1984) Temperature-sensitive mutants of avian erythroblastosis virus: surface expression of the *erbB* product correlates with transformation. *Cell 36,* 963–72.

Beug, H., Kahn, P., Doederlein, G., Hayman, M. J. & Graf, T. (1985) Characterization of hematopoietic cells transformed in vitro by AEV-H, a v-*erbB*-containing avian erythroblastosis virus. In: *Modern Trends in Human Leukemia VI,* eds. Neth, A., Gallo, R. C., Greaves, M. & Janka, T., pp. 290–7, Springer Verlag, Heidelberg.

Beug, H., Kitchener, G., Doederlein, G., Graf, T. & Hayman, M. J. (1980) Mutant of avian erythroblastosis virus defective for erythroblast transformation: deletion in the *erb* portion of p75 suggests function of the protein in leukemogenesis. *Proc. Natl. Acad. Sci. USA 77,* 6683–6.

Beug, H., Palmieri, S., Freudenstein, C., Zentgraf, H. & Graf, T. (1982) Hormone-dependent terminal differentiation *in vitro* of chicken erythroleukemia cells transformed by *ts* mutants of avian erythroleukemia virus. *Cell 28,* 907–19.

Debuire, B., Henry, C., Benaissa, M., Biserte, G., Claverie, J., Saule, S., Martin, P. & Stehelin, D. (1984) Sequencing the *erbA* oncogene of avian erythroblastosis virus reveals a new type of oncogene. *Science 224,* 1456–9.

Frykberg, L., Palmieri, S., Beug, H., Graf, T., Hayman, M. J. & Vennström, B. (1983) Transforming capacities of avian erythroblastosis virus mutants deleted in the *erbA* or *erbB* oncogenes. *Cell 32,* 227–38.

Green, S., Walter, P., Kumar, V., Krust, A., Bornert, J., Argos, P. & Chambon, P. (1986) Human oestrogen receptor cDNA: sequence, expression and homology to v-*erbA*. *Nature 320,* 134–9.

Kahn, P., Frykberg, L., Brady, C., Stanley, I. J., Beug, H., Vennström, B. & Graf, T. (1986) V-*erbA* cooperates with sarcoma oncogenes in leukaemic cell transformation. *Cell 45,* 357–64.

Royer-Pokora, B., Grieser, S., Beug, H. & Graf, T. (1979) Mutant avian erythroblastosis virus with restricted target cell specificity. *Nature 282,* 750–2.

Schwarz, D., Tizard, R. & Gilbert, W. (1983) Nucleotide sequence of Rous Sarcoma Virus. *Cell 32,* 853–69.

Sealy, L., Privalsky, M. L., Moscovici, G., Moscovici, C. & Bishop, J. M. (1983) Site-specific mutagenesis of avian erythroblastosis virus: *erbB* is required for oncogenicity. *Virology 130,* 155–78.

Ullrich, A., Loussens, L., Hayflick, T., Dull, T., Tam, A., Lee, J., Yarden, Y., Libermann, T., Schlessinger, J., Downward, J., Mayes, E., Whittle, N., Waterfield, M. & Seeberg, P. (1984) Human epidermal growth factor receptor cDNA sequence. *Nature 309,* 418–25.

Vennström, B., Fanshier, L., Moscovici, C. & Bishop, J. M. (1980) Molecular cloning of the avian erythroblastosis virus genome and recovery of oncogenic virus by transfection of chicken cells. *J. Virol. 36,* 575–85.

Weinberger, C., Hollenberg, S. M., Rosenfeld, M. G. & Evans, R. M. (1985) Domain structure of human glucocorticoid receptor and its relationship to the v-*erbA* oncogene product. *Nature 318,* 670–2.

DISCUSSION

RAPP: Where did the cDNAs come from, which cell types?

VENNSTRÖM: We made the RNA from total chicken embryos.

RAPP: And you do know there the distribution of c-*erb*A?

VENNSTRÖM: I have looked at c-*erb*A expression in eight different types of cell lines or tissues, and it is low in all of these. Five–ten copies of RNA per cell.

CORY: What happens if you deliberately reduce the steroid concentration in the growth medium?

VENNSTRÖM: We have not done that experiment.

SHERR: The differentiation in your systems is dependent on erythropoietin?

VENNSTRÖM: It is like this: If you have erythroblasts transformed by a ts mutant of v-*erb*B and incubate the cells at a non-permissive temperature, then the differentiation to erythrocytes is dependent on erythropoetin. However, the spontaneous differentiation we see in the absence of v-*erb*A expression is independent of erythropoietin, suggesting that the activity of v-*erb*B replaces the activity mediated by the erythropoetin receptor, but that is speculation.

SHERR: And similarly, self-renewal then would be equally independent?

VENNSTRÖM: Yes.

HANAHAN: Can you complement the v-*erb*A with the cDNA of the cellular gene of c-*erb*A?

VENNSTRÖM: We have not done that experiment yet. We will do it.

HANAHAN: Could you speculate about what amino acid changes in the v-*erb*A are doing, based on homology?

VENNSTRÖM: Not really. The viral *erb*A has N- and C-terminal deletions and a number of amino acid changes scattered throughout the gene, and it does not tell us very much.

SCHLESSINGER: It is probably belonging to the family of the steroid receptor or maybe DNA binding proteins. The homology in the binding regions is rather weak. If we take the EGF receptor and the insulin receptor it appears that there is a significant homology in the binding domain, but there is really no homology between insulin and EGF at all. So, maybe the ligand for the c-*erb*A may be very different from steroids.

VENNSTRÖM: Yes, I agree. It is quite possible that it is not a true steroid hormone at all, that is the ligand. But on the other hand, when you compare the estrogen and the glucocorticoid receptors, then you see that the ligand binding domains are similar in overall structure, but quite distinct in amino acid sequence. A hydrophobic pocket might be all that is required for hormone binding.

FORCHHAMMER: Yesterday, Dr. Bravo mentioned that he had manipulated some other pathways with TPA or ionophores. Have you tried if you could dispense for *erb*A by manipulating *erb*B plus some inhibitors or stimulators?

VENNSTRÖM: No, we have not done that.

Mechanism of Transformation by the *fms* Oncogene and the Relationship of its Product to the CSF-1 Receptor

Charles J. Sherr, Carl W. Rettenmier & Martine F. Roussel

The c-*fms* proto-oncogene encodes a cell surface glycoprotein which is related, and probably identical, to the receptor for the mononuclear phagocyte colony stimulating factor, CSF-1 (M-CSF) (Sherr *et al.* 1985). Recombination between feline leukemia virus (FeLV) and c-*fms* proto-oncogene sequences in cat cellular DNA led to the formation of the McDonough strain of feline sarcoma virus (SM-FeSV) (McDonough *et al.* 1971) which transforms fibroblasts in culture and induces fibrosarcomas in domestic cats (Sarma *et al.* 1972). Since the c-*fms* gene appears to function in regulating the growth and differentiation of normal macrophages and their progenitors, transduction of c-*fms* sequences in a retrovirus must have altered the receptor gene to confer new properties of fibroblast transformation. Studies of the structure and expression of the v-*fms* gene and its c-*fms* progenitor should therefore provide insights about critical alterations in receptor genes that subvert their normal functions and generate inappropriate signals for cell growth and malignant transformation.

1. RELATIONSHIP OF THE v-*fms* AND c-*fms* GENE PRODUCTS

a. *The v-*fms *oncogene*

The v-*fms* oncogene was transduced into the open reading frame of the FeLV *gag* gene so that SM-FeSV encodes a glycosylated, 180 kilodalton (kd) polyprotein (gP180*gag-fms*) consisting of 459 aminoterminal *gag*-coded amino acids fused to 975 amino acids encoded by v-*fms* (Barbacid *et al.* 1980, Ruscetti *et al.* 1980, Van de Ven *et al.* 1980). The polyprotein is synthesized on membrane-bound

---

Department of Tumor Cell Biology, St. Jude Children's Research Hospital, 332 North Lauderdale, Memphis, Tennessee 38105, U.S.A.

VIRAL CARCINOGENESIS, Alfred Benzon Symposium 24,
Editors: N. O. Kjeldgaard, J. Forchhammer, Munksgaard, Copenhagen 1987.

polyribosomes as an integral transmembrane glycoprotein and is cotranslationally glycosylated (Anderson *et al.* 1982, 1984, Rettenmier *et al.* 1985b). Proteolysis of the polyprotein near the *gag-fms* junction generates an aminoterminal 55 kd *gag*-coded polypeptide (p55*gag*) and a membrane-bound 120 kd glycoprotein (gp120v-*fms*) encoded exclusively by v-*fms* sequences. A comparison of the v-*fms* (Hampe *et al.* 1984) and human c-*fms* nucleotide sequences (Coussens *et al.* 1986, Wheeler *et al.* 1986) suggested that the site of recombination between FeLV and the cat c-*fms* gene occurred in a region of the proto-oncogene encoding the 5' untranslated region of c-*fms* mRNA. Thus, gp180*gag-fms* contains the complete aminoterminal domain of the c-*fms* gene product, and proteolytic cleavage within the polyprotein occurs at a site immediately downstream of the c-*fms*-coded "signal peptide" sequence (Wheeler *et al.* 1986).

The v-*fms*-coded glycoprotein, gp120v-*fms*, becomes oriented in the membrane of the endoplasmic reticulum (ER) with its glycosylated aminoterminal domain (ca. 450 amino acids) in the ER cisternae and its carboxylterminal domain (ca. 400 amino acids) in the cytoplasm. The latter domain has homology to prototypic members of the tyrosine kinase gene family (Hampe *et al.* 1984) and encodes the expected enzymatic activity (Barbacid & Lauver 1981). The structure of the v-*fms*-coded kinase domain is somewhat unusual in that amino acid sequences flanking the binding site for ATP are separated from the remaining region of tyrosine kinase homology in the distal carboxyl terminus by a 72 amino acid "spacer region". This unusual structure of the kinase domain has recently been predicted for two other gene products, including the receptor for the platelet-derived growth factor (PDGF) (Yarden *et al.* 1986) and another feline oncogene, v-*kit* (Besmer *et al.* 1986). This feature distinguishes this group of kinases from other known members of the tyrosine kinase gene family (including v-*src*, v-*abl*, v-*fes*/*fps*, v-*ros*, the receptors for epidermal growth factor (EGF) and insulin, etc.). Unlike transformants generated by many of the latter viral oncogene products, cells transformed by SM-FeSV do not show an overall increase in phosphotyrosine; moreover, in immune complex kinase reactions, the v-*fms* gene products are their own best substrates (Barbacid & Lauver 1981) and do not readily phosphorylate admixed polypeptides such as immunoglobulin, casein, or angiotensin (C. W. Rettenmier, unpublished data).

Transport of the v-*fms*-coded glycoprotein to the cell surface is required for transformation (Roussel *et al.* 1984, Nichols *et al.* 1985). During transport through the ER-Golgi complex, gp120v-*fms* undergoes modification of its asparagine (N)-linked oligosaccharide chains, so that the mature cell surface form of the glycoprotein is increased in apparent molecular mass (gp140v-*fms*). The glycoprotein becomes oriented at the cell surface with its glycosylated aminoterminal domain outside the cell and its kinase domain at the inner face of the plasma membrane (Rettenmier *et al.* 1985b). Once expressed at the cell surface, gp140v-*fms* segregates into clathrin-coated pits, and is sequestered into endosomes and degraded (Manger *et al.* 1984). Mutations within the polypeptide chain that inhibit Golgi transport (Roussel *et al.* 1984) or drugs that concordantly inhibit oligosaccharide modification and surface localization (Nichols *et al.* 1985) abrogate transformation even in the face of wild type levels of kinase activity. Thus, it appears that the relevant physiologic targets of v-*fms*-coded tyrosine kinase reside at the plasma membrane.

b. *The c-*fms *gene product*

The c-*fms* proto-oncogene is expressed at high levels in placenta (Muller *et al.* 1983a, b) and in mononuclear phagocytes (Sariban *et al.* 1985, Sherr *et al.* 1985, Woolford *et al.* 1985, Rettenmier *et al.* 1986). The c-*fms* gene product was first identified in feline splenocytes, and was shown to be a ca. 170 kd glycoprotein with an associated tyrosine-specific protein kinase activity (Rettenmier *et al.* 1985a). Using antisera to a v-*fms*-coded polypeptide expressed in bacteria, a closely related glycoprotein of 165 kd was precipitated from murine macrophage cell lines and was demonstrated to be the receptor for CSF-1 (Sherr *et al.* 1985). Moreover, gp140$^{v\text{-}fms}$ expressed at the cell surface of SM-FeSV transformed cells is able to specifically bind CSF-1 (Sacca *et al.* 1986, and see below). Taken together, these findings indicated that c-*fms* and the gene encoding the CSF-1 receptor are related and possibly identical genes. However, because the amino acid sequence of the purified CSF-1 receptor has not been determined, there is as yet no formal proof that the two genes are one and the same.

The structure of the human c-*fms* gene has been recently predicted from its cDNA sequence (Coussens *et al.* 1986). The amino acid sequences of the feline v-*fms* and human c-*fms* gene show an overall homology of 84%, with extracellular sequences displaying a lower level of homology (75%) than cytoplasmic sequences within the kinase domain (95%). Since the sequence of the feline c-*fms* gene is unknown, the amino acid differences between v-*fms* and the human c-*fms* gene product could reflect species differences between the cat and human genes as well as alterations in coding sequences that followed viral transduction. The two molecules differ significantly at their extreme carboxyl termini in a region distal to the tyrosine kinase domain. The v-*fms*-coded polypeptide is 29 amino acids shorter than the predicted c-*fms* gene product, and the carboxyl terminal 11 amino acids of the v-*fms* product show no homology to the 40 carboxyl terminal amino acids of the c-*fms*-coded polypeptide. In agreement with the predicted amino acid sequences in this region, monoclonal antibodies to epitopes in the extreme v-*fms*-coded carboxyl terminus do not react with c-*fms*-coded molecules (Furman *et al.* 1986). The unique region of the human c-*fms* carboxyl terminus includes a single tyrosyl residue which, by analogy to sites of preferred tyrosine phosphorylation in the EGF receptor (Downward *et al.* 1984a), may represent a regulatory phosphorylation site within the c-*fms* gene product (Coussens *et al.* 1986).

Although c-*fms* expression appears to be restricted to mononuclear phagocytes in adult tissues, expression of the gene in placenta connotes an alternative function of the gene outside the context of hematopoiesis. Placental expression is probably not due to the presence of macrophages, since choriocarcinoma cell lines derived from placental trophoblasts express c-*fms* as well (Muller *et al.* 1983b). Comparative biochemical studies of the c-*fms* gene product on human peripheral blood monocytes and choriocarcinoma cell lines showed that the polypeptides on both classes of cells were indistinguishable; moreover, choriocarcinoma cell lines were found to bind human urinary CSF-1 (Rettenmier *et al.* 1986). Thus, it seems that CSF-1 may have a unique function in placental development. Consistent with this prediction, the levels of uterine murine CSF-1 rise precipitously during gestation, and are highest at parturition (Bradley *et al.* 1971, E. R. Stanley and A. Bartocci, personal communication).

2. THE MONONUCLEAR PHAGOCYTE COLONY STIMULATING FACTOR, CSF-1

Murine CSF-1 is a 40–70 kd homodimer composed of two ca. 14 kd polypeptide chains covalently linked by disulfide bonds (Stanley & Heard 1977, Das & Stanley 1982). The variation in molecular weight of the mature secreted growth factor may reflect variable glycosylation or, alternatively, different modes of post-translational, proteolytic processing of the polypeptide chains. The human CSF-1 gene has been molecularly cloned (Kawasaki *et al.* 1985), and its sequence predicts a polypeptide of 26 kd. It seems likely that the molecule is proteolytically cleaved at its carboxyl terminus to yield a shortened polypeptide that forms dimers prior to secretion from the cell. CSF-1 is synthesized by fibroblasts including stromal cells in the bone marrow (Tushinski *et al.* 1982). Although the mRNA coding for secreted CSF-1 is only 1.6 kb in length, larger mRNAs up to

4.5 kb long have been detected, and it is possible that other biologically active forms of the molecule are synthesized (Kawasaki *et al.* 1985).

CSF-1 mediates its biological effects through its interaction with the CSF-1 receptor expressed on macrophages and their precursors (Guilbert & Stanley 1980, Byrne *et al.* 1981, Stanley & Guilbert 1981). The synthesis of the CSF-1 receptor is one of the earliest markers of commitment of hematopoietic progenitors to the mononuclear phagocyte lineage. In the absence of other colony stimulating factors, CSF-1 stimulates committed precursors of mature mononuclear phagocytes to proliferate and differentiate. CSF-1 also appears to act on earlier progenitors that respond to other hemopoietins. Expression of CSF-1 receptors increases as mononuclear phagocyte precursors differentiate and is highest on mature monocytes and macrophages (Bartelmez & Stanley 1985). The growth factor is necessary for the survival of post-mitotic macrophages and may "prime" the mature cells to carry out their normal differentiated functions (Tushinski *et al.* 1982).

3. ROLE OF THE v-*fms* AND c-*fms* GENES IN TRANSFORMATION

a. *Fibroblast transformation by v-*fms

SM-FeSV transforms immortalized fibroblast cell lines in culture and induces fibrosarcomas when inoculated into cats (McDonough *et al.* 1971, Sarma *et al.* 1972). Since the aminoterminal, ligand-binding domain of the CSF-1 receptor was retained in the recombination event that generated the v-*fms* oncogene (see above), cells transformed by SM-FeSV were tested for their ability to bind purified murine CSF-1 (Sacca *et al.* 1986). All SM-FeSV-transformed cell lines tested, regardless of their species of origin, expressed specific binding sites for CSF-1 at their cell surface, and chemical crosslinking experiments established that the growth factor bound to gp140v-*fms* molecules on the plasma membrane. Moreover, a monoclonal antibody to a v-*fms*-coded epitope was able to interfere with the binding of CSF-1 to gp140v-*fms* on the surface of SM-FeSV-transformed cells. In control experiments, binding was not observed using cells transformed by feline sarcoma viruses containing the v-*fes* oncogene. Although murine CSF-1 binds to murine macrophages with high affinity, the binding of the murine growth factor to $gp140^{v\text{-}fms}$ was of considerably lower affinity. This appears to be due to the interspecies interaction of murine CSF-1 with the feline receptor, since the binding of murine CSF-1 to normal feline peritoneal exudate macrophages was of similar affinity to that observed with SM-FeSV-transformed cells. Thus, unlike the v-*erb*B oncogene, which encodes a truncated form of the EGF receptor lacking a ligand-binding domain (Downward *et al.* 1984b, Ullrich *et al.* 1984), the v-*fms* gene encodes a receptor-like molecule that retains its ability to specifically bind CSF-1.

The fact that fibroblasts produce CSF-1 raises the possibility that transformation by this virus involves the transduction of a receptor-like molecule into cells that produce the growth factor. Indeed, established cell lines, such as mouse NIH-3T3 or rat NRK that are susceptible to transformation by SM-FeSV, were found to produce CSF-1. However, SM-FeSV transformants do not depend upon an exogenous source of CSF-1 for growth. Moreover, neutralizing antibodies to CSF-1 or monoclonal antibodies to gp140v-*fms* that interfere with binding to the growth factor have no demonstrable effect on the transformed phenotype or on the ability of the transformed cells to form colonies in semi-solid medium (Sacca *et al.* 1986). Although these experiments did not formally exclude the possibility of an intracellular interaction between CSF-1 and the v-*fms*-coded receptor in transformed cells, the experiments suggest that transformation by SM-FeSV is probably not mediated by a simple autocrine mechanism.

In macrophage membrane preparations incubated with ^{32}P-labeled ATP *in vitro*, phosphorylation of the murine CSF-1 receptor on tyrosine is enhanced in the presence of the growth factor (Sherr *et al.* 1985). In parallel experiments performed with membranes from SM-FeSV-transformed fibroblasts, *in vitro* phosphorylation of the v-*fms*-coded glycoproteins was constitutive and was unaffected by the presence of high concentrations of murine CSF-1 (Sacca *et al.* 1986). The simplest interpretation is that the v-*fms* gene product is active as a tyrosine kinase in the absence of exogenous CSF-1 and acts as a promiscuous enzyme. Possibly, truncation of the carboxyl terminus of the v-*fms*-coded polypeptide removes a negative regulatory site of tyrosine phosphorylation (Coussens *et al.* 1986) as has been demonstrated for pp60c-*src* (Courtneidge 1985, Iba *et al.* 1985, Cooper *et al.* 1986).

b. *Transformation of hematopoietic cells*

In spite of the relationship of the v-*fms* gene product to the CSF-1 receptor, SM-FeSV has not been shown to transform hematopoietic cells or to produce hematopoietic malignancies in animals. It should be possible to infect CSF-1-dependent macrophage cell lines with SM-FeSV and test whether insertion of a v-*fms* gene renders the cells independent of the growth factor or tumorigenic in animals. If v-*fms* encodes an unregulated tyrosine kinase, it might be possible to demonstrate such effects in macrophages or, possibly, in other hematopoietic cell lines dependent on other CSFs. As a precedent, the v-*abl* oncogene has been shown to convert interleukin-3-dependent myeloid cells to factor independence (Oliff *et al.* 1985, Pierce *et al.* 1985, Cook *et al.* 1985). Analogous effects on avian erythroid cells were previously observed using other oncogenes of the tyrosine kinase gene family (Adkins *et al.* 1984, Kahn *et al.* 1984).

If v-*fms* were found to alter the growth of hematopoietic target cells, the results would suggest that critical rearrangements of the c-*fms* gene affecting its tyrosine kinase domain might, in turn, contribute to leukemia. To date, no c-*fms* rearrangements have been documented in any malignancy. However, the c-*fms* gene has been mapped to human chromosome 5q33.2–33.3 (Groffen *et al.* 1983, Le Beau *et al.* 1986b), a frequent site of interstitial deletion in a variety of hematopoietic disorders, including refractory anemia, myelodysplasias, and therapy-related acute myelogenous leukemias (AML) (reviewed in Wisniewski & Hirschhorn 1983, Le Beau *et al.* 1986a). Indeed, patients with the 5q- refractory anemia syndrome are frequently hemizygous at the c-*fms* locus (Nienhuis *et al.* 1985). Recently, the CSF-1 gene was also assigned to chromosome 5 at band q33.1 and was found to be rearranged in a single case of AML (Le Beau et al. 1986a). Moreover, the gene encoding the granulocyte-macrophage colony stimulating factor (GM-CSF) is linked to c-*fms* and CSF-1 at bands 5q23–31 (Huebner *et al.* 1985, Le Beau *et al.* 1986b). These results suggest that a family of genes regulating hematopoiesis are closely linked on human chromosome 5 and raise the possibility that interstitial deletions affecting chromosome 5q might affect one or more of these loci.

4. RELATIONSHIP OF THE CSF-1 AND PDGF RECEPTORS

Yarden and coworkers (1986) recently found that the PDGF receptor is structurally related to the c-*fms* gene product and that the PDGF receptor gene maps adjacent to c-*fms* on human chromosome 5q31–32. These observations suggest that the CSF-1 and PDGF receptors evolved from the same ancestral gene. It is intriguing that PDGF-like molecules can be produced by macrophages and activated monocytes (Shimokado *et al.* 1985, Martinet *et al.* 1986) and interact with receptors on mesenchymal cells, whereas CSF-1 is produced by fibroblasts and binds to receptors on mononuclear phagocytes. Both growth factors are disulfide-linked dimers that appear to undergo proteolytic processing prior to secretion. In a physiologic sense, the two growth factor receptor systems appear to have diverged in such a way as to provide mirror image signals between mesenchymal and hematopoietic cells. In accord with this hypothesis, the two receptors may each act through second messengers generated by phospholipase C hydrolysis of phosphatidylinositol-4,5-diphosphate (Habenicht *et al.* 1981, Jackowski *et al.* 1986). It seems evident that further studies of this gene family will yield additional information about such diverse processes as hematopoiesis, wound healing, inflammation, and aberrant receptor signals which contribute to neoplasia.

REFERENCES

Adkins, B., Leutz, A. & Graf, T. (1984) Autocrine growth induced by *src*-related oncogenes in transformed chicken myeloid cells. *Cell 39,* 439–45.

Anderson, S. J., Furth, M., Wolff, L., Ruscetti, S. K. & Sherr, C. J. (1982) Monoclonal antibodies to the transformation-specific glycoprotein encoded by the feline retroviral oncogene v-*fms*. *J. Virol. 44,* 696–702.

Anderson, S. J., Gonda, M. A., Rettenmier, C. W. & Sherr, C. J. (1984) Subcellular localization of glycoproteins encoded by the viral oncogene v-*fms*. *J. Virol. 51,* 730–41.

Barbacid, M. & Lauver, A. V. (1981) Gene products of McDonough feline sarcoma virus have an *in vitro*-associated protein kinase that phosphorylates tyrosine residues: Lack of detection of this enzymatic activity *in vivo*. *J. Virol. 40,* 812–21.

Barbacid, M., Lauver, A. V. & Devare, S. G. (1980) Biochemical and immunological characterization of polyproteins coded for by the McDonough, Gardner-Arnstein, and Snyder-Theilen strains of feline sarcoma virus. *J. Virol. 33,* 196–207.

Bartelmez, S. H. & Stanley, E. R. (1985) Synergism between hemopoietic growth factors (HGFs) detected by their effects on cells bearing receptors for a lineage specific HGF: Assay of hemopoietin-1. *J. Cell Physiol. 122,* 370–8.

Besmer, P., Murphy, J. E., George, P. C., Qiu, F., Bergold, P. J., Lederman, L., Snyder, H. W., Jr., Brodeur, D., Zuckerman, E. E. & Hardy, W. D. (1986) A new acute transforming feline retrovirus and relationship of its oncogene v-*kit* with the protein kinase gene family. *Nature 320,* 415–21.

Bradley, T. R., Stanley, E. R. & Summer, M. A. (1971) Factors from mouse tissues stimulating colony growth of mouse bone marrow cells *in vitro*. *Aust. J. Exp. Biol. Sci. 49,* 595–603.

Byrne, P. V., Guilbert, L. J. & Stanley, E. R. (1981) Distribution of cells bearing receptors for a colony-stimulating factor (CSF-1) in murine tissues. *J. Cell Biol. 91,* 848–53.

Cook, W. D., Metcalf, D., Nicola, N. A., Burgess, A. W. & Walker, F. (1985) Malignant transformation of a growth factor-dependent myeloid cell line by Abelson virus without evidence of an autocrine mechanism. *Cell 41,* 677–83.

Cooper, J. A., Gould, K. L., Cartwright, C. A. & Hunter, T. (1986) Tyr^{527} is phosphorylated in $pp60^{c\text{-}src}$: Implications for regulation. *Science 231,* 1431–3.

Courtneidge, S. A. (1985) Activation of the $pp60^{c\text{-}src}$ kinase by middle T binding or by dephosphorylation. *EMBO J 4,* 1471–7.

Coussens, L., Van Beveren, C., Smith, D., Chen, E., Mitchell, R. L., Isacke, C. M., Verma, I. M. & Ullrich, A. (1986) Structural alteration of viral homologue of receptor proto-oncogene *fms* at carboxyl terminus. *Nature 320,* 277–81.

Das, S. K. & Stanley, E. R. (1982) Structure-function studies of a colony stimulating factor (CSF-1). *J. Biol. Chem. 257,* 13679–84.

Downward, J., Parker, P. & Waterfield, M. D. (1984a) Autophosphorylation sites on the epidermal growth factor receptor. *Nature 311,* 483–5.

Downward, J., Yarden, Y., Mayes, E., Scrace, G., Totty, N., Stockwell, P., Ullrich, A., Schlessinger, J. &

Waterfield, M. D. (1984b) Close similarity of epidermal growth factor receptor and v-*erb*B oncogene protein sequences. *Nature (London) 307,* 521–7.

Furman, W. L., Rettenmier, C. W., Chen, J. H., Roussel, M. F., Quinn, C. O. & Sherr, C. J. (1986) Antibodies to distal carboxylterminal epitopes in the v-*fms*-coded glycoprotein do not cross-react with the c-*fms* gene product. *Virology 152,* 432–45.

Groffen, J., Heisterkamp, N., Spurr, N., Dana, S., Wasmuth, J. J. & Stephenson, J. R. (1983) Chromosomal localization of the human c-*fms* oncogene. *Nucleic Acid Res. 11,* 6331–9.

Guilbert, L. J. & Stanley, E. R. (1980) Specific interaction of murine colony-stimulating factor with mononuclear phagocytic cells. *J. Cell. Biol. 85,* 153–9.

Habenicht, A. J. R., Glomset, J. A., King, W. C., Nist, C., Mitchell, C. D. & Ross, R. (1981) Early changes in phosphatidylinositol and arachidonic acid metabolism in quiescent Swiss 3T3 cells stimulated to divide by platelet-derived growth factor. *J. Biol. Chem. 256,* 12329–35.

Hampe, A., Gobet, M., Sherr, C. J. & Galibert, F. (1984) The nucleotide sequence of the feline retroviral oncogene v-*fms* shows unexpected homology with oncogenes encoding tyrosine-specific protein kinases. *Proc. Natl. Acad. Sci. USA 81,* 85–9.

Huebner, K., Isobe, M., Croce, C. M., Golde, D. W., Kaufman, S. E. & Gasson, J. C. (1985) The human gene encoding GM-CSF is at 5q21-q32, the chromosome region deleted in the $5q^-$ anomaly. *Science 230,* 1282–5.

Iba, H., Cross, F., Garber, E. A. & Hanafusa, H. (1985) Low level of cellular protein phosphorylation by nontransforming overproduced $p60^{c\text{-}src}$. *Mol. Cell. Biol. 5,* 1058–66.

Jackowski, S., Rettenmier, C. W., Sherr, C. J. & Rock, C. O. (1986) A guanine nucleotide-dependent phosphatidylinositol-4,5-diphosphate-specific phospholipase C in cells transformed by the v-*fms* and v-*fes* oncogenes. *J. Biol. Chem. 261,* 4978–85.

Kahn, P., Adkins, B., Beug, H. & Graf, T. (1984) *src-* and *fps*-containing avian sarcoma viruses transform chicken erythroid cells. *Proc. Natl. Acad. Sci. USA 81,* 7122–6.

Kawasaki, E. S., Ladner, M. B., Wang, A. M., Van Arsdell, J., Warren, M. K., Coyne, M. Y., Schweickart, V. L., Lee, M. T., Wilson, K. J., Boosman, A., Stanley, E. R., Ralph, P. & Mark, D. F. (1985) Molecular cloning of a complementary DNA encoding human macrophage-specific colony stimulating factor (CSF-1). *Science 230,* 291–6.

Le Beau, M. M., Pattenati, M. J., Lemons, R. S., Diaz, M. O., Westbrook, C. A., Larson, R. A., Sherr, C. J. & Rowley, J. D. (1986a) Assignment of the *GM-CSF, CSF-1* and *FMS* genes to human chromosome 5 provides evidence for linkage of a family of genes regulating hematopoiesis and for their involvement in the deletion (5q) in myeloid disorders. *Cold Spring Harbor Symp. Quant. Biol. 51* (in press).

Le Beau, M. M., Westbrook, C. A., Diaz, M. O., Larson, R. A., Rowley, J. D., Gasson, J. C., Golde, D. W. & Sherr, C. J. (1986b) Evidence for the involvement of GM-CSF and c-*fms* in the deletion (5q) in myeloid disorders. *Science 231,* 984–7.

Manger, R., Najita, L., Nichols, E. J., Hakomori, S.-I. & Rohrschneider, L. (1984) Cell surface expression of the McDonough strain of feline sarcoma virus *fms* gene product (gp140*fms*). *Cell 39,* 327–37.

Martinet, Y., Bitterman, P. B., Mornex, J.-F., Grotendorst, G. R., Martin, G. R. & Crystal, R. G. (1986) Activated human monocytes express the c-*sis* protooncogene and release a mediator showing PDGF-like activity. *Nature 319,* 158–60.

McDonough, S. K., Larsen, S., Brodey, R. S., Stock, N. D. & Hardy, Jr., W. D. (1971) A transmissible feline fibrosarcoma of viral origin. *Cancer Res. 31,* 953–6.

Müller, R., Slamon, D. J., Adamson, E. D., Tremblay, J. M., Muller, D., Cline, M. J. & Verma, I. M. (1983a) Transcription of c-*onc* genes c-ras^{ki} and c-*fms* during mouse development. *Mol. Cell. Biol. 3,* 1062–9.

Müller, R., Tremblay, J. M., Adamson, E. D. & Verma, I. M. (1983b) Tissue and cell type specific expression of two human c-*onc* genes. *Nature 304,* 454–6.

Nichols, E. J., Manger, R., Hakomori, S., Herscovics, A. & Rohrschneider, L. R. (1985) Transformation by the v-*fms* oncogene product: role of glycosylational processing and cell surface expression. *Mol. Cell Biol. 5,* 3467–75.

Nienhuis, A. W., Bunn, H. F., Turner, P. H., Gopal, T. V., Nash, W. G., O'Brien, S. J. & Sherr, C. J. (1985) Expression of the human c-*fms* proto-oncogene in hematopoietic cells and its deletion in the $5q^-$ syndrome. *Cell 42,* 421–8.

Oliff, A., Agranovsky, O., McKinney, M. D., Murty, V. V. V. S. & Banchwitz, R. (1985) Friend murine leukemia virus-immortalized myeloid cells are converted into tumorigenic cells lines by Abelson leukemia virus. *Proc. Natl. Acad. Sci. USA 82,* 3306–10.

Pierce, J. H., Di Fiore, P. P., Aaronson, S. A., Potter, M., Pumphrey, J., Scott, A. & Ihle, J. N. (1985) Neoplastic transformation of mast cells by Abelson MuLV: Abrogation of IL-3 dependence by a nonautocrine mechanism. *Cell 41,* 685–93.

Rettenmier, C. W., Chen, J. H., Roussel, M. F. & Sherr, C. J. (1985a) The product of the c-*fms* proto-oncogene: a glycoprotein with associated tyrosine kinase activity. *Science 228,* 320–2.

Rettenmier, C. W., Roussel, M. F., Quinn, C. O., Kitchingman, G. R., Look, A. T. & Sherr, C. J. (1985b) Transmembrane orientation of glycoproteins encoded by the v-*fms* oncogene. *Cell 40,* 971–81.

Rettenmier, C. W., Sacca, R., Furman, W. L., Roussel, M. F., Holt, J. T., Nienhuis, A. W., Stanley, E. R. & Sherr, C. J. (1986) Expression of the human c-*fms* proto-oncogene product (CSF-1 receptor) on peripheral blood mononuclear cells and choriocarcinoma cells lines. *J. Clin. Invest. 77,* 1740–6.

Roussel, M. F., Rettenmier, C. W., Look, A. T. & Sherr, C. J. (1984) Cell surface expression of v-*fms*-coded glycoproteins is required for transformation. *Mol. Cell. Biol. 4,* 1999–2009.

Ruscetti, S. K., Turek, L. P. & Sherr, C. J. (1980) Three independent isolates of feline sarcoma virus code for three distinct *gag*-X polyproteins. *J. Virol. 35,* 259–64.

Sacca, R., Stanley, E. R., Sherr, C. J. & Rettenmier, C. W. (1986) Specific binding of the mononuclear phagocyte colony stimulating factor, CSF-1, to the product of the v-*fms* oncogene. *Proc. Natl. Acad. Sci. USA 83,* 3331–5.

Sariban, E., Mitchell, T. & Kufe, D. (1985) Expression of the c-*fms* proto-oncogene during human monocytic differentiation. *Nature 316,* 64–6.

Sarma, P. S., Sharar, A., & McDonough, S. (1972) The SM strain of feline sarcoma virus. Biologic and antigenic characterization of virus. *Proc. Soc. Exp. Biol. Med. 140,* 1365–8.

Sherr, C. J., Rettenmier, C. W., Sacca, R., Roussel, M. F., Look, A. T. & Stanley, E. R. (1985) The c-*fms* proto-oncogene product is related to the receptor for the mononuclear phagocyte growth factor, CSF-1. *Cell 41,* 665–76.

Shimokado, K., Raines, E. W., Madtes, D. K., Barrett, T. B., Benditt, E. P. & Ross, R. (1985) A significant part of macrophage-derived growth factor consists of at least two forms of PDGF. *Cell 43*, 277–86.

Stanley, E. R. & Guilbert, L. J. (1981) Methods for the purification, assay, characterization and target cell binding of a colony stimulating factor (CSF-1). *J. Immunol. Methods 42,* 253–84.

Stanley, E. R. & Heard, P. M. (1977) Factors regulating macrophage production and growth. Purification and some properties of the colony stimulating factor from medium conditioned by mouse L cells. *J. Biol. Chem. 252,* 4305–12.

Tushinski, R. J., Oliver, I. T., Guilbert, L. J., Tynan, P. W., Warner, J. R. & Stanley, E. R. (1982) Survival of mononuclear phagocytes depends on a lineage-specific growth factor that the differentiated cells selectively destroy. *Cell 28,* 71–81.

Ullrich, A., Coussens, L., Hayflick, J. S., Dull, T. J., Gray, A., Tam, A. W., Lee, J., Yarden, Y., Libermann, T. A., Schleissinger, J., Downward, J., Mayes, E. L. V., Whittle, N., Waterfield, M. D. &

Seeburg, P. H. (1984) Human epidermal growth factor receptor cDNA sequence and aberrant expression of the amplified gene in A431 epidermoid carcinoma cells. *Nature 309,* 418–25.

Van de Ven, W. J. M., Reynolds, F. H., Jr., Nalewaik, R. P. & Stephenson, J. R. (1980) Characterization of a 170,000-dalton polyprotein encoded by the McDonough strain of feline sarcoma virus. *J. Virol. 35,* 165–75.

Wheeler, E. F., Roussel, M. F., Hampe, A., Walker, M. H., Fried, V. A., Look, A. T., Rettenmier, C. W. & Sherr, C. J. (1986) The aminoterminal domain of the v-*fms* oncogene product includes a functional signal peptide that directs synthesis of a transforming glycoprotein in the absence of FeLV *gag* sequences. *J. Virol. 59,* 224–33.

Wisniewski, L. P. & Hirschhorn, K. (1983) Acquired partial deletions of the long arm of chromosome 5 in hematologic disorders. *Am. J. Hematol. 15,* 295–310.

Woolford, J., Rothwell, V. & Rohrschneider, L. (1985) Characterization of the human c-*fms* gene product and its expression in cells of the monocyte-macrophage lineage. *Mol. Cell Biol. 5,* 3458–66.

Yarden, Y., Escobedo, J. A., Kuang, W. J., Yang-Feng, T. L., Daniel, T. O., Tremble, P. M., Chen, E. Y., Ando, M. E., Harkins, R. N., Francke, U., Fried, V. A., Ullrich, A. & Williams, L. T. (1986) The receptor for PDGF: Structural homologies define a family of closely related growth factor receptors. *Nature 323,* 226–32.

DISCUSSION

SCHLESSINGER: Would TPA modulate the kinase activity of *fms* in a similar way to its effect on the EGF receptor?

SHERR: Yes, TPA has been shown to down-regulate the CSF-1 receptor, and when I say down-regulate, I mean to reduce drastically and immediately the number of binding sites for CSF-1 on the macrophage cell surface.

SCHLESSINGER: Do you have an equivalent site as *thr* 654 of EGF-receptor?

SHERR: No. So, the phenomena are analogous, but the presumed target for protein kinase C would have to be different.

BALTIMORE: You probably screened quite a number of different ligands and membranes and whatever for activation for the phospholipase C. Is there some kind of generality?

SHERR: We have tested several receptor systems, and the difficult ones we have ignored, but *fms* is the home town oncogene so we are concentrating on that. Charles Rock and Suzanne Jackowski are independently working very hard on the enzyme in other systems, and I can describe some of their results. They have worked on platelets, for example, and have tried to stimulate the enzyme with alpha thrombin which is supposed to give a good PI response. That works very well. You get an immediate alpha thrombin response in the absence of calcium using permeabilized platelet membranes. It is instantaneous. The enzyme activity comes on as quickly as you can measure it.

BALTIMORE: Is that GTP-dependent?

SHERR: That is an interesting question. If you add a lot of alpha thrombin, the answer is no. But you can also activate the enzyme by GTP alone. At suboptimal concentrations, alpha thrombin and GTP seem additive in that system, and we are not sure why. Moreover, there is a component of the enzyme activity which appears to be calcium-sensitive. In the presence of added calcium you see an overshoot in the maximum activity. That suggests that there is a calcium-activated phospholipase that is participating as a component of this reaction. The simplest idea is that calcium generated by IP3 activates the cytosolic phospholipase, and you begin to assay that as a contaminant of the membrane preparation, but that certainly is not a clear result. Moreover, I emphasize that Phil Majerus and our own group have been discussing for a year whether we are looking at one or two enzymes. He might argue that the cytosolic enzyme becomes associated with the membrane in the presence of GTP and becomes hormone-responsive.

BALTIMORE: It is homologous to a protein kinase C argument.

SHERR: Yes, and there is no way around that. I cannot exclude that the cytosolic and membrane enzymes are not the same enzyme. But to me it is a lot simpler to propose that they are different. They are different in all their properties.

SCHLESSINGER: So calcium is also required for phospholipase C, and not only for kinase C?

SHERR: No, it is not. This is a very difficult point. Our interpretation is that there is membrane-bound phospholipase C which is GTP regulated. We know also that there is a cytosolic phospholipase C, which is GTP-independent. At physiological concentrations of calcium, and by that I mean that you do not add any, you can permeabilize membranes, add PIP2, and convert it to IP3 with the membrane-bound enzyme. What I am suggesting is that the calcium signal generated through IP3 is also known to activate the cytosolic enzyme, and it has an effect on its substrate specificity. The membrane-bound enzyme is PIP2-specific, whereas the cytosolic enzyme recognizes not only PIP2, but PIP and PI. So, one possibility is that calcium changes the substrate specificity of the cytosolic enzyme. Diacylglycerol is also generated by this enzyme, and activates protein kinase C. One of the enzymes that degrades IP3 to IP2, so-called 5′ phosphatase, is a substrate for protein kinase C, and the active form of that enzyme is the phosphorylated one. One way to attenuate a calcium signal may be to activate 5′ phosphatase through protein kinase C. I am contending that this is a very complex metabolism, and we have not really thought through all these possibilities.

BALTIMORE: In the protein kinase C system the activation removes the enzyme physically from the cytosol and binds it to the membrane. So the question would be, do you see a loss in soluble phospholipase C?

SHERR: I do not know the answer to that.

SCHLESSINGER: May I add that calcium is not necessarily recruited from intracellular compartments, as for EGF it was shown that it may stimulate the entry of calcium into the cell.

SHERR: Yes, Rodrigo Bravo made this point yesterday, that you could use an ionophore to provide calcium extracellularly.

BRUGGE: I have a question for either you or Dr. Schlessinger: Have any natural conditions been identified under which protein kinase C is activated, i.e. exclusive of diacylglycerol homologues or TPA?

SCHLESSINGER: PDGF stimulates kinase C which will reduce the affinity of EGF receptor. Bombesin has a similar effect.

BRUGGE: Does it cause the internalization?

SCHLESSINGER: Well, it mainly affects affinity.

BRUGGE: I just wonder whether the TPA effects on internalization are really mediated by phosphorylation of protein kinase C.

SCHLESSINGER: I think that the main effect of TPA is to reduce the ligand affinity.

VAN DER EB: Has it been possible to make a C terminal small truncation in the c-*fms* gene and check the transforming capability?

SHERR: Of course it is an obvious experiment. Two groups have already tried and reported different results. One group has suggested that the truncation is activating. Actually, what has been done is to make an exchange between the C-termini of c-*fms* and v-*fms*, and to place the c-*fms* C-terminus on v-*fms*. You have to keep in mind that v-*fms* is feline and that the source of c-*fms* cDNA is human, so that is a complication. One group has suggested that that these chimeric constructs are non-transforming, whereas another has reported that they transform, but give small foci. We have not completed our own experiments but are in the process of testing many constructs, and I have no definitive data, so I cannot tell you who is right and who is wrong. I can guarantee that somebody is right, and somebody is wrong.

I think one of the other points that I will make is that some of the vectors that we have made can be used to address what I think is an important question in the context of some of these discussions, and that is to insert the v-*fms* gene product into hematopoietic cells. The implication of some of this work is that genetic rearrangements affecting c-*fms in situ* might have an effect on hematopoiesis. That is the basis of our interest in chromosome 5 deletions, but there is in fact no data that the v-*fms* gene product has any effect on hematopoiesis, and the virus has never been shown to transform hematopoietic cells or produce hematological tumors in cats. So, now that we have some constructs that are quite useful in this regard, we are inserting the v-*fms* gene product into CSF-1 dependent macrophages to see if we can convert such cells to growth factor independence. In principle, these experiments are quite interesting, because they also allow us to look at trans-modulation between the v-*fms* gene product, which we think is a constitutive kinase, and the c-*fms* gene product, which is regulated by CSF-1.

Acute Transformation by Simian Sarcoma Virus is Mediated by an Externalized PDGF-like Growth Factor

Bengt Westermark[1], Christer Betsholtz[1], Ann Johnsson[2] & Carl-Henrik Heldin[2,3]

The discovery that the oncogenes of acutely transforming retroviruses are derived from cellular gene sequences (reviewed by Bishop 1983) raised important questions about the transforming activity of the virally-encoded proteins in relation to the physiological function of the normal cellular homologues. A clue to this enigma was provided by the finding of a near identity in amino acid sequence between a region in the transforming protein $p28^{sis}$ of simian sarcoma virus (SSV) and the B-chain of human platelet-derived growth factor (PDGF) (Waterfield *et al.* 1983, Doolittle *et al.* 1983). This finding strongly suggests that SSV transformation is mediated by a growth factor, encoded by the v-*sis* oncogene. Subsequent to this observation, the normal cellular homologues of several retroviral oncogenes have been identified as structural genes for proteins which are proven, or thought, to play a role in mitogenesis. These findings have generated the idea that retroviral oncogene products subvert the mitogenic pathway at key regulatory points and thereby elicit an uncontrolled growth stimulus (Heldin & Westermark 1984).

The purpose of the present communication is to review recent data on the structural and functional properties of SSV and its oncogene product. The data that will be presented favor the hypothesis that SSV transformation of cells in culture is caused by an externalized PDGF-like growth factor that stimulates cell proliferation by an autocrine mechanism.

STRUCTURAL ASPECTS OF THE v-*sis* GENE AND ITS TRANSLATION PRODUCT

The replication-defective simian sarcoma virus and its helper virus, simian sarcoma-associated virus (SSAV), were isolated from a fibrosarcoma of a pet woolly

---

[1]Dept. of Pathology, University Hospital, S-751 85 Uppsala, [2]Dept. of Medical and Physiological Chemistry, Biomedical Center, S-751 23 Uppsala, and [3](Present address): Uppsala Branch of the Ludwig Institute for Cancer Research, Biomedical Center, S-751 23 Uppsala, Sweden.

VIRAL CARCINOGENESIS, Alfred Benzon Symposium 24,
Editors: N. O. Kjeldgaard, J. Forchhammer, Munksgaard, Copenhagen 1987.

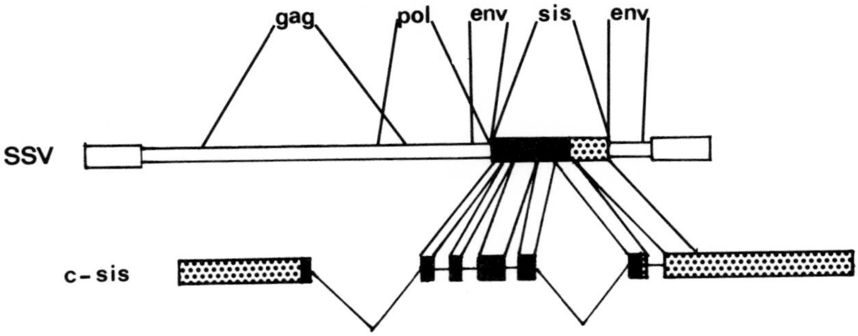

Fig. 1. Schematic representation of the structural relationship between the human PDGF B-chain gene (c-*sis*) and the SSV genome. Coding sequences of c-*sis* exons and homologous sequences in v-*sis* are represented by filled boxes and noncoding portions by dotted boxes. Note that introns are not drawn to scale. The 5′ and 3′ long terminal repeats of SSV are represented by open boxes. (Data compiled from Devare *et al.* 1983, Johnsson *et al.* 1984, Josephs *et al.* 1984, Collins *et al.* 1985, Rao *et al.* 1986).

monkey (Theilen *et al.* 1971, Wolfe *et al.* 1971). Analysis of molecularly cloned SSV proviral DNA has shown that the v-*sis* gene is integrated within a partially deleted helper virus *env* gene (Devare *et al.* 1983) (c.f. Figure 1). The nucleotide sequence of the open reading frame of v-*sis* predicts a 28 kDa primary translation product denoted $p28^{sis}$ (Devare *et al.* 1983), the mid-portion of which is virtually identical to the B-chain of human PDGF (c.f. Figure 2). Molecular cloning of v-*sis*-related cDNA and genomic DNA sequences has led to the identification of the human c-*sis* gene as the structural gene for a PDGF B-chain precursor (Josephs *et al.* 1984, Johnsson *et al.* 1984, Chiu *et al.* 1984, Collins *et al.* 1985, Rao *et al.* 1986). The precursor molecule has the properties of a preproprotein, i.e. it has a strectch of hydrophobic amino acid residues in the amino terminus characteristic of a signal peptide sequence, as well as additional amino and carboxy terminal sequences that are not matched by the mature product. As can be seen from Figure 1, the v-*sis* gene excludes the 5′ part of the c-*sis* gene that contains the coding sequence for the signal peptide of the B-chain precursor. However, recent studies have shown that translation of the v-*sis* transcript is probably initiated within the 5′ flanking *env* segment, that evidently compensates for the loss of the cellular signal sequence (King *et al.* 1985); the primary translation product is thus an *env-sis* fusion protein (Figure 2).

STRUCTURE OF PDGF

Human PDGF consists of dimers of A-chains and B-chains (Johnsson *et al.* 1982) that are about 60% homologous in their amino acid sequences and encoded

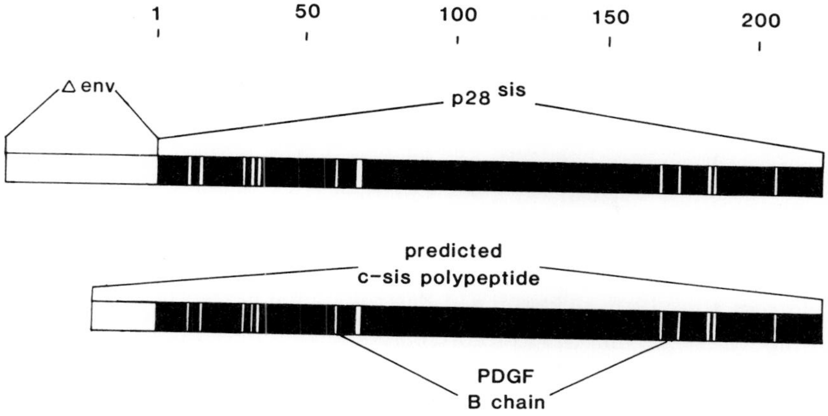

Fig. 2. Schematic representation of the structural relationship between the transforming gene product of SSV and the human PDGF B-chain precursor. Note that the amino terminal part of the viral gene product is encoded by sequences derived from the partially deleted helper virus *env* gene. Black boxes represent homologous sequences. (Data compiled from Devare *et al.* 1983, Josephs *et al.* 1984, Johnsson *et al.* 1984, Collins *et al.* 1985, King *et al.* 1985).

by different genes (Johnsson *et al.* 1984, Betsholtz *et al.* 1986a). A dimeric structure of PDGF is required for its mitogenic activity. In view of the structural homology between the subunits, it may be assumed that any dimeric combination of chains (A-A, B-B, A-B) should have PDGF-like biological activity. This notion is supported by experimental data. Thus, human PDGF, as purified from platelets, is most likely assembled as an A-B heterodimer (Heldin *et al.*, unpublished observation) whereas the osteosarcoma- (Heldin *et al.* 1986) and melanoma-derived growth factors (Westermark *et al.* 1986) are homodimers of A-chains. Moreover, pig PDGF consists of only one type of chain, which is homologous to the human B-chain (Stroobant & Waterfield 1984).

METABOLIC PROCESSING AND BIOLOGICAL FUNCTION OF THE v-*sis* GENE PRODUCT

Experiments on the metabolic processing of $p28^{sis}$ in SSV-transformed cells (Robbins *et al.* 1983) have provided evidence for an intimate structural relationship between the processed v-*sis* gene product and PDGF. Rapidly after synthesis, $p28^{sis}$ is dimerized and proteolytically trimmed in both ends. The apparently stable end product is a homodimer of M_r 24 000, that is identical or nearly identical to a putative PDGF B-chain homodimer and thus is closely related to pig PDGF. Therefore, although it has not been formally proven, it is reasonable to believe that the M_r 24 000 v-*sis* product is fully active as a PDGF-receptor agonist. Indeed, SSV-transformed cell lysates and conditioned media contain a PDGF-like growth factor activity that can be neutralized by PDGF antibodies (Deuel *et al.* 1983, Johnson *et al.* 1985a). The secreted activity binds to the PDGF receptor and stimulates its inherent tyrosine kinase activity (Johnsson *et al.* 1985a). Moreover, exposure of human fibroblasts to conditioned medium from SSV-transformed cells leads to a rapid reorganization of actin and induction of membrane ruffling of the same kind as that seen after exposure to PDGF (Mellström *et al.* 1983, Johnsson and Betsholtz, unpublished observation).

Although the primary *env-sis* translation product fulfills the structural requirements for a secretory product, the protein does not behave as such. The major part of the mature v-*sis* product remains cell-associated (Robbins *et al.* 1985) and the amount of PDGF-like activity that is released to the bulk medium in cultures of SSV-transformed cells is minute (Johnsson *et al.*, unpublished observation). In part, this may be due to a rapid binding and internalization of the secreted factor. The fact that the PDGF receptors on SSV-transformed cells are extensively down-regulated (Garret *et al.* 1984, Johnsson *et al.* 1986, Betsholtz *et al.* 1986b) indeed suggests that the juxtacellular concentration of the PDGF-like factor is much higher than that in the bulk medium. It is pertinent to point out that the morphology of SSV-transformed foci is in accordance with the notion that transformation is mediated via a locally acting PDGF-like growth factor. The foci appear as very distinct and sharply demarcated bundles or whirls of densely packed cells in parallel arrays (Johnsson *et al.* 1985b). Such a morphology does not seem to be consistent with an extensive paracrine action of a secreted growth factor.

EVIDENCE THAT SSV TRANSFORMATION IS SOLELY MEDIATED BY A PDGF-LIKE FACTOR THAT INTERACTS WITH PDGF RECEPTORS AT THE CELL SURFACE

If the v-*sis* gene product functions as a PDGF-like growth factor in SSV transforamtion one should expect the virus to exclusively transform cells that carry PDGF receptors. This question was addressed by Leal *et al.* (1985) who found a strict correlation between SSV transformation and binding of ^{125}I-labeled PDGF. Thus, vascular endothelial cells and epithelial cells are resistant to SSV transformation whereas various kinds of fibroblasts and vascular smooth muscle cells are susceptible. Observations made in our laboratory confirm these findings (Betsholtz, unpublished observations). The tissue tropism of SSV *in vivo* is also in accordance with the view that the virus only transforms PDGF-responsive cells. In newborn marmosets, SSV induces fibromas, fibrosarcomas or gliomas (Deinhardt 1980), i.e. tumors that originate from cells that carry receptors for PDGF (Heldin *et al.* 1981). The proposed model of SSV transformation further requires that the v-*sis* gene product be co-compartmentalized with the PDGF receptor, which is an integral membrane protein with an extracellular ligand binding portion (Heldin *et al.* 1983). Circumstantial evidence in favor of this notion is provided by the finding that deletion of the *env*-derived signal sequence abolishes the transforming capacity of the gene product (Hannink & Donoghue 1984, King *et al.* 1985).

The finding that the v-*sis* gene product does not behave as a classical secretory product has generated some controversy regarding its subcellular site of action. However, strong evidence in favor of the model that SSV transformation is mediated by a PDGF-like growth factor that interacts with cell surface PDGF receptors has been obtained by experiments showing that specific and nonspecific PDGF antagonists revert the transformed phenotype of SSV-transformed fibroblasts in culture. Thus, PDGF antibodies attenuate focus formation, block the proliferation in serum-free medium and revert the transformed morphology of human foreskin fibroblasts, transformed by SSV (Johnsson *et al.* 1985b). These findings are of primary importance since they show that the transforming activity of SSV is immunologically related to PDGF and that its site of action must be on the exterior of the cell since immunoglobulins, for obvious reasons, cannot pass the plasma membrane. An even more efficient but less specific means of interfering with the putative autocrine loop in SSV-transformed cells was employed by Betsholtz *et al.* (1986b) who took advantage of the previous finding that the highly charged, polycyclic compound suramin blocks PDGF binding and displaces receptor-bound PDGF (Williams *et al.* 1984). Suramin was found to efficiently revert the transformed morphology and block the proliferation in serum-free medium of SSV-transformed fibroblasts without having any such effect on cells transformed by Rous sarcoma virus or Kirsten murine sarcoma virus. The effect of suramin was entirely reversible; after drug removal, the cells reattained their transformed phenotype after 1–2 days.

The model that SSV transformation is solely mediated by a PDGF-like growth factor predicts that the phenotypic properties of SSV-transformed cells should be identical to those of nontransformed cell continuously exposed to PDGF. This was shown to be the case in a study by Johnsson *et al.* (1986). Thus, it was found that SSV induces growth in semi-solid media to about the same extent as PDGF added to control cells grown under the same conditions. Moreover, serial passage

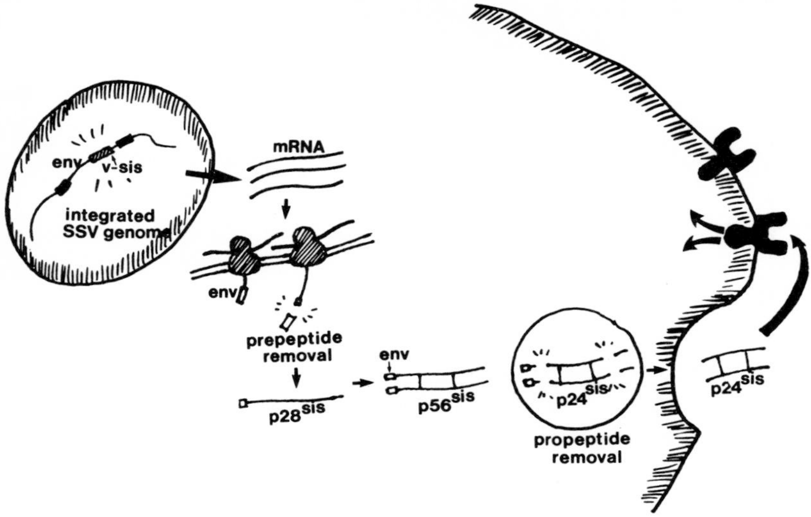

Fig. 3. Mechanism of SSV-transformation. The transforming protein of SSV is synthesized as an *env-sis* fusion product. The signal peptide, encoded by *env*-derived sequences, is cleaved off at the endoplasmic reticulum. The resulting p28^{sis} molecule is dimerized and trimmed to a 24 kDa dimer that is externalized and binds to the PDGF receptor at the cell surface and thereby stimulates cell proliferation by an autocrine mechanism. (Model based on data presented by Robbins *et al.* 1983, 1985, King *et al.* 1985, Johnsson *et al.* 1985, 1986a, b, Betsholtz *et al.* 1986b). (See text for details).

of SSV-transformed human fibroblasts showed that SSV is not an immortalizing virus; rather, the life span of SSV-transformed cells is identical to that of their nontransformed counterparts. Interestingly, at the end of their life span, the SSV-transformed cells become large and flat with an extremely high cytoplasmic/nuclear ratio and are morphologically indistinguishable from nontransformed cells of comparable passage level. Such senescent SSV-transformed cells retain the SSV genome and continue to produce PDGF-like activity but are apparently refractory to its mitogenic activity. In this context it should be mentioned that senescent normal fibroblasts bind PDGF but do not respond to the growth factor (Paulsson *et al.*, in preparation). Most likely, therefore, normal and SSV-transformed fibroblasts senesce by the same mechanism, possibly involving a block in the postreceptor pathway.

In conclusion, we suggest a very simplistic model for SSV transformation of cultured cells: the v-*sis* gene product is processed to a PDGF receptor agonist

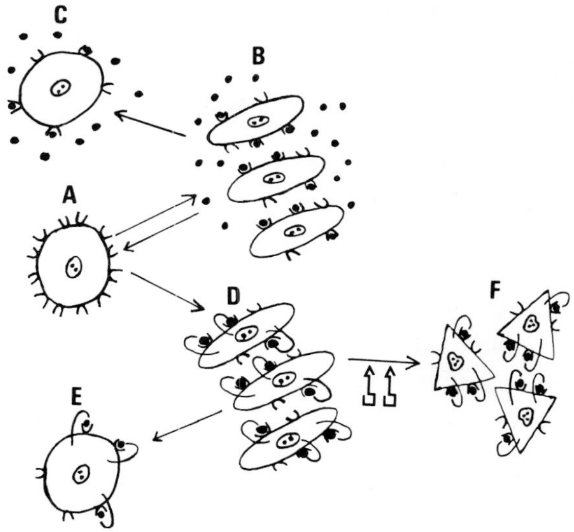

Fig. 4. Hypothetical relationship between PDGF stimulation, SSV transformation *in vitro* and formation of SSV-induced malignant tumors *in vivo*. The cell in A represents a quiescent fibroblast that in B is continuously exposed to PDGF (dots). This results in a blockade and down-regulation of the PDGF receptors and cell division. After repeated cell cycles, the cell in C has become senescent and refractory to PDGF although it retains cell surface PDGF receptors. In D, cells are SSV-transformed. The transforming gene product (dots) has the same biological function as exogenous PDGF, i.e. blocks and down-regulates the PDGF receptors and stimulates cell proliferation, but by an autocrine mechanism. In E, the SSV-transformed cell has become senescent after multiple cell cycles and is refractory to the endogenous growth factor as is the cell in C to exogenous PDGF. *In vivo*, SSV causes a polyclonal expansion by the same mechanism as *in vitro* (D). A small (monoclonal?) subpopulation undergoes secondary genetic changes and develops the fully malignant phenotype in F.

that is externalized and stimulates cell proliferation by an autocrine mechanism. This model is schematically depicted in Figures 3 and 4.

TUMORIGENICITY OF SSV *IN VIVO*

Although the molecular mechanism of SSV transformation *in vitro* may be on the verge of being elucidated, a critical question that remains to be answered concerns the oncogenic potential of the virus. Deinhardt and collaborators have shown that SSV, when injected together with its helper virus, is a *bona fide* tumor virus since it induces malignant tumors such as fibrosarcomas and glioblastomas (Deinhardt 1980). However, since the phenotypic and genotypic characteristics of SSV-induced tumors have not been subjected to a detailed analysis, any discussion on the relationship between *in vitro* transformation and *in vivo* tumorigenicity has to be speculative. We consider it unlikely that the endogenous production of a PDGF-like growth factor alone is sufficient to induce and maintain a fully malignant phenotype *in vivo*. Rather, we would suggest that an autocrine response by a polyclonal population of SSV-infected cells may be the first step in neoplastic conversion that is complemented by secondary changes in the host genome, perhaps occurring only in a minor or even in a clonal subpopulation (Figure 4). Clearly, the helper virus may play a part in such a secondary event. In order to shed some light on the problem, a careful analysis of the molecular biology of SSV-induced tumors should be undertaken. This may be especially important since the system seems to be well suited for the study of the role of autocrine growth stimulation in tumorigenesis.

ACKNOWLEDGMENTS

Our own work cited in the text was supported by grants from the Swedish Cancer Society and the Swedish Department of Agriculture.

REFERENCES

Betsholtz, C., Johnsson, A., Heldin, C.-H. & Westermark, B. (1986b) Efficient reversion of SSV-transformation and inhibition of growth factor-induced mitogenesis by suramin. *Proc. Natl. Acad. Sci. USA. 83,* 6440–4.

Betsholtz, C., Johnsson, A., Heldin, C.-H., Westermark, B., Lind, P., Urdea, M. S., Eddy, R., Shows, T. B., Philpott, K., Mellor, A. L., Knott, T. J. & Scott, J. (1986a) cDNA sequence and chromosomal localization of human platelet-derived growth factor A-chain and its expression in tumour cell lines. *Nature 320*, 695–9.

Bishop, J. M. (1983) Cellular oncogenes and retroviruses. *Ann. Rev. Biochem. 52*, 301–54.

Chiu, I. M., Reddy, E. P., Givol, D., Robbins, K. C., Tronick, S. R. & Aaronson, S. A. (1984) Nucleotide sequence analysis identifies the human c-*sis* proto-oncogene as a structural gene for platelet-derived growth factor. *Cell 37*, 123–9.

Collins, T., Ginsburg, D., Boss, J. M., Orkin, S. H. & Pober, J. S. (1985) Cultured human endothelial cells express platelet-derived growth factor B chain: cDNA cloning and structural analysis. *Nature 316*, 748–50.

Deinhardt, F. (1980) The biology of primate retroviruses. In: *Viral Oncology*, ed. Klein, G., pp. 359–398. Raven Press, New York.

Deuel, T. F., Huang, J. S., Stroobant, P. & Waterfield, M. (1983) Expression of a platelet-derived growth factor-like protein in simian sarcoma virus transformed cells. *Science 221*, 1348–50.

Devare, S. G., Reddy, E. P., Law, J. P., Robbins, K. C. & Aaronson, S. A. (1983) Nucleotide sequence of the simian sarcoma virus genome; demonstration that its acquired cellular sequences encode the transforming gene product $p28^{sis}$. *Proc. Natl. Acad. Sci. USA 80*, 731–5.

Doolittle, R. F., Hunkapiller, M. W., Hood, L. E., Devare, S. G., Robbins, K. C., Aaronson, S. A. & Antoniades, H. N. (1983) Simian sarcoma virus oncogene, v-sis, is derived from the gene (or genes) encoding a platelet-derived growth factor. *Science 221*, 275–7.

Garrett, J. S., Coughlin, S. R., Niman, H. L., Tremble, P. M., Giels, G. M. & Williams, L. T. (1984) Blockade of autocrine stimulation in simian sarcoma virus-transformed cells reverses down-regulation of platelet-derived growth factor receptors. *Proc. Natl. Acad. Sci. 81*, 7466–70.

Hannink, M. & Donoghue, D. J. (1984) Requirement for a sequence in biological expression of the v-sis oncogene. *Science 230*, 1197–9.

Heldin, C.-H., Ek, B. & Rönnstrand, L. (1983) Characterization of the receptor for platelet-derived growth factor on human fibroblasts: Demonstration of an intimate relationship with a 185,000 dalton substrate for the platelet-derived growth factor-stimulated kinase. *J. Biol. Chem. 258*, 10054–61.

Heldin, C.-H., Johnsson, A., Wennergren, S., Wernstedt, C., Betsholtz, C. & Westermark, B. (1986) A human osteosarcoma cell line secretes a growth factor structurally related to a homodimer of PDGF A chains. *Nature 319*, 511–4.

Heldin, C.-H., Westermark, B. & Wasteson, Å. (1981) Specific receptors for platelet-derived growth factor on cells derived from connective tissue and glia. *Proc. Natl. Acad. Sci. USA 78*, 3663–8.

Heldin, C.-H. & Westermark, B. (1984) Growth factors: Mechanism of action and relation to oncogenes. *Cell 37*, 9–20.

Johnsson, A., Betsholtz, C., Heldin, C.-H. & Westermark, B. (1985b) Antibodies against platelet-derived growth factor inhibit acute transformation by simian sarcoma virus. *Nature 317*, 438–40.

Johnsson, A., Betsholtz, C., Heldin, C.-H. & Westermark, B. (1986) The phenotypic characteristics of simian sarcoma virus-transformed human fibroblasts suggest that the v-sis gene product acts solely as a PDGF receptor agonist in cell transformation. *EMBO J. 5,* 1535–41.

Johnsson, A., Betsholtz, C., von der Helm K., Heldin, C.-H. & Westermark, B. (1985a) Platelet-derived growth factor agonist activity of a secreted form of the v-sis oncogene product. *Proc. Natl. Acad. Sci. USA 82*, 1721–5.

Johnson, A., Heldin, C.-H., Wasteson, Å., Westermark, B., Deuel, T. F., Huang, J. S., Seeburg, D. H., Gray, E., Ullrich, A., Scrace, G., Stroobant, P. & Waterfield, M. D. (1984) The c-sis gene encodes a precursor of the B chain of platelet-derived growth factor. *EMBO J. 3*, 921–8.

Johnsson, A., Heldin, C.-H., Westermark, B. & Wasteson, Å. (1982) Platelet-derived growth factor: identification of constituent polypeptide chains. *Biochem. Biophys. Res. Commun. 104*, 66–74.

Josephs, S. F., Guo, C., Ratner, L. & Wong-Staal, F. (1984) Human protooncogene nucleotide sequences corresponding to the transforming region of simian sarcoma virus. *Science 223*, 487–90.

King, C. R., Giese, N. A., Robbins, K. C. & Aaronson, S. A. (1985) In vitro mutagenesis of the v-sis transforming gene defines functional domains of its growth factor-related product. *Proc. Natl. Acad. Sci. USA 82*, 5295–9.

Leal, F., Williams, L. T., Robbins, K. C. & Aaronson, S. A. (1985) Evidence that the v-sis gene product transforms by interaction with the receptor for platelet-derived growth factor. *Science 230*, 327–30.

Rao, C. D., Igarashi, H., Chiu, I.-M., Robbins, K. C. & Aaronson, S. A. (1986) Structure and sequence of the human c-sis/platelet-derived growth factor 2 (SIS/PDGF2) transcriptional unit. *Proc. Natl. Acad. Sci. USA 83*, 2392–6.

Robbins, K. C., Antoniades, H. N., Devare, S. G., Hunkapiller, M. W. & Aaronson, S. A. (1983) Structural and immunological similarities between simian sarcoma virus gene product(s) and human platelet-derived growth factor. *Nature 305*, 605–8.

Robbins, K. C., Leal, F., Pierce, J. H. & Aaronson, S. A. (1985) The v-sis/PDGF2 transforming gene product localizes to cell membranes but is not a secretory protein. *EMBO J. 4*, 1783–92.

Stroobant, P. & Waterfield, M. D. (1984) Purification and properties of porcine platelet-derived growth factor. *EMBO J. 3*, 2963–7.

Theilen, G. H., Gould, D., Fowler, M. & Dungworth, D. L. (1971) C-type virus in tumor tissue of a woolly monkey with fibrosarcoma. *J. Natl. Cancer Inst. 47*, 881–99.

Waterfield, M. D., Scrace, T., Whittle, N., Stroobant, P., Johnsson, A., Wasteson, Å., Westermark, B.,

Heldin, C.-H., Huang, J. S. & Deuel, T. F. (1983) Platelet-derived growth factor is structurally related to the putative transforming protein $p28^{sis}$ of simian sarcoma virus. *Nature 304*, 35–9.

Westermark, B., Johnsson, A., Paulsson, Y., Betsholtz, C., Heldin, C.-H., Herlyn, M., Rodeck, U. & Koprowski, H. (1986) Human melanoma cell lines of primary and metastastic origin express the genes encoding the chains of platelet-derived growth factor and produce a PDGF-like growth factor. *Proc. Natl. Acad. Sci. USA* (in press).

Williams, L. T., Tremble, P. M., Lavin, M. F. & Sunday, M. E. (1984) Platelet-derived growth factor receptors form a high affinity state in membrane preparations. *J. Biol. Chem. 259*, 5287–94.

Wolfe, L. G., Deinhardt, F., Theilen, G. H., Rabin, H., Kawakami, T. & Bustad, L. K. (1971) Induction of tumors in marmoset monkeys by simian sarcoma virus, type 1 (Lagothrix): A preliminary report. *J. Natl. Cancer Inst. 47*, 1115–20.

DISCUSSION

VENNSTRÖM: According to Anita Robert's results you would need PDGF, TGF beta and TGF alpha for anchorage-independent growth of normal fibroblasts. Do you have any indications that SSV-transformed cells also make TGF alpha and beta?

WESTERMARK: No, they might do that as perhaps cells that are stimulated by PDGF. We do not know. Once we have reported the finding that both PDGF and SSV cause anchorage independence, but we have not pursued it any more.

VAN DER EB: Are the SSV-transformed cells still dependent on insulin or EGF? Or are they completely independent of any growth factor?

WESTERMARK: That may be a complicated question, because when we grow these cells in a very rich medium, specifically designed for human fibroblasts, which contains no growth factors, just with a balanced composition of amino acids and salts and other nutrients, these cells do not require insulin, when they are normal and non-transformed. They grow happily in PDGF alone or in EGF alone, and if we add insulin to the cells, nothing happens with the regard to their DNA synthesis.

VAN DER EB: Human fibroblasts excrete insulin or EGF themselves apparently?

WESTERMARK: That we do not know, but they may do that. We have taken IGF-1 antibodies or IGF-1 receptor antibodies and added to the cells, but those antibodies would not block the growth stimulation by PDGF. So, if they do, perhaps they do not require it.

RIGBY: Can you just clarify the status of the SSV-transformed human fibroblasts. Do they become senescent just like regular fibroblasts. They are transformed but not immortalized?

WESTERMARK: Yes.

SHERR: The gibbon ape leukemia virus is quite oncogenic and induces disease after short latency. There are well-documented cases of transmission between animals. So, I think that argues that the SSAV helper virus had an oncogenic capability that could function in the context of a secondary event in the SSV model system.

WESTERMARK: The question is whether it functions only by transmission in the way that you recruit new cells all the time, so you get an expansion of the tumor because of recruitment of cells. In that case you would have a polyclonal tumor.

SHERR: What was the latency in Deinhardt's original experiments, do you know?

GALLO: A few months.

CORY: I assume that the tumors that you get in the animals are tumorigenic again when passaged?

WESTERMARK: That has not been done. Many of these studies are just anecdotal. Real, full papers have not been published. Of course one did not know anything about the molecular biology at that time, so that is why I think it is important to go back to the monkeys and do the experiments again and look with all the new tools and all the new concepts of what has happened.

CORY: So it is conceivable that you are just getting proliferation?

WESTERMARK: Yes, but there is one thing that really argues against that. Because, if you infect the brain with SSV, and you get a proliferation of cells, you would get something that histopathologically would look like gliosis, glia cells would grow. But these tumors are, according to a very experienced neuropathologist, bona fide glioblastomas with all the hallmarks of a human glioblastoma. I think that argues against that.

GALLO: I think the same is true with the fibrosarcomas also. They are real malignancies.

WESTERMARK: Yes, but fibrosarcomas are even in man a little bit more difficult to diagnose. The cells are elongated and look sort of normal, but in the brain there is no question about it.

SCHLESSINGER: What is know about the effect of suramin on the binding of EGF and PDGF?

WESTERMARK: It is a charged hydrophobic 14 000 mol.weight polycyclic compound that interacts wth proteins non-specifically and probably does not pass the plasma membrane, because it would kill the cell. It is used for the treatment of certain parasitic disorders. It has been used, for instance, by Brown and Goldstein to inhibit LDL binding to the LDL receptor. It is very useful if you think that you have the ligand receptor binding, because you are sure that it would inhibit that. But it is useless if you want some specificity.

A Personal Perspective

David Baltimore

To sum up the Symposium David Baltimore gave an account of viral carcinogenesis as "A Personal Perspective". This was introduced by the following summary:

The field of viral carcinogenesis has provided an extraordinary range of understanding at the genetic level, but like all research in carcinogenesis, it has left open questions of mechanism. There is a great lacuna in our knowledge about cellular growth regulation. How mitogenic signals are transduced from the cell surface to critical places in the interior of the cell remains a mystery. We are quite sure the growth regulation involves changes in transcription, but we are still in the earliest stages of understanding how mammalian transcription is regulated. The power of genetics has not run its course: there is especially hope that genetic lesions in simple systems will provide clues to the pathways of growth regulation. It does seem evident, however, that the hard work of the biochemist is going to play an increasingly important role during the next phase of research.

In his presentation Dr. Baltimore described the structure of the c-*abl* genes and continued with a survey of the regulatory proteins implicated in the control of activity of the immunoglobulin heavy chain and kappa light chain enhancers.

The following discussion of Dr. Baltimore's presentation is pertinent mainly to these aspects.

Whitehead Institute for Biomedical Research, Nine Cambridge Center. Cambridge, Massachusetts 02142, USA.

VIRAL CARCINOGENESIS, Alfred Benzon Symposium 24,
Editors: N. O. Kjeldgaard, J. Forchhammer, Munksgaard, Copenhagen 1987.

DISCUSSION

YANIV: It is certain that the idea of having common motifs in regulatory elements of different genes is suggested in many systems in higher eukaryotes. But why does yeast behave differently, having specific signals and specific proteins for inducible genes or a family of inducible genes?

BALTIMORE: Yeast does not have the requirement of having multiple types of differentiated cells. That may be the reason.

YANIV: One more comment, in bacteriophages and adenovirus or herpesvirus you have a series of factors that control a series of events in a linear fashion. In T4 you have immediately early, early, delayed early, and late genes. So, I think we should not exclude such a mechanism during cell differentiation.

BALTIMORE: I do not want to exclude anything. What I am really trying to do is to generalize from our experience, and it seems to me, from most people's experience ... I tended to say, until recently at least, that the fact that we find the same motif in the enhancers that control differentiated genes, and also in viruses, is because viruses have to grow on a lot of cells, so they have to have an agglomeration of all sorts of specific sites. But with the increasing difficulty of proving that anything is specific, I am beginning to doubt that that is true. And the fact that the more that one looks at things, the more one sees commonality, leads me to wonder if there really is going to be specificity. There are very few systems in which we understand the individual motifs as well as we do in the immunoglobulins.

HANAHAN: You spoke of the complexity dilemma, with respect to the amount of specific information required for each regulated gene, such that it has its specific transacting factors (which in turn require other trans-acting factors in order to be expressed, and so on). If all the DNA binding proteins are resident in an average cell, does that necessarily get you around this problem in any significant way?

BALTIMORE: No, it causes a different problem. It gets you around that problem. The problem is: how do you turn beta-galactosidase induction into a permanent induction that is independent of beta-galactosides, and people puzzled over that one for as many years as they have looked at inducible enzymes in bacteria. There are lots of potential solutions that have been proposed. I think none is satisfactory. The only solution to that problem which I find intriguing at the moment – and it is solely, I guess, because it is new and involves protein kinases – is the paper by Mary Kennedy and somebody else in "Cell" about two months ago, in which they show that a protein kinase, which is ordinarily a calcium-activated protein kinase, becomes calcium-independent if it is phosphorylated on a few out of what is a very large number of potential sites that can be phosphorylated on that enzyme. She points out quite rightly that this is a perfect model for turning a transient event into a stable event if you can build into it some kind of self-replication. The self-replication would be that this protein kinase phosphorylates new molecules of itself up to the level where it becomes calcium-independent, so that a single pulse of calcium changes the transcriptional program of the cell by that kind of model.

HANAHAN: But does that not in some sense just shift the problem? You have a lymphoid cell that has a very specific character whatever the nature of the developmental pathway is, it turns on a whole battery of genes that make that cell what it is. Then, as you should say, once you get an event that turns on all those genes the problem is solved. But meanwhile you just switch the whole problem of tissue-specific regulation back to that event, whatever it is.

BALTIMORE: No, it does not. It explains a part of the puzzle. It does not explain it all. I certainly agree with that. Because then you have to go back into what are classically called determination events in differentiation, and whether those are the same sort of things we are seeing here, or something qualitatively different. I do not know. What I do suspect, however, is that in transient transfection experiments or even in stable transfection experiments we may not be looking at those events, and therefore we may simply have no way of asking the question at the moment, and perhaps only through transgenic mice can you look at the question.

ROIZMAN: I have two comments and a question. The comments really relate to the fact that there are a large number of herpes virus genes which can be deleted. These viruses will grow perfectly well in transformed cells, but when you put them in primary cells or in animals, they will not grow. What I am wondering about is whether, by analogy, in these cell lines which you tested are bags of factors. The factors are in these cells only because they are already transformed cells; but in a natural lineage, those factors may not be present.

BALTIMORE: The question that one might logically want to ask, and a question which worries me, is whether tissues from animals have all of these factors the way cell lines do, and I do not know that. We have not done that experiment, largely because, as somebody was pointing out the other day, a given tissue has so many kinds of cells in it that it is not an awfully precise experiment. You get a negative answer and you are very worried. But we have really not done that, and we are going to start on a fairly systematic look at questions like that. That is a real worry that what we are looking at are immortalized lines, and that they are not a perfect representation of the native state.

ROIZMAN: The second comment before I get to my question is an observation that the major regulatory protein of herpes simplex has a binding site, and it has a strong positional effect. In genes that it normally inhibits, the position of the binding site is within the leader sequence or the capping site. Where it induces, the location of the binding site is way upstream, and the question is, do you know anything about binding sites or can you relate binding sites to specific effects of induction?

BALTIMORE: Only in one case, and that is with the upstream octamer of the kappa promoter. If you move it further upstream it no longer activates. It does not inhibit, but it no longer activates. So, a part of a reason for octamer specificity seems to be its closeness to the starting site of transcription. There are also a whole lot of complicated experiments from Strasbourg, which have not yet resolved themselves, that relate to the heavy-chain enhancers having both negative and positive effects. How much of that can be ascribed to distances and relationship between sites, and how much of it is absolute, I think is still an open question.

ROIZMAN: My last question has to do with your differentiation of proteins B1 and B3. Your B3 is a different form of B1. How do you explain the DNA band shift in mobility if B3 is in fact a different protein. Do you have an estimation of its affinity relative to B1?

BALTIMORE: No, we cannot get reliable affinity measurements from crude extract binding. You can play around and pretend to do it, but it is so imperfect that we have usually avoided the question although on occasions people try to get answers. We do not see any particular, obvious difference that would tell us anything. But you are right about saying that, because there are two different bands there, there must be two different proteins, and our experience is – and this is an interesting aspect of this assay which I do not understand – that when a given protein is binding to a DNA of quite variable size, the band that is formed almost always has very close to the identical mobility, so that we can cut the DNA in many ways, and all of the complex bands will be in almost identical positions, whereas the free DNA will run with various mobilities.

ROIZMAN: Almost, no perfect.

MATTHIAS: I have one comment and at least one question. The comment extends Hanahan's remark. I think it is a matter of combination how things will work eventually. We have taken the reverse approach, and we tested directly a variety of enhancer subsegments after polymerization. While the SV40 core sequence is active in almost every cell type, the region of the SphI site has a marked B-cell preference. This indicates that a motif can have a different cell type specificity than the enhancer it comes from, and the combination of the various motifs will have the cell type specificity one knows. So, it could mean that in cell type A the SphI region will not be relevant, but it is also possible that overlapping with the SphI region is another motif which will be relevant in that cell type. I think we should always keep in mind that motifs can be overlapping and that only the combination of all motifs will determine the phenotype of the whole enhancer.

My question is, do you get the very same band shift if you look at the octamer sequence from the kappa promoter and from the heavy-chain enhancer, if you do that side by side? My feeling was that, when you talked about the octamer, that was all done with the octamer from the promoter region.

BALTIMORE: No, we have done more of the same work with the octamer sequence in the mu enhancer. And you get two bands, they have the same relationship to each other, they compete out …

MATTHIAS: You do not see any cell type specificity there either?

BALTIMORE: No, we see the same cell type specificity, i.e. the upper band shows no specificity and the lower band shows the lymphoid specificity.

MATTHIAS: And do you have any experiments in the human enhancer?

BALTIMORE: No.

MATTHIAS: We have done some and we do not get any shift with the equivalent fragment. It is interesting to see that there is a one nucleotide difference within the octamer sequence, and it raises the question whether this region is functional at all in the human enhancer.

BALTIMORE: I guess that I have once noticed that there was a one nucleotide difference and was worried about it.

CORY: I am wondering about the changes that may be occurring to the B-site factor after induction with LPS or TPA and then later on in a differentiated cell. Do you have any hint that that protein in itself is being changed in any way and is it changed differently on LPS stimulation than in a mature B cell?

BALTIMORE: No, the bands co-migrate, so there is no hint from that. That would be the only way we get evidence because we have no way of assaying the protein before it is activated. We just get a blank lane. We know nothing about that protein until it becomes a DNA binding protein.

You can make many guesses about what might be happening. The obvious guess is that the factor is being phosphorylated and changing its conformation. I do not like that. I guess it is more emotional than rational, but I would rather imagine that the binding site is actually covered by an inhibitor protein, and that what the phosphorylation does is to remove the inhibitor protein, thus releasing the binding site.

CORY: It has to be a different change in the LPS-inducible cells than in the more permanently differentiated cells.

BALTIMORE: No, I do not think it has to be a different one. All we know is that something has to stabilize it. And I do not know what that is or how it works. You are absolutely right that a B cell and a transiently induced 70Z cell are different in some fundamental way that allows the B cell to continue to have this factor and the 70Z cell seems to lose it.

DYNAN: There is a tendency to represent these binding phenomena on a slide as a colored ball sitting on a line of sequence. In reality, there are mechanistic processes that we know relatively little about that are set in motion by the binding of the factors. I would hope to elicit your comments or thoughts on some of the enigmatic aspects of these factors, the bidirectionality, the flexibility in positioning, and the occurrence of multiple copies of the binding site in a single control region.

BALTIMORE: You are still hoping! That, as far as I am concerned, is the most difficult and enigmatic part of all the transcriptional control, how an enhancer can work at a number of kb away. Actually, once you get enough kb's away it does not really matter what direction you are in, so it is really the distance which is the most remarkable aspect of the ability of the enhancers to function. In fact, if the distance gets long enough, *cis* and *trans* become equivalent. I am not sure that they are not equivalent. There are some phenomena, in particular associated with immunoglobulin, the lack of need for immunoglobulin enhancers, that in certain kinds of cell situations may mean that there is transactivation from other enhancers. I do not have anything useful to say except that I like the idea that DNA bends around and brings together the proteins bound to the enhancer with the promoter region. The fact that I like it does not make it true. And without having a set of purified factors in a really well-defined system I do not think that we are going to find out.

DYNAN: In the promoter, the spacing between sites is sometimes only 10–20 bp. It is much harder to imagine a loop forming within this short distance than it is to imagine a large loop between enhancer and promoter.

BALTIMORE: It is, in that case, much harder to imagine than the large loop, and certainly in the experiment that Chambon's lab did, where they looked at distance relationship, the evidence for quite close approximation of proteins in enhancer, promoter and TATA box was very impressive, and should not be forgotten. That looks more like a site-to-site interaction than it does of bending because of distances involved. The octamer in the promoter can be in either direction. It works perfectly in either direction. It is naturally found in one direction in heavy-chain and in the other direction in the light chain.

YANIV: In our laboratory we study the expression of the albumin gene, a liver-specific protein. We identified two upstream elements conserved among albumin genes from chicken to man, but absent in other known liver-specific genes. Liver nuclei contain factors binding to these upstream sequences; little or none of these factors is present in other organs. Still, when we examined tissue culture cells of different histotypic origin we found in most cells proteins that bind to one of those sequences but giving a slightly different footprint. A family of factors that bind to the same sequence may exist or alternatively these are different forms of the same factor. A specific differentiated cell can use a single member of a family of factors, a specifically modified form or a unique novel factor to regulate tissue-specific expression. At the end, we have probably a combinatory mechanism that uses some ubiquitous factors, some specifically modified forms of frequent factors and some tissue-specific factors.

BALTIMORE: Once we know enough genes, we will be able to say the extent to which there are very specific factors, and the extent to which it is complementary in some sense. You are quite right in saying that there is likely to be almost anything you can imagine. Every imaginable control mechanism probably plays some role in differentiation.

Subject Index